P9-CAD-918

SOFTWARE
FOR
AMATEUR RADIO

No. 1560
$21.95

SOFTWARE FOR AMATEUR RADIO

BY JOE KASSER, G3ZCZ

FIRST EDITION

FIRST PRINTING

Library of Congress Cataloging in Publication Data

Kasser, Joe.
Software for amateur radio.

Includes index.
1. Amateur radio stations—Data processing. I. Title.
TK9956.K27 1984 621.3841'66 84-8706
ISBN 0-8306-0360-3
ISBN 0-8306-0260-7 (pbk.)

This book is dedicated to the memory of my late brother.

Acknowledgments

I would like to thank Snow Micro Systems Inc. for permission to publish listings of the software that they are distributing in machine-readable form. Thanks are also due to Sharon Properzio for her help in the typing of the manuscript.

Contents

SOFTWARE
FOR
AMATEUR RADIO

Introduction

A few of the radio club members had been playing with computers for some months now. Some had built computers. Others had purchased computers and were busy this week building hardware and software for the contest that began that morning.

Fred was not a dyed-in-the-wool contester, although he liked to get on during a contest to work a few new countries. He had dabbled in various other aspects of amateur radio and was now thinking about joining the fun by setting a computer and using it with his amateur radio station. He was on his way to the Club station to see what the "experts" had built.

As Fred came close to the building, he noticed that there seemed to be a lot more antennas than usual. The new ones were mostly simple dipoles, which the regular contesters wouldn't touch with a 30-foot bargepole, so he wondered what was going on.

Inside, several new operating positions were set up, each with computer terminals and a club member wearing headphones and engrossed in what he was doing. Fred wondered what was going on, but he knew better than to interrupt the contest operators. Fred spotted Marc, who did not seem to be operating. Instead, he was sitting in front of an area with a lot of TV sets in front of him, and he was making notes on a yellow tablet. Some TVs contained summary data of bands, contacts made, and the current score. Others showed graphic displays of the USA with the different states and sections colored in red, yellow, or green, and, as he watched, two states changed from yellow to red and others changed from red to green and back to yellow. Fred was fascinated. He had never seen anything quite like it.

Marc looked up and grinned. "Go take a look in the corner over there," he suggested before he bent over his work again.

Fred found himself at an unmanned operating table. There was a computer and an HF rig, with sounds of Morse code coming from the speaker. As he listened, he heard what seemed to be Marc's callsign making contacts, but there was no operator in sight.

The whole set-up like something out of a science fiction movie. Was this what the Pentagon War Room looked like? Fred impatiently waited for a chance to ask Marc a few questions.

As he stood there watching, Marc finally put his pen down and came across the room. "Fred, I'm due for a short break now, so come and get a cup of coffee with me and I'll fill you in on what's going on."

They went over to the refreshment area and Fred said, "Well, I hardly know where to begin. I knew you people were playing with computers, but this is fantastic."

"Not really," replied Marc. "What we have done is to take a number of microcomputers and modified them and their software to work in the contest environment. The programs that they are running are all standard, or logical extensions of standard programs."

"Is that operating position working without an operator?" asked Fred.

"Yes," replied Marc, "That's one of our experimental programs. It's running my callsign and I want to see how it does. If it does well enough I may even send in an entry for it."

"Great!" said Fred. "Where'd you get the idea?"

"Mostly from two books that we have been using as a guide," said Marc.

"Which books?" asked Fred "I'd like to read them."

"Well, the first book is *Microcomputers in Amateur Radio* by G3ZCZ. It's published by TAB BOOKS Inc. (No.1305). It deals with the basic applications of the microcomputer in the ham station and forms the basis of our construction project. The second book is *Software for Amateur Radio* by the same author, which begins where the first one left off. It describes different programs that can be used for logging, contests, record-keeping, awards, OSCAR satellite-tracking, and antenna direction and distance computations. There's also a contest simulation in the form of a game that we found very useful for developing tactics for use in the contest. We also set up the operating programs to have similar dialogs so that we could use the game program as a training aid.

"What about those maps?" asked Fred, pointing at the TV screens where the colors were slowly changing from yellow to red to green or yellow again.

"That's our spotter network," replied Marc. "We are operating in the 3-transmitter class, so we can only have three stations active at any one time. The problem is knowing which is the best band to be on. Bill, Pat, and some of the SWL members of the club are over there tuning the bands. When they hear a new one calling CQ, they enter it's callsign and the frequency into their terminals. That's when the display changes from yellow to red. The operators use that to help them decide whether to join the pile up or to continue calling CQ. If it's a rare one they'll go for him; otherwise they usually stay where they are. Anyhow, when they contact an area it changes color to green. If they don't work it, the computer changes the color back to yellow after five minutes. The spotters keep updating the screens and I can see which band we should be transmitting on." Marc glanced back at the TV's.

"Oops," he exclaimed, "looks like we should change from 10 Meters to 15 Meters about now." As he said that, they saw Mike signalling to the operating stations to make the change.

"Is that program in the book?" asked Fred "It looks very complicated."

"No," replied Marc. "We modified one of the graphics packages that came with the computer by crossing it with two parts of different programs in the books."

"Does that mean that the programs in

the book need lots of work before they will run?" asked Fred.

"No, we have rather an extreme case here. The programs in the book are written in BASIC and are for use in the average ham station, and can be used with little or no modification other than possibly converting them to a different dialect of BASIC."

"Dialects?" echoed Fred. "Does a computer language have dialects?"

"Sort of. The programs in this book are written in either MICROSOFT or Northstar BASIC, which have minor differences in the way they do several things.

Microcomputers in Amateur Radio discusses the basic concepts involved. This book is the second in the series and presents software to be used in the amateur radio station in the number crunching and data processing modes. Each program is explained line by line so that anyone can figure out what is going on. In fact, putting these programs on a system and customizing them to add those tiny bells and whistles is a good way to learn BASIC. There are many different ways of doing different things, and although the same operation may crop up in more than one program, the technique used in one program may differ from that used to perform the same operation in a later program."

"Sounds good," said Fred "What about dialects?"

"Well, when you look at the listings you will note that the ":" or the "\" are used between statements on the same line. There are also differences in string handling, print statements, and disk operations. For example both versions of BASIC allow a short form of the PRINT statement. In one, it is a "!"; in the other it is a "?". There are also differences in the way functions are defined, and most important there is a major difference in the way the logic flow operates."

"For instance?" asked Fred.

"Well, consider the following statements" said Marc.

```
900 IF X = 1 THEN Y = 1 : GOTO 1000
910 REM CONTINUE HERE.
```

MICROSOFT Basic performs the test on X in line 900. If it fails the program flow branches to line 910. Northstar basic on the other hand performs the same test on X and if it fails does not set the value of Y to 1. It then continues after the separator. Thus MICROSOFT will only branch to line 1000 if the test is valid, whereas Northstar will always branch to line 1000 no matter what the value of X. This is the most difficult change to make because if it is fouled up the logic flows are wrong and the programs do not perform as expected."

"Sounds difficult," said Fred.

"Not really," Marc replied. "It's like anything else. It's simple once you get the hang of it. Most of the differences are explained in more detail in the manual that comes with the computer. Here—you can take these books if you want. Just return them next week, OK? I've got to go on duty again."

Marc put his cup down, and returned to the control point. Fred wandered around looking at the various displays, shaking his head.

"I can't believe the whole thing," he said to himself. He sat down, opened the books, and began to skim them. Within seconds he was engrossed. Slowly and surely it began to make sense. Fred was still sitting there hours later when the contest ended.

The sudden transition at the end of the contest brought Fred back to earth.

"How did it go?" he asked Marc.

"It's difficult to say. We have to check everything, but our qso rate is up on last year and the score is definately much better. After we've run the dupe program and double checked the log we'll be sure. Still, that chore will be a breeze. No matter what, we had a lot of fun both putting the equipment together and operating the contest, and now we'll take our equipment back to our homes and put them to use in the regular station. Using computers and software for amateur radio is great!"

Also by the Author from TAB Books Inc.

1305 Microcomputers in Amateur Radio

Computer Aided Design and Circuit Analysis

A microcomputer is an excellent tool to use when you design and analyze electronic circuits, analyze a circuit without using soldering irons, parts substitution boxes, multimeters, or oscilloscopes. This chapter describes only a few of the many ways that the computer can be used for circuit design and analysis. If a formula can be found to describe the operation of the circuit, the computer can analyze it. The depth of the analysis and the complexity of the circuit determine the sophistication of the program needed to perform the analysis. For professional applications, programs costing thousands of dollars perform sophisticated circuit analysis. For amateur applications, you can write simple programs yourself that analyze the circuits you use. To help you get started, a few examples are shown in this chapter.

BATTERY-CHARGING CIRCUIT

The computer can be used to examine the behavior of a battery charging circuit. The relationship between the parts of the circuit, shown in Fig. 1-1, conforms to Ohm's law and can be summarized as follows:

$$V_1 = [I \times R] + V_2$$

$$\text{or} \quad I = \frac{V_1 - V_2}{R}$$

Thus, the value of current for different input voltages and currents can be calculated. If V_2 is assumed to be equal to the voltage of the nickel-cadmium batteries being charged, and V_1 is the output voltage of the dc power supply, the value of charging current for different resistances for each value of input voltage can be calculated. Here's a simple program written in BASIC

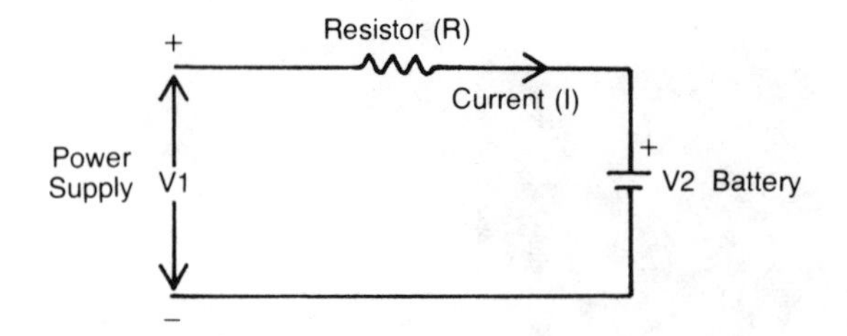

Fig. 1-1. Battery-charging circuit.

that prints the value of I (current) when the amount of resistance (R) in the circuit is 100 ohms.

```
10  V1 = 6 : v2 = 3.75
20  R = 100
30  I = (V1 − V2)/R
40  PRINT I,R
```

It works, but a pocket calculator could do the same thing.

The advantages of using a computer become evident when we start changing some of the variables. For example, we can consider the value of I for different values of R between 10 and 100 ohms by changing line 20 into a FOR/NEXT loop as follows:

```
10  V1 = 6 : V2 = 3.75
20  For R = 10 to 100 step 10
30  I = (V1 − V2)/R
40  PRINT I,R
50  NEXT
```

This prints the value of current (I) for resistance values of 10, 20, 30, 40, 50, 60, 70, 80, 90, and 100 ohms.

These programs assume that the input voltage is 6 volts. If the value of input voltage in line 10 is changed to 9, 12, or 15 volts (common dc power supply output voltages), the corresponding values of I and R can be obtained.

If a second FOR/NEXT loop for values of V_1 is set up, a table of the current values obtained using different voltages and resistances can be printed, as shown in Fig. 1-2.

TUNED CIRCUITS

Tuned circuits are used in oscillators and amplifiers. The resonant frequency of a tuned circuit is given by the formula

$$L = \frac{1}{2\pi}\sqrt{LC}$$

The most common type of variable-frequency oscillator (VFO) is one in which the inductance (L) is fixed and the capacitance is varied to change the frequency. A small program can be written to print out frequencies (F) as a function of capacitances (C) for particular inductances (L). One such program is shown in Fig. 1-3. The program requests an inductance (L) for a given run and then lists the frequencies that result

```
20 REM charging resistor calculator
30 V2 = 3.75
40  LPRINT "INPUT VOLTAGE",6,9,12,15
50 LPRINT
60 LPRINT "RESISTANCE",,,"CURRENT (mA)"
70 FOR R = 10 TO 100 STEP 10
80 LPRINT R,
90 FOR J = 1 TO 4
100 READ V1
110 V = V1 - V2
120 I = V / R
130 I1 = I * 1000
140 LPRINT I1,
150 NEXT
160 RESTORE
170 LPRINT
180 NEXT
190 DATA 6,9,12,15
```

INPUT VOLTAGE	6	9	12	15
RESISTANCE			CURRENT (mA)	
10	225	525	825	1125
20	112.5	262.5	412.5	562.5
30	75	175	275	375
40	56.25	131.25	206.25	281.25
50	45	105	165	225
60	37.5	87.5	137.5	187.5
70	32.1429	75	117.857	160.714
80	28.125	65.625	103.125	140.625
90	25	58.3333	91.6667	125
100	22.5	52.5	82.5	112.5

Fig. 1-2. Resistance charging values program.

```
1010 REM COMPUTE FREQUENCY OF AN OSCILLATOR
20 X = 50
30 Y = 500
40 Z = 25
50 INPUT "L = (uH) " ; L
60 L = L * .000001
70  LPRINT "L (uH) =" ; L * 1E+06
80 LPRINT : LPRINT
90 LPRINT "F (Mhz)","C (pF)"
100 FOR C1 = X TO Y STEP Z
110  C = C1 * 1E-12
120 F = 1 / (2*3.1419*(SQR(L*C)))
130 F1 = F * .000001
140 C2 = C * 1E+12
150 LPRINT F1,C2
160 NEXT
170 LPRINT CHR$(12)
180 GOTO 50
```

Fig. 1-3. Computing oscillator frequency program.

when capacitances between 50 and 500 pF (in steps of 25 pF) are used. Figure 1-4 shows sample listings for inductances of 0.5 and 1.0 μH.

L (uH) = .5		L (uH) = 1	
F (Mhz)	C (pF)	F (Mhz)	C (pF)
31.8278	50	22.5057	50
25.9873	75	18.3758	75
22.5057	100	15.9139	100
20.1297	125	14.2338	125
18.3758	150	12.9937	150
17.0127	175	12.0298	175
15.9139	200	11.2528	200
15.0038	225	10.6093	225
14.2338	250	10.0649	250
13.5714	275	9.59646	275
12.9937	300	9.18791	300
12.4839	325	8.82745	325
12.0298	350	8.50636	350
11.6219	375	8.21791	375
11.2528	400	7.95697	400
10.9169	425	7.7194	425
10.6093	450	7.5019	450
10.3263	475	7.30182	475
10.0649	500	7.11694	500

Fig. 1-4. Frequencies for L = 0.5 μH to 1.0 μH.

Most transceivers mix a VFO output with a crystal-controlled oscillator to get the injection frequency for the various amateur bands. These injection frequencies can be calculated using the program listed in Fig. 1-5. The program contains the band edge frequencies in data statements (lines 10-50). It first lists a header for the table to come (line 65), then, for each band (line 70), it reads the band edge frequencies (line 80) and prints the possible injection frequencies (line 90). There are two possible injection frequencies because the VFO can be above or below the i-f; namely, the injection frequency can be added to or subtracted from the i-f frequency to obtain the amateur band frequency. A listing generated by the program example is shown in Fig. 1-6. The negative values of the low oscillator columns for the 1.8, 3.5 and 7 MHz bands indicate that the tuning is reversed. For example, the 5.0 to 5.5 MHz range appears in the low OSC entries for the 3.5/4.0 MHz band. Using 9.0 MHz, a very common intermediate frequency for transceivers, we find that:

$$9.0 \text{ MHz} - 5.0 \text{ MHz} = 4.0 \text{ MHz}$$

$$9.0 \text{ MHz} - 5.5 \text{ MHz} = 3.5 \text{ MHz}$$

```
10 REM FREQUENCY PLOTTING ROUTINE
20 DATA 1.8,2, 3.5,4, 7,7.5
30 DATA 10,10.5,14,14.5,18,18.5
40 DATA 21,21.5,24,24.5
50 DATA 28,28.5,28.5,29,29,29.5,29.5,30
60 DATA 50,50.5
70 I = 9 : REM 9 BANDS
80 LPRINT "BAND","       LOW OSC",,"      HIGH OSC"
90 FOR X = 1 TO 13
100 READ F1,F2
110 LPRINT F1, F1-I,F2-I,F1+I,F2+I
120 NEXT
```

Fig. 1-5. Transceiver injection frequency calculator.

BAND	LOW OSC		HIGH OSC	
1.8	-7.2	-7	10.8	11
3.5	-5.5	-5	12.5	13
7	-2	-1.5	16	16.5
10	1	1.5	19	19.5
14	5	5.5	23	23.5
18	9	9.5	27	27.5
21	12	12.5	30	30.5
24	15	15.5	33	33.5
28	19	19.5	37	37.5
28.5	19.5	20	37.5	38
29	20	20.5	38	38.5
29.5	20.5	21	38.5	39
50	41	41.5	59	59.5

Fig. 1-6. Injection frequencies table.

A VFO frequency of 5.0 MHz corresponds to an actual frequency of 4.0 MHz, and 5.5 MHz corresponds to 3.5 MHz. On the 14/14.5 MHz range, however, the mixing is additive and 5.0 MHz corresponds to 14.0 MHz while 5.5 MHz corresponds to 14.5 MHz.

VFO's with frequencies between 10.5 and 27.5 MHz can be used as injection frequencies. If a linear or reasonably linear tuning range can be found, it may be possible to build one VFO, switching in fixed values of capacitance to get 500 kHz of tuning at the different bands. This is the bandset and bandspread technique commonly used on HF multiband portable-radio receivers.

Another program can test this concept. If we return to the basic resonant frequency program and listings of Figs. 1-3 and Fig. 1-4, and plot a graph of frequency against capacitance as shown in Fig. 1-7, the relationship can be displayed in a different manner.

The graph of Fig. 1-7 shows that the rate of change of frequencies is nonlinear, and that the curve for an inductance (L) of 0.5 μH is more linear than that for 1.0 μH. The slope of the curve can be calculated using the formula:

$$\text{Slope} = \frac{\text{Change in Frequency (F)}}{\text{Change in Capacity (C)}},$$

The program listed in Fig. 1-8 can tabulate the slope of the curve, and list frequencies and corresponding capacitances. Listings for various inductances are presented in Fig. 1-9.

Figure 1-9A shows that an inductance of 0.5 μH meets the criteria for the oscillator circuit identified by the frequency calculator program of Fig. 1-5. It also shows that the tuning rate is different on each band. It takes a total change of about 50 pF to cover 10 to 10.5 MHz, but only requires a change of less than 10 pF to cover 20 to 20.5 MHz. This analysis does not take anything into consideration apart from the effects of inductance and capacitance, but it would seem that one inductance cannot give a fairly constant tuning rate on each band. If an inductance of 0.5 μH is used for frequencies between 10 and 13 MHz, the oscillator provides coverage of the 1.5/2.0 MHz, 3.5/4.0 MHz, and 21.0/21.5 MHz bands. A different inductance is needed to provide sufficient bandspread for all amateur bands.

Let's consider what happens if the inductance (L) is increased. If the inductance is changed to 2.0 μH, the circuit has the characteristics shown in Fig. 1-9B. This circuit gives a reasonably linear tuning range of 5 to 5.5 MHz with a change in capacitance of about 100 pF. If the inductance is 16 μH, the characteristics of the circuit become those shown in Fig. 1-9C. This circuit makes a good in-band VFO for 160 Meters (1.8 to 2.0 MHz) because the band is covered in a 100 pF range and the tuning rate is slow and very linear.

If the inductance is set to 180 μH, a tuning range of 530 kHz to 1.7 MHz is achieved, as shown in Fig. 1-9D. This figure clearly shows why the tuning on broadcast band radios is very nonlinear.

We've only discussed the VFO aspects of a multiband HF radio transmitter or receiver here; most of the other problems involved have been left out. Our goal has been to obtain a reasonably linear tuning range over a wide range of capacitance. Of course, it's much easier to get a linear tuning range if the amount of capacitance change is kept small.

CRYSTAL FREQUENCIES

The computer can also be used to calculate crystal frequencies. Let's consider the case of a two meter FM receiver and calculate the transmit and receive frequencies for the US repeater channels between 146.61 MHz and 147.00 MHz using the program listed in Fig. 1-10.

The channel spacing of 30 kHz (line 20) and the offset of 600 kHz (line 40) are set up first. F1, F2, and F4 are the transmit frequencies of crystals in the 18, 12, and 8 MHz

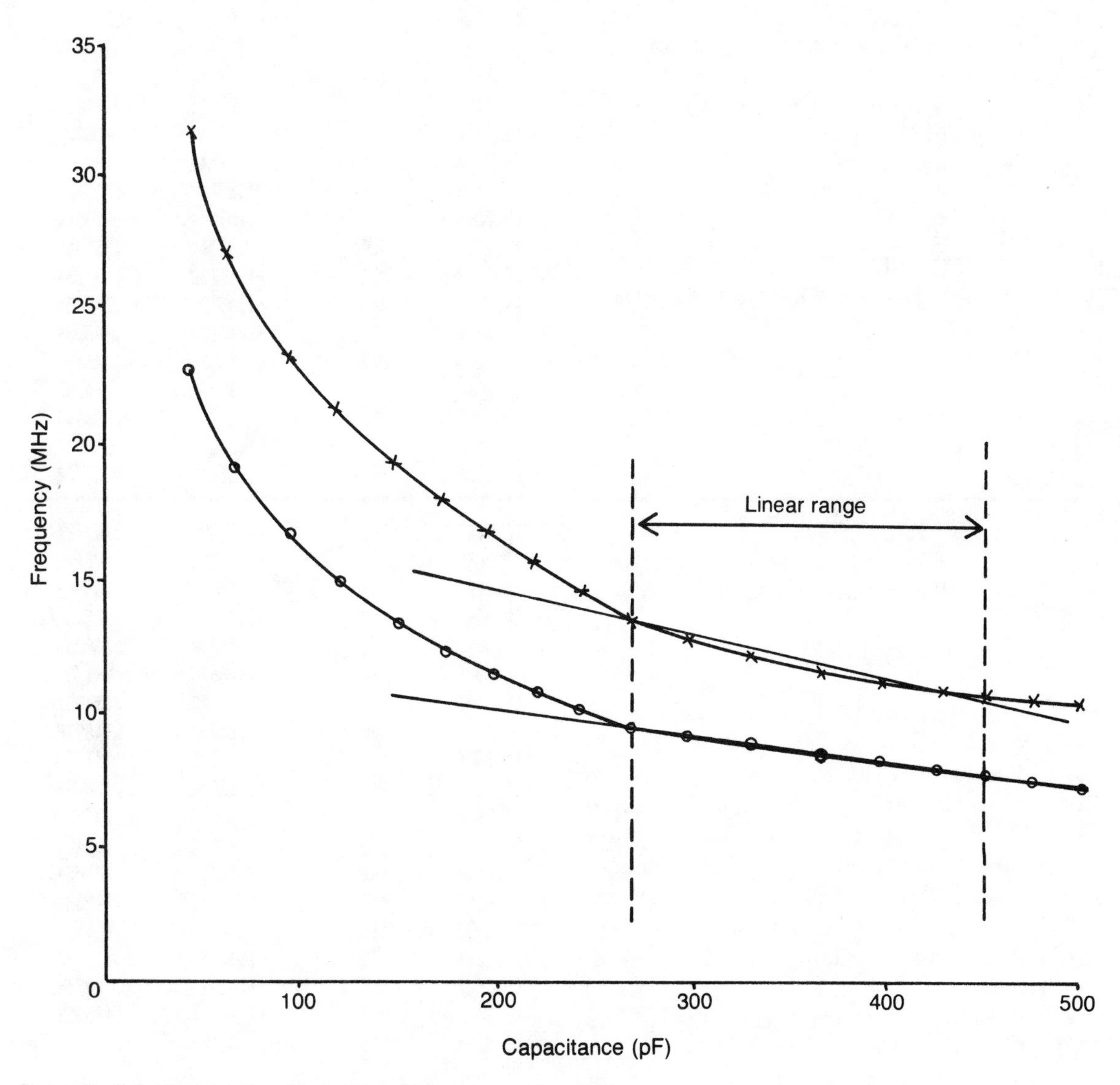

Fig. 1-7. Graphical relationship between frequency and capacitance.

```
10 REM COMPUTE OSCILLATOR FREQUENCY
20 X = 50  : REM MINIMUM VALUE OF CAPACITANCE
30 Y = 500 : REM MAXIMUM VALUE OF CAPACITANCE
40 Z = 10  : REM STEP
50 INPUT "L (uH) ";L
60 C3 = 0 : F3 = C3 : F4 = F3
70 LPRINT "L (uH) = ";L
80 L = L * .000001 : REM CONVERT uH TO H
90 LPRINT
100 LPRINT "F (Mhz)", "C (pF)", "SLOPE","DELTA F"
110 FOR C1 = X TO Y STEP Z
120 C = C1 * 1E-12 : REM CONVERT pF TO F
130 REM
140 F = 1 / ( 2 * 3.1419 * (SQR( L * C )))
150 IF C1 = X THEN F5 = F
160 F1 = F * .000001
170 F4 = F3 - F1
180 C2 = C * 1E+12
190 M = (F3 - F2) / (C3 - C2)
200 IF M = 0 THEN F4 = 0 : REM FUDGE FIRST VALUE
210 LPRINT F1,C2,M,F4
220 F3 = F1
230 C3 = C2
240 NEXT
250 F5 = F5 / F
260 LPRINT
270 LPRINT "TUNING RATIO IS";F5
```

Fig. 1-8. Program to calculate incremental changes in frequency.

ranges (lines 80 and 100). The receiver frequency is F3 (line 90). The computation is controlled by the FOR/NEXT loop setup (lines 70 to 120) and the listing is generated (line 110), as shown in Fig. 1-11. Figure 1-11 shows that the change of fundamental frequency is small. This explains why the crystals have to be tweaked up on frequency. Since the crystal frequency is dependent on the capacitance of the circuit, a crystal marked *146.85 MHz* may generate 146.82 MHz if it's installed in a transmitter with a different value of resonant capacitance in the oscillator circuit.

A

L (uH) = .5

F (Mhz)	C (pF)	SLOPE	DELTA F
31.8278	50	0	0
29.0548	60	-3.18278	2.77305
26.8995	70	-2.90548	2.15529
25.1621	80	-2.68995	1.73733
23.7231	90	-2.51622	1.43905
22.5057	100	-2.37231	1.2174
21.4584	110	-2.25057	1.04734
20.5448	120	-2.14584	.91354
19.7388	130	-2.05448	.806004
19.0208	140	-1.97388	.718039
18.3758	150	-1.90208	.644943
17.7923	160	-1.83758	.583508
17.2611	170	-1.77923	.53126
16.7748	180	-1.72611	.486282
16.3273	190	-1.67748	.447434
15.9139	200	-1.63273	.41342
15.5304	210	-1.59139	.383521
15.1734	220	-1.55304	.357057
14.8398	230	-1.51734	.333534
14.5274	240	-1.48398	.312437
14.2338	250	-1.45274	.293539
13.9575	260	-1.42338	.276392
13.6965	270	-1.39575	.260907
13.4497	280	-1.36965	.246804
13.2158	290	-1.34497	.233939
12.9937	300	-1.32158	.222122
12.7824	310	-1.29937	.211295
12.5811	320	-1.27824	.201324
12.389	330	-1.25811	.19206
12.2054	340	-1.2389	.183564
12.0298	350	-1.22054	.17563
11.8615	360	-1.20298	.168256
11.7002	370	-1.18615	.161386
11.5452	380	-1.17002	.154981
11.3962	390	-1.15452	.148978
11.2528	400	-1.13962	.14337
11.1148	410	-1.12528	.138053
10.9816	420	-1.11148	.133137
10.8532	430	-1.09816	.128429
10.7292	440	-1.08532	.124034
10.6093	450	-1.07292	.119881
10.4933	460	-1.06093	.115963
10.3811	470	-1.04933	.112237
10.2724	480	-1.03811	.108703
10.167	490	-1.02724	.105359
10.0649	500	-1.0167	.102178

TUNING RATIO IS 3.16227

Fig. 1-9. Tables of changes in frequency for fixed values of inductance.

B

L (uH) = 2

F (Mhz)	C (pF)	SLOPE	DELTA F
15.9139	50	0	0
14.5274	60	-1.59139	1.38655
13.4497	70	-1.45274	1.07764
12.5811	80	-1.34497	.868681
11.8615	90	-1.25811	.71951
11.2528	100	-1.18615	.608715
10.7292	110	-1.12528	.523653
10.2724	120	-1.07292	.456783
9.86939	130	-1.02724	.403001
9.5104	140	-.986939	.358994
9.18791	150	-.95104	.322496
8.89616	160	-.918791	.291742
8.63056	170	-.889616	.265607
8.38738	180	-.863057	.243176
8.16367	190	-.838738	.223716
7.95697	200	-.816367	.206699
7.7652	210	-.795697	.191761
7.58667	220	-.77652	.178539
7.4199	230	-.758667	.166766
7.26369	240	-.74199	.156209
7.11694	250	-.726369	.146751
6.97873	260	-.711694	.138215
6.84827	270	-.697873	.130454
6.72486	280	-.684827	.123411
6.6079	290	-.672486	.116961
6.49684	300	-.66079	.111061
6.39118	310	-.649684	.105656
6.29054	320	-.639118	.100646
6.19449	330	-.629054	.0960474
6.10272	340	-.619449	.0917735
6.0149	350	-.610272	.0878148
5.93077	360	-.601492	.0841355
5.85008	370	-.593077	.0806856
5.77259	380	-.585008	.0774903
5.69809	390	-.577259	.0744972
5.62642	400	-.569809	.0716691
5.55739	410	-.562642	.0690341
5.49083	420	-.555739	.0665612
5.4266	430	-.549083	.0642285
5.36459	440	-.54266	.0620103
5.30465	450	-.536459	.0599403
5.24667	460	-.530465	.0579815
5.19056	470	-.524667	.0561051
5.1362	480	-.519056	.054358
5.08352	490	-.51362	.0526862
5.03243	500	-.508352	.0510888

TUNING RATIO IS 3.16228

C

L (uH) = 16

F (Mhz)	C (pF)	SLOPE	DELTA F
5.62642	50	0	0
5.1362	60	-.562642	.490219
4.7552	70	-.51362	.381004
4.44807	80	-.47552	.307126
4.19369	90	-.444808	.254385
3.97848	100	-.419369	.215208
3.79334	110	-.397848	.185144
3.63184	120	-.379334	.161498
3.48936	130	-.363184	.142483
3.36244	140	-.348936	.126919
3.24842	150	-.336244	.11402
3.14527	160	-.324842	.103151
3.05135	170	-.314527	.0939145
2.96539	180	-.305136	.0859673
2.8863	190	-.296539	.0790918
2.81321	200	-.28863	.0730832
2.74541	210	-.281321	.0677977
2.68229	220	-.274541	.0631228
2.62333	230	-.268229	.0589576
2.5681	240	-.262333	.0552316
2.51621	250	-.25681	.0518904
2.46735	260	-.251621	.0488601
2.42123	270	-.246735	.0461223
2.3776	280	-.242123	.0436292
2.33625	290	-.23776	.0413547
2.29698	300	-.233625	.0392666
2.25963	310	-.229698	.0373516
2.22404	320	-.225963	.0355835
2.19008	330	-.222404	.0339611
2.15764	340	-.219008	.0324469
2.12659	350	-.215764	.0310471
2.09684	360	-.21266	.0297465
2.06832	370	-.209684	.0285268
2.04092	380	-.206832	.0273969
2.01458	390	-.204092	.0263388
1.98924	400	-.201458	.0253414
1.96483	410	-.198924	.0244046
1.9413	420	-.196483	.0235356
1.9186	430	-.19413	.0227029
1.89667	440	-.19186	.0219263
1.87548	450	-.189667	.0211924
1.85498	460	-.187548	.0204994
1.83514	470	-.185498	.019841
1.81592	480	-.183514	.0192114
1.7973	490	-.181592	.0186297
1.77923	500	-.17973	.0180625

TUNING RATIO IS 3.16228

Fig. 1-9. Tables of changes in frequency for fixed values of inductance. (continued)

L (uH) = 180

D

F (Mhz)	C (pF)	SLOPE	DELTA F
1.67748	50	0	0
1.53132	60	-.167748	.146157
1.41772	70	-.153132	.113595
1.32616	80	-.141772	.0915614
1.25032	90	-.132616	.0758449
1.18615	100	-.125032	.0641626
1.13095	110	-.118615	.0552009
1.08281	120	-.113095	.0481478
1.04033	130	-.108281	.0424787
1.00248	140	-.104033	.0378439
.968492	150	-.100248	.0339916
.937738	160	-.0968492	.0307535
.90974	170	-.0937738	.0279987
.884108	180	-.0909741	.0256317
.860527	190	-.0884108	.0235806
.838738	200	-.0860527	.0217892
.818525	210	-.0838738	.0202134
.799705	220	-.0818525	.0188196
.782127	230	-.0799705	.0175778
.76566	240	-.0782127	.0164678
.750191	250	-.076566	.0154688
.735623	260	-.0750191	.0145683
.721871	270	-.0735623	.0137519
.708863	280	-.0721871	.0130078
.696534	290	-.0708863	.0123287
.684827	300	-.0696534	.011707
.673691	310	-.0684827	.0111362
.663081	320	-.0673691	.0106098
.652957	330	-.0663081	.0101243
.643283	340	-.0652957	9.67378E-03
.634027	350	-.0643283	9.25642E-03
.625159	360	-.0634029	8.86792E-03
.616653	370	-.0625159	8.50588E-03
.608485	380	-.0616653	.0081681
.600633	390	-.0608485	7.85196E-03
.593077	400	-.0600633	7.55537E-03
.5858	410	-.0593077	7.27761E-03
.578784	420	-.05858	7.01541E-03
.572015	430	-.0578784	.0067696
.565477	440	-.0572015	6.53786E-03
.559159	450	-.0565477	6.31833E-03
.553048	460	-.0559159	6.11097E-03
.547132	470	-.0553048	5.91546E-03
.541403	480	-.0547132	5.72914E-03
.53585	490	-.0541403	5.55289E-03
.530465	500	-.053585	5.38522E-03

TUNING RATIO IS 3.16228

Fig. 1-9. Tables of changes in frequency for fixed values of inductance. (continued)

```
10 REM 2 METER FREQUENCY TEST
20 I = .03
30 REM I = CHANNEL SPACING
40 O1 = .6 : REM 600 kHz LOW
50 LPRINT "2M FREQ","REPEATER INPUTS ( - 600 kHz)"
60 LPRINT,"TX FUND","TX FUND","TX FUND","RX FUND"
70 FOR F = 146.61 TO 147 STEP I
80 F1 = (F - O1)/8 : F2 = (F - O1)/12
90 F3 = (F - 10.7)/3
100 F4 = (F - O1)/18
110 LPRINT F,F1,F2,F4,F3
120 NEXT
```

Fig. 1-10. 2 meter fundamental frequency calculator.

2M FREQ	REPEATER INPUTS (- 600 kHz)			
	TX FUND	TX FUND	TX FUND	RX FUND
146.61	18.2513	12.1675	8.11167	45.3033
146.64	18.255	12.17	8.11333	45.3133
146.67	18.2588	12.1725	8.115	45.3233
146.7	18.2625	12.175	8.11667	45.3333
146.73	18.2662	12.1775	8.11833	45.3433
146.76	18.27	12.18	8.12	45.3533
146.79	18.2737	12.1825	8.12167	45.3633
146.82	18.2775	12.185	8.12333	45.3733
146.85	18.2812	12.1875	8.125	45.3833
146.88	18.285	12.19	8.12667	45.3933
146.91	18.2887	12.1925	8.12833	45.4033
146.94	18.2925	12.195	8.13	45.4133
146.97	18.2962	12.1975	8.13167	45.4233
147	18.3	12.2	8.13333	45.4333

Fig. 1-11. Listing of transmit and receive frequency crystals.

UPLINK	XTALS		
145.85	18.2313	12.1542	8.10278
145.87	18.2338	12.1558	8.10389
145.89	18.2363	12.1575	8.105
145.91	18.2388	12.1592	8.10611
145.93	18.2413	12.1608	8.10722
145.95	18.2438	12.1625	8.10833
145.97	18.2463	12.1642	8.10945
145.99	18.2488	12.1658	8.11056

Fig. 1-12. Listing of fundamental frequencies producing 145.85-146 MHz.

```
10 REM OSCAR UPLINK FREQUENCY PLOTTER
20 I = .02
30 REM GAP BETWEEN SAMPLES
40 LPRINT " UPLINK",,"XTALS"
50 LPRINT "-----------------------------------------------"
60 FOR F = 145.85 TO 146 STEP I
70 F1 = F/8 : F2 = F/12 : F3 = F/18
80 LPRINT F,F1,F2,F3
90 NEXT
```

Fig. 1-13. Program to generate fundamental frequencies producing 145.85-146 MHz.

Since frequencies can be pulled a little, let's consider the OSCAR satellite Mode A/J uplink band. Figure 1-12 shows a listing of the fundamental crystal frequencies that can be used to generate signals between 145.85 and 146 MHz. Crystals at 18.24 MHz, 12.16 MHz, or 8.107 MHz, could probably be tweaked to cover most of the passband. The program that generated the table is shown in Fig. 1-13. Figures 1-14 and 1-15 show a similar table and program for examining the 435 to 436 MHz passband to be used on Mode B and in future OSCAR satellites. (Note the "round-off error" at the 436 MHz end of the table.)

```
10 REM OSCAR MODE B UPLINK FREQUENCY PLOTTER
30 I = .02
40 REM GAP BETWEEN SAMPLES
50 LPRINT "UPLINK",,"XTALS"
60 FOR F = 435 TO 436 STEP I
70 F1 = F/24 : F2 = F/36 : F3 = F/(18*3)
80 LPRINT F,F1,F2,F3
90 NEXT
```

Fig. 1-14. Listing of fundamental frequencies producing 435-436 MHz.

UPLINK		XTALS	
435	18.125	12.0833	8.05556
435.02	18.1258	12.0839	8.05593
435.04	18.1267	12.0844	8.0563
435.06	18.1275	12.085	8.05667
435.08	18.1283	12.0856	8.05704
435.1	18.1292	12.0861	8.05741
435.12	18.13	12.0867	8.05778
435.14	18.1308	12.0872	8.05815
435.16	18.1317	12.0878	8.05852
435.18	18.1325	12.0883	8.05889
435.2	18.1333	12.0889	8.05926
435.22	18.1342	12.0894	8.05963
435.24	18.135	12.09	8.06
435.26	18.1358	12.0906	8.06037
435.28	18.1367	12.0911	8.06074
435.3	18.1375	12.0917	8.06111
435.32	18.1383	12.0922	8.06148
435.34	18.1392	12.0928	8.06185
435.36	18.14	12.0933	8.06222
435.38	18.1408	12.0939	8.06259
435.4	18.1417	12.0944	8.06296
435.42	18.1425	12.095	8.06333
435.44	18.1433	12.0956	8.0637
435.46	18.1442	12.0961	8.06407
435.48	18.145	12.0967	8.06444
435.5	18.1458	12.0972	8.06481
435.52	18.1467	12.0978	8.06518
435.54	18.1475	12.0983	8.06555
435.56	18.1483	12.0989	8.06592
435.58	18.1492	12.0994	8.06629
435.6	18.15	12.1	8.06666
435.62	18.1508	12.1005	8.06703
435.64	18.1517	12.1011	8.0674
435.66	18.1525	12.1017	8.06777
435.68	18.1533	12.1022	8.06814
435.7	18.1542	12.1028	8.06851
435.72	18.155	12.1033	8.06888
435.74	18.1558	12.1039	8.06925
435.76	18.1566	12.1044	8.06962
435.78	18.1575	12.105	8.06999
435.8	18.1583	12.1055	8.07036
435.82	18.1591	12.1061	8.07073
435.84	18.16	12.1067	8.0711
435.86	18.1608	12.1072	8.07147
435.88	18.1616	12.1078	8.07184
435.9	18.1625	12.1083	8.07221
435.92	18.1633	12.1089	8.07258
435.94	18.1641	12.1094	8.07295
435.96	18.165	12.11	8.07332
435.979	18.1658	12.1105	8.07369
435.999	18.1666	12.1111	8.07407

Fig. 1-15. Program to generate fundamental frequencies producing 435-436 MHz.

Logging

Logging is one area in which the computer shines. A log contains all sorts of useful information. The usual information—date, band, and callsign—can be augmented by any other data you care to add—name and town of the other operator, or which antenna you used. This log information can be stored on disk and accessed in many different ways.

A basic logging program performs the following functions:

- □ Allocates disk space for logs.
- □ Allows data to be entered into the log.
- □ Allows data to be read from the log.
- □ Allows entries in the log to be edited.
- □ Allows different logs to be combined into one.
- □ Allows logs to be deleted from the disk.
- □ Allows logs to be renamed.
- □ Allows logs to be recovered if a problem occurs while information is being entered.
- □ Allows QSL cards to be printed using the information in the log. All these operations can be done by one large program or by a number of small programs that are used together. Since microcomputers often have limited amounts of memory, using a number of programs with the overlay approach is presented in this chapter.

Logs can be stored on disks either as random files or as sequential files. Random files are faster to access, but use up a fixed amount of disk space for each entry no matter how large that particular entry is. Sequential files take longer to access, especially if they are full, but they generally use up less space per entry because the space in each entry is determined by the size of that entry. A sequential access technique is used in these programs, because the increased access time is not a problem here, and the more efficient data storage is a distinct advantage.

This chapter presents a logging "package" that consists of a number of separate programs and an executive or system program that allows each operator to be performed by typing in its name.

The package uses a common subroutine library that is added to each program as needed. This way it can be debugged only once, and then used over and over again in different programs.

Having the system program means that when the program name is entered, the SYSTEM program automatically loads and runs that program. Also, the program allows the automatic selection of a disk drive for command execution. Each disk drive is searched sequentially (beginning with the first drive) until a program in BASIC is found. Then each program is loaded and executed. This way, if the first drive fails for some reason, the system continues to operate from the second, third, and fourth drives (if they are present).

STORING LOGS ON A FLOPPY DISK

When space is allocated for a log using Northstar BASIC, two files are created. One is the actual log file that carries the data, and the second is a pointer file that begins with the prefix "*." For example, a log file for the 1981 CQWPX contest could have been called "CQWPX81." Its corresponding pointer file would automatically be named "*CQWPX81." The pointer file is a minimum size, and only contains the total number of entries in the main log file.

Logs can be stored on a floppy disk in either sequential or random formats. The random format allows access to any record in a minimum time frame, but uses up more space on the disk than the sequential method. Since floppy disks only contain between 100K and 1M bytes of data the sequential storage technique was chosen for use here although response time when using large log data files will be slow. A sample log data file listing is shown in Fig. 2-1.

THE LOGGING PACKAGE

The package contains a number of commands as shown in Fig. 2-2. Each command is actually the name of a program. A system program (listed in Fig. 2-3) loads and executes the named command. This technique allows any BASIC program (Type 2 file) to be executed by name in the same way (as far as the user is concerned) as CP/M commands. A menu feature to show the user which commands and data files are on which disk is handled by HELP FILES.

SYSTEM. The SYSTEM program listed in Fig. 2-3 outputs a prompt character (>) to the console and awaits an input. The input is assumed to be a Type 2 file or BASIC language program. The FOR . . . NEXT loop (lines 60 through 80) scans the directory on each floppy disk sequentially until a match is found. When one is found it is CHAINED (i.e., loaded and executed). If one is not found, an error message is displayed (line 80) and the SYSTEM file is reloaded from a disk. Extensive error checking is employed to make sure that if Drive 1 fails the programs will still continue.

FILES. The FILES program listed in Fig. 2-4 asks which drive to interrogate. The directory file "<*>" is opened and each entry is read. If the entry is a type 3 file (by convention, a data file that can be accessed by BASIC programs) its name is displayed on the screen. If it is not a type 3 file, the next entry is read in. Again, extensive error-checking is performed in this program.

Any file whose name begins with an asterisk (*) is not listed. This allows invis-

STATION LOG G3ZCZ/4X PAGE 1 FILE 1981/2

QSO #	DATE	TIME	RAND	STATION	S	R	MODE	PWR	QSL	COMMENTS
1	30 Aug 81	1429	20.	UK3XAM	59	58	SSB	200	--	
2	30 Aug 81	1436	20.	UK5MCU	57	59	SSB	200	--	
3	30 Aug 81	1440	20.	LZ1XI	59	59	SSB	200	--	
4	30 Aug 81	1443	20.	I5TDJ	59	59	SSB	200	--	Pete
5	30 Aug 81	1509	20.	DK9ZP	59	59	SSB	200	--	
6	30 Aug 81	1519	20.	SM0AVK	59	59	SSB	200	--	
7	30 Aug 81	1629	15.	YB2DI	53	52	SSB	200	--	Java
8	30 Aug 81	1819	20.	GJ3VLX	59	59	SSB	200	--	(G3VLX)
9	30 Aug 81	1823	20.	F0GMM	59	59	SSB	200	--	(DF7NJ)
10	30 Aug 81	1913	2.	4Z4ZB	59	59	FM	2	--	R9
11	30 Aug 81	1939	15.	CT2CR	57	57	SSB	200	--	
12	31 Aug 81	0510	20.	ZL3HH	59	55	SSB	200	--	
13	31 Aug 81	2030	20.	FG7TD/FS7	55	55	SSB	200	--	Q=F6AZN
14	31 Aug 81	0749	10.	UA4CCF	57	59	SSB	200	--	
15	31 Aug 81	0811	10.	UA3DQE	53	59	SSB	200	--	
16	31 Aug 81	0834	10.	SN0WPC	59	59	SSB	200	--	Q=SP3AUZ
17	31 Aug 81	0917	2.	4X6DD	59	59	FM	2	--	
18	31 Aug 81	0925	10.	DL5RBG	32	53	SSB	200	--	
19	31 Aug 81	0920	10.	DL2NAI	53	55	SSB	200	--	Israel
20	31 Aug 81	0946	10.	UA9WGJ	55	56	SSB	200	--	
21	31 Aug 81	1013	15.	YU4EBL	59	59	SSB	200	--	
22	31 Aug 81	1055	15.	W4YLU	55	55	SSB	200	--	
23	31 Aug 81	1153	10.	SK0EJ	53	58	SSB	200	--	
24	31 Aug 81	1230	10.	EA7CEL	53	53	SSB	200	--	
25	1 Sep 81	1251	10.	DH1EAG	53	52	SSB	200	--	
26	1 Sep 81	1330	20.	I0UXK	59	59	SSB	200	--	
27	1 Sep 81	1354	10.	YO3AJN	53	53	SSB	200	--	
28	1 Sep 81	1743	2.	4X4IL	59	59	FM	2	--	
29	1 Sep 81	1754	2.	4Z4AQ/P	59	59	FM	2	--	R9
30	1 Sep 81	2210	20.	K3STM	59	55	SSB	200	--	patch Marc
31	2 Sep 81	1615	2.	4Z4ZB	59	59	FM	2	--	
32	2 Sep 81	1622	2.	4X6CQ	59	59	FM	2	--	
33	2 Sep 81	1622	2.	4Z4JS	59	59	FM	2	--	
34	2 Sep 81	1859	15.	HB9XH	55	57	SSB	200	--	
35	3 Sep 81	1114	10.	DJ1ZU	55	58	SSB	200	--	
36	4 Sep 81	1301	10.	PA3AJG	57	55	SSB	200	--	
37	4 Sep 81	1321	10.	PA3BNT	52	51	SSB	200	--	
38	4 Sep 81	1331	10.	G4BYK	53	55	SSB	200	--	
39	4 Sep 81	1357	10.	SM6JNT	53	54	SSB	200	--	
40	4 Sep 81	1402	10.	UQ2GHT	55	59	SSB	200	--	
41	4 Sep 81	1406	10.	UA4HFG	55	57	SSB	200	--	
42	4 Sep 81	1812	20.	DL0RC/P	55	55	SSB	200	--	1

Fig. 2-1. Sample log listing.

```
43  4 Sep 81 1836  15.   DF9ZP/P    58 58  SSB  200  --  2
44  7 Sep 81 1249  10.   DL1YAL     55 59  SSB  200  --
45  7 Sep 81 1331  10.   PA6KEI     55 55  SSB  200  --  (PA3BKX)
46  7 Sep 81 1340  20.   SV0BL      59 59  SSB  200  --  Q=K9QXY
47  7 Sep 81 1459  20.   UA3AOV     59 59  SSB  200  --
48  7 Sep 81 1536  10.   YU3DBA     57 59  SSB  200  --
49  7 Sep 81 1748  20.   HA4XH      59 59  SSB  200  --
50  7 Sep 81 1754   2.   4Z4ZB      59 59  FM     2  --  R9
51  7 Sep 81 1757   2.   4X4JW      59 59  FM     2  --
52  7 Sep 81 1803   2.   4Z4MK      59 59  FM    30  --  S17
53  7 Sep 81 2217  20.   W2VOX      59 59  SSB  200  --
54  7 Sep 81 2217  20.   4Z4US      59 59  SSB  200  --
55 11 Sep 81 1405   2.   4X6CQ      59 59  FM    30  --  R9
56 11 Sep 81 1408   2.   4Z4ZB      59 59  FM    30  --
57 11 Sep 81 1410  10.   Y38UL      55 58  SSB  200  --
58 12 Sep 81 1705  15.   YU2CKH     59  0  SSB  200  --  5937
59 12 Sep 81 1707  15.   OE6HZG     55  0  SSB  200  --  59343
60 12 Sep 81 1709  15.   HA5KKG     59  0  SSB  200  --  59449
61 12 Sep 81 1711  15.   OK3JW      59  0  SSB  200  --  5927
62 12 Sep 81 1711  15.   OE3JSA     59  0  SSB  200  --  59117
63 12 Sep 81 1712  15.   G4CVZ      59  0  SSB  200  --  59117
64 12 Sep 81 1713  15.   G2CKQ      59  0  SSB  200  --  5902
65 12 Sep 81 1719  20.   LZ2TG      59  0  SSB  200  --  5918
66 12 Sep 81 1720  20.   Y46XF      55  0  SSB  200  --  5976
67 12 Sep 81 1723  20.   EA2ABJ     59  0  SSB  200  --  5931
68 12 Sep 81 1724  20.   UA3FT      55  0  SSB  200  --  5906
69 12 Sep 81 1727  20.   EA2YD      55  0  SSB  200  --  5955
70 12 Sep 81 1731  15.   GM5AIW     59  0  SSB  200  --  5958
71 12 Sep 81 1736  15.   SP6IXF     59  0  SSB  200  --  5942
72 12 Sep 81 1738  15.   I5RCR      59  0  SSB  200  --  5913
73 12 Sep 81 1746  15.   UK2BBB     59  0  SSB  200  --  59893
74 12 Sep 81 1753  15.   DK3GI      59  0  SSB  200  --  591056
75 12 Sep 81 1756  15.   OK1AGN     59  0  SSB  200  --  59503
76 12 Sep 81 1755  15.   OK2BSA     59  0  SSB  200  --  59024
77 12 Sep 81 1757  15.   HA7TM      59  0  SSB  200  --  5907
78 12 Sep 81 1802  15.   HA7KPL     59  0  SSB  200  --  59563
79 12 Sep 81 1803  15.   Y44ZI/P    59  0  SSB  200  --  59861
80 12 Sep 81 1804  15.   SP8GTS     59  0  SSB  200  --  59123
81 12 Sep 81 1807  20.   I0JU       59  0  SSB  200  --  5913
82 12 Sep 81 1808  20.   PA0QX      59  0  SSB  200  --  5920
83 12 Sep 81 1809  20.   YU1KQ      59  0  SSB  200  --  59119
84 12 Sep 81 1811  20.   IX1BGJ     59  0  SSB  200  --  59331
85 12 Sep 81 1929  20.   YO6AWR/P   59  0  SSB  200  --  59233
86 12 Sep 81 1930  20.   UK5MAF     59  0  SSB  200  --  59876
87 12 Sep 81 1931  15.   UK5MAF     53  0  SSB  200  --  59877
```

Fig. 2-1. Sample log listing. (continued)

```
 88 12 Sep 81 1935  20.  YU4EAW      59   0  SSB  200  --  59468
 89 12 Sep 81 1939  20.  DJ9MT       59   0  SSB  200  --  59452
 90 12 Sep 81 1941  20.  HG5A        59   0  SSB  200  --  591113
 91 13 Sep 81 1629  15.  CT2CQ       59   0  SSB  200  --  59698
 92 13 Sep 81 1633  15.  OK3CFP      59   0  SSB  200  --  59125
 93 13 Sep 81 1636  15.  UK2BCR      59   0  SSB  200  --  59470
 94 13 Sep 81 1641  15.  HA5KKC/7    59   0  SSB  200  --  591549
 95 13 Sep 81 1642  15.  SP8ECV      59   0  SSB  200  --  59325
 96 13 Sep 81 1809  15.  DF0DX       59   0  SSB  200  --  571244
 97 13 Sep 81 1813  15.  SP2PDI      59   0  SSB  200  --  591629
 98 13 Sep 81 1816  15.  SN0WPC      59   0  SSB  200  --  591298
 99 13 Sep 81 1817  15.  DJ9KH       59   0  SSB  200  --  591192
100 13 Sep 81 1819  15.  DL6WT       59   0  SSB  200  --  59936
101 13 Sep 81 1840  15.  DA2DC       57   0  SSB  200  --  59806
102 13 Sep 81 1905  15.  G3IOR       59  59  SSB  200  --  <*>
103 13 Sep 81 1910  15.  WA2LQQ      41  11  SSB  200  --  <*>
104 13 Sep 81 1947  15.  4X4GI       59  59  SSB  200  --  <*>
105 13 Sep 81 2014   2.  4X4GI       59  59  FM    30  --  R7
106 14 Sep 81 0326  20.  K3STM       57  53  SSB  200  --
107 14 Sep 81 0355  20.  WA4WTG      57  53  SSB  200  --
108 14 Sep 81 1650   2.  4X6AS       59  59  FM    30  --  R9
109 16 Sep 81 1515  15.  JA3BOA      59  55  SSB  200  --
110 16 Sep 81 2300  20.  WA2UND/MM3  53  53  SSB  200  --
111 16 Sep 81 2300  20.  W2VLX       55  59  SSB  200  --
112 16 Sep 81 2300  20.  K3STM       59  55  SSB  200  --
113 17 Sep 81 0735  20.  4N2DX       59  59  SSB  200  --  <YU2DX>
114 17 Sep 81 0751  10.  IZ5ARI      59  59  SSB  200  --  Q=I5HCH
115 17 Sep 81 2155   2.  4Z4US       59  59  FM     2  --  R9
116 17 Sep 81 2200  20.  K3STM       59  55  SSB  200  --  QSP Fay's parents
117 17 Sep 81 2200  20.  4X4GI       53  53  SSB  200  --
118 18 Sep 81 1152   2.  4X4IK       59  59  FM     2  --
119 19 Sep 81 1714   2.  N4DII/4X    59  59  FM     2  --
120 19 Sep 81 1716   2.  4Z4ZA       59  59  FM     2  --
121 19 Sep 81 1717   2.  4X4IO       59  59  FM     2  --  QSP de K3STM
122 19 Sep 81 1742  15.  DF5CW       59  59  SSB  200  --
123 20 Sep 81 0450  20.  4Z4US       59  59  SSB  200  --
124 20 Sep 81 0508  20.  K3JW        59  56  SSB  200  --  QSP to Fay's parents
125 20 Sep 81 1900   2.  GI4FUM/4X   59  59  FM    30  --  S20
126 20 Sep 81 1910  15.  F8ZS        55  55  SSB  200  --
127 24 Sep 81 0418  20.  K3JW        59  55  SSB  200  --
128 24 Sep 81 0420  20.  4X4US       59  59  SSB  200  --
129 24 Sep 81 0420   2.  4Z4ZB        0   0  FM     2  --
130  1 Oct 81 1925   2.  ZC4ESB       0   0  FM     2  --  <ZC4NB> R9
131  1 Oct 81 1932   2.  ZR6AX/4X     0   0  FM     2  --
```

Fig. 2-1. Sample log listing. (continued)

```
STATION LOG  G3ZCZ/4X            PAGE 1            FILE 1981/2
QSO #    DATE     TIME BAND     STATION     S   R  MODE  PWR  QSL  COMMENTS
   10 30 Aug 81 1913   2.      4Z4ZB       59  59  FM      2  --   R9
   17 31 Aug 81 0917   2.      4X6DD       59  59  FM      2  --
   28  1 Sep 81 1743   2.      4X4IL       59  59  FM      2  --
   29  1 Sep 81 1754   2.      4Z4AQ/P     59  59  FM      2  --   R9
   31  2 Sep 81 1615   2.      4Z4ZB       59  59  FM      2  --
   32  2 Sep 81 1622   2.      4X6CQ       59  59  FM      2  --
   33  2 Sep 81 1622   2.      4Z4JS       59  59  FM      2  --
   50  7 Sep 81 1754   2.      4Z4ZB       59  59  FM      2  --   R9
   51  7 Sep 81 1757   2.      4X4JW       59  59  FM      2  --
   52  7 Sep 81 1803   2.      4Z4MK       59  59  FM     30  --   S17
   54  7 Sep 81 2217  20.      4Z4US       59  59  SSB   200  --
   55 11 Sep 81 1405   2.      4X6CQ       59  59  FM     30  --   R9
   56 11 Sep 81 1408   2.      4Z4ZB       59  59  FM     30  --
  104 13 Sep 81 1947  15.      4X4GI       59  59  SSB   200  --   (*)
  105 13 Sep 81 2014   2.      4X4GI       59  59  FM     30  --   R7
  108 14 Sep 81 1650   2.      4X6AS       59  59  FM     30  --   R9
  113 17 Sep 81 0735  20.      4N2DX       59  59  SSB   200  --   (YU2DX)
  115 17 Sep 81 2155   2.      4Z4US       59  59  FM      2  --   R9
  117 17 Sep 81 2200  20.      4X4GI       53  53  SSB   200  --
  118 18 Sep 81 1152   2.      4X4IK       59  59  FM      2  --
  119 19 Sep 81 1714   2.      N4DII/4X    59  59  FM      2  --
  120 19 Sep 81 1716   2.      4Z4ZA       59  59  FM      2  --
  121 19 Sep 81 1717   2.      4X4IO       59  59  FM      2  --   QSP de K3STM
  123 20 Sep 81 0450  20.      4Z4US       59  59  SSB   200  --
  125 20 Sep 81 1900   2.      GI4FUM/4X   59  59  FM     30  --   S20
  128 24 Sep 81 0420  20.      4X4US       59  59  SSB   200  --
  129 24 Sep 81 0420   2.      4Z4ZB        0   0  FM      2  --
  131  1 Oct 81 1932   2.      ZR6AX/4X     0   0  FM      2  --
```

Fig. 2-1. Sample log listing. (continued)

ible files (like the log pointer files) to be present.

HELP. The HELP program shown in Fig. 2-5 is almost identical to the files program except that the names of type 2 files (BASIC language programs) are displayed on the screen. Again, extensive error-trapping and checking is employed utilizing the built-in error-handling features of Northstar BASIC.

LOG. The LOG program shown in Fig. 2-6 is the initial entry into the package. It clears the screen (line 20), sending the three most common screen clear characters to the console. Then the sign-on message is sent and the SYSTEM program is chained. LOG contains the same error-handling routines as the previous programs.

Most of the remaining programs in the package utilize a common subroutine library. This technique minimizes errors in development because programs only have to be typed and debugged once—not once per program. Also, updates and changes are easier. If a change is made and debugged in one program, it can be appended to all the others without further testing. The disad-

The following commands are the system commands:

SYSTEM (invisible)	System program
FILES	Lists all data files to the console
HELP	Lists all command files to the console

FILES HELP perform the directory read functions. The following are additional commands from the logging package:

STNINFO	Updates the STNDATA file in the package
NEWLOG	Allocates space on a floppy disk for a log data file
LOGDEL	Deletes a log data file from a floppy disk
LOGEDIT	Edits the contents of a log data file
LOGENTER	Enters information into a log data file
LOGMERGE	Joins two log data files
LOGPRINT	Prints the contents of a log data file
LOGRENAME	Renames a log data file
LOGRESTR	Restores a crashed log data file
QSLPRINT	Prints QSL labels from the contents of a log data file

Fig. 2-2. Commands used in logging package.

vantage is that there may be lines of code in a program that are not used within that program. Since the programs are not memory-limited, the extra lines are inconsequential.

LOGLIB. The subroutine package contains the routines to set up variables, to define constants, to perform disk I/O, and other tasks that are done by more than one program in the logging package.

The listings presented in this chapter terminate at line 5000. The subroutines are appended to each individual program prior to execution and the whole is saved on disk as one program. The deletion in the listings makes the programs more readable by leaving out the common modules after describing them once.

The subroutine library shown in Fig. 2-7 is interfaced to the applications programs by the jump table (lines 5000 to 5050). This slows the execution time down marginally, but allows changes to be made to the in-

```
10 REM SYSTEM PROGRAM VERSION 790702
20 ERRSET 110,E1,E2
30 REM COPYRIGHT SNOW MICRO SYSTEMS INC. 1979
40 REM BY JOE KASSER JUNE 1979
45 PRINT
50 DIMX$(15) \ INPUT">",C$
60 FOR I=1 TO 4\ I$=STR$(I) \X$=C$+","+I$(2,2)
70 IF FILE(X$)=2 THEN EXIT 100
80 NEXT\ PRINT"cannot find ",C$
90 GOTO 170
100 CHAIN X$
110 ERRSET 110,E1,E2
120 IF E2=15 THEN 170
130 IF E1<>70 THEN 140 ELSE IF I>4 THEN 90 ELSE 80
140 IF E1<>180 THEN150 ELSE IF I>4 THEN 210 ELSE 190
150 IF E2<>8 THEN 160 ELSE PRINT"HARD DISC ERROR"\GOTO
    170
160 PRINT"ERROR ",E2," AT LINE ",E1 \GOTO 170
170 FOR I=1 TO 4
180 IF FILE("SYSTEM,"+STR$(I))=2 THEN EXIT 200
190 NEXT \ GOTO 210
200 CHAIN "SYSTEM,"+STR$(I)
210 PRINT"Put a system disc in any drive, then hit
    'RETURN'"
220 INPUT" ",A$ \ GOTO 170
```

Fig. 2-3. SYSTEM program.

```
10 REM VERSION 790822
20REM FILES\- LISTS TYPE THREE FILES AND SIZES IN BLOCKS
30 ERRSET 320,E1,E2
40 INPUT'Which drive ? ',D$\IFD$=''THEN 40
60 PRINT
70 PRINT 'Data files on drive ',D$,' are :- '
80 DIM L$(1028)
90 DIM A(8)
100 PRINT
110 OPEN #0,'(*),'+D$
120 FOR I=0 TO 63
130 L$=''
140 READ#0%I*16,&A(0),&A(1),&A(2),&A(3),&A(4),&A(5),&
A(6),&A(7)
150 READ#0,&J1,&J2,&J3,&J4,&J5
160 READ #0%I*16+12,&T
170 IF A(0)=32 THEN 290
180 IF T>128 THEN T=T - 128
190 IF T <> 3 THEN 290
200 FOR K=0 TO 7
210 L$=L$+CHR$(A(K))
220 NEXT K
230 IF L$(1,1)='*' THEN 290
240 IF L$(1,3) = '(*)' THEN 290
250 Q= Q + 1
260 PRINTL$,TAB(12), \ J3=J3+256*J4
270 IF J5<128 THEN PRINT' S  ', ELSE PRINT' D  ',
280 IFJ5<128 THEN PRINTJ3 ELSE PRINT2*J3
290 NEXT I
300 IF Q = 0 THEN PRINT 'NO DATA FILES PRESENT ON DISC'
310 CLOSE #0 \ GOTO 350
320 ERRSET 320,E1,E2
330 IF E1<>360 THEN 340 ELSE IF I>4 THEN 380 ELSE 370
340 IF E2<>8 THEN 345 ELSE PRINT'HARD DISC ERROR'\GOTO
 350
345 PRINT'ERROR ',E2,' AT LINE ',E1
350 FOR I=1 TO 4 \ I$=STR$(I)
360 IF FILE('SYSTEM,'+I$)=2 THEN EXIT 400
370 NEXT
380 PRINT'Put a system disc in a drive, hit 'RETURN'
390 INPUT' ',A$ \ GOTO 350
400 CHAIN'SYSTEM,'+I$
```

Fig. 2-4. FILES program.

ternal structure without affecting any of the application programs that use routines in the package. Constants and variables are set up, as well as the function (FNT) which converts the time of delay (FuNction Time) to a string to allow formatted printing in the logs (lines 5060 to 5100). The log looks much better with time printed in four digits whether it is 0005 or 2315 hours.

The data that has been or is about to be stored in a log data file is printed (lines 5110 to 5170). A title line on the designated list device (A) is printed next (lines 5110 to 5170). Northstar BASIC allows the output to be sent to different devices by defining the device number in the print statement (i.e., PRINT #A). If no device number is specified in the print statement, device 0 is assumed. Most Northstar BASIC users set up the console as device 0 for the sake of convenience.

```
10 REM HELP VER 790814
20 ERRSET 300,E1,E2
30 INPUT 'Which drive ? ',D\IF D<1 OR D>4 THEN 30
40 PRINT \ D$=STR$(D)
50 PRINT 'COMMANDS ON DRIVE ',D$,' ARE :- '
60 DIM L$(1028)
70 DIM A(8)
80 PRINT
90 OPEN #0,'(*),'+D$
100 FOR I=0 TO 63
110 L$=''
120 READ#0%I*16,&A(0),&A(1),&A(2),&A(3),&A(4),&A(5),&A(6),&A(7)
130 READ #0%I*16+12,&T
140 IF A(0)=32 THEN 220
150 IF T>128 THEN T=T - 128
160 IF T <> 2 THEN 220
170 FOR K=0 TO 7
180 L$=L$+CHR$(A(K))
190 NEXT K
200 IF L$(1,6) = 'SYSTEM' THEN 220
210 PRINT L$
220 NEXT I
230 CLOSE #0
240 FOR I=1 TO 4 \ I$=STR$(I)
250 IF FILE('SYSTEM,'+I$)=2 THEN EXIT 290
260 NEXT
270PRINT'Put a system disc in any drive, then hit 'RETURN''
280 INPUT' ',A$\ GOTO 240
290 CHAIN 'SYSTEM,'+I$
300 ERRSET 300,E1,E2
310 IF E2=15 THEN 240 \REM ^C ABORT
320 IF E1<>90 THEN 330 ELSE PRINT'DIRECTORY ERROR'\ GOTO 240
330 IF E1<>250 THEN 330 ELSE IF I>4 THEN 270 ELSE 260
340 IF E2<>7 THEN 350 ELSE  PRINT'FILE ERROR' \GOTO 240
350 IF E2<>8 THEN 360 ELSE PRINT'HARD DISC ERROR'\GOTO 240
360 PRINT'ERROR ',E2,' AT LINE ',E1\ GOTO 240
```

Fig. 2-5. HELP program.

```
10 ERRSET 9060,E1,E2
20 PRINT CHR$(12),CHR$(26),CHR$(27)\REM CLRS
30 PRINT'G3ZCZ AMATEUR RADIO PACKAGE VERSION 2.0'
40 PRINT
60 PRINT'(C) Copyright Snow Micro Systems Inc. 1979'
70 PRINT
9000 FOR I=1 TO 4
9010 IF FILE('SYSTEM,'+STR$(I))=2 THEN EXIT 9030
9020 NEXT \ GOTO 9040
9030 CHAIN 'SYSTEM,'+STR$(I)
9040 PRINT'Put a system disc in any drive, then hit 'RETURN''
9050 INPUT' ',A$ \ GOTO 9000
9060ERRSET 9060,E1,E2
9070 IF E2=15THEN9000
9080 IFE1=9010THENIFI<1ORI>4THEN9040ELSE9020
9090 IFE2<>7THEN9100ELSEPRINT'FILE ERROR'\GOTO9000
9100 IFE2<>8THEN9110ELSEPRINT'HARD DISC ERROR'\GOTO9000
9110 PRINT'ERROR ',E2,' AT LINE ',E1\GOTO9000
```

Fig. 2-6. LOG program.

The user is asked to allocate the output device (lines 5230 to 5250). Since Northstar BASIC only allows device numbers 0–7, the

```
5000 REM SUBROUTINE PACKAGE VER 791008
5005 GOTO 5550 \ REM GET STNDATA
5010 GOTO 5060 \ REM SET UP VARIABLES & CONSTANTS
5015 GOTO 5180 \ REM PRINT TITLE LINE
5020 GOTO 5110 \ REM PRINT LINE OF DATA
5025 GOTO 5600 \ REM READ ENTRY FROM DISC
5030 GOTO 5500 \ REM PRINT HEADING
5035 GOTO 5620 \ REM WRITE ENTRY TO  DISC
5040 GOTO 5230 \ DETERMINE OUTPUT DEVICE
5045 GOTO 5640 \ REM RETURN TO SYSTEM
5050 GOTO 5260 \ REM DETERMINE NAME OF LOG FILE & GET N
5060 DIMX$(64) \ P1=66\P2=1
5065DIMM1$(36)\M1$='JanFebMarAprMayJunJulAugSepOctNovDec'
5070 T0 = 9 \ T1=T0+ 4 \ T2= T1+ 3 \ T3 = T2 +3
5075 T4=T3+10\T5=T4+10\T6=T5+4\T7=T6+5\T8=T7+4\T9=T8+6
5080 DEF FNT$(V)
5085 V1=INT(V/1000)\V2=INT((V-V1*1000)/100)
5090 V3=INT((V-V1*1000-V2*100)/10)\V4=INT(V-V1*1000-V2*100-V3*10)
5095 RETURN CHR$(48+V1)+CHR$(48+V2)+CHR$(48+V3)+CHR$(48+V4)
5096 FNEND
5100 RETURN
5110 PRINT #A,%5I,I,TAB(6),%2I,D1,TAB(T0),M1$((D2-1)*3+1,(D2-1)*3+3),
5112 PRINT #A,TAB(T1),%2I,D3,TAB(T2),
5115 T$=FNT$(T)
5120 PRINT #A,T$,TAB(T3),%Z8F3,F,TAB(T4),C$,TAB(T5),%3I,S,TAB(T6),
5130 PRINT #A,%3I,R,TAB(T7),M$,TAB(T8),%4I,P,TAB(T9),
5140 IF Q1=1 THEN PRINT#A,'S', ELSE PRINT#A,'-',
5150 IF Q2=1 THEN PRINT#A,'R', ELSE PRINT#A,'-',
5160 PRINT#A,'  ',X$
5165 P1=P1-1\IFP1>2THEN5170 ELSE P1=66\P2=P2+1\!#A\!#A'-'\!#A\GOSUB5500
5170 RETURN
5180 PRINT#A,'QSO #',TAB(5),'   DATE',TAB(T1+3),'TIME',TAB(T3),
5190 PRINT#A,' BAND',TAB(T4),'STATION',TAB(T5),' S',TAB(T6),
5200 PRINT#A,' R',TAB(T7-1),'MODE',TAB(T8),' PWR',TAB(T9),
5210 PRINT#A,'QSL  COMMENTS'
```

Fig. 2-7. LOGLIB program.

user input is tested (line 5240) to see that it is valid. If a non-valid number has been entered, the request is repeated.

The name of the Log file to be accessed is requested (lines 5260 to 5360) and determines how many entries (N) are in the file. Error-checking also appears here (lines 5275, 5325 and 5327).

An identifier is printed on the top line of each page (lines 5500 to 5510), and column headings are printed (line 5180).

The standard station data is picked up from the *STNDATA file (lines 5550 to 5595).

A log entry can be read from (lines 5600 to 5610), or written to lines 5620 to 5630 the log data file.

The package is returned to the system mode by searching the disks sequentially from 1 to 4 to find and then chain the

```
5220 RETURN
5230 INPUT'Which output device (0-7) ? ',A
5240 IF A>7 OR A<0 THEN 5230
5250 RETURN
5260 INPUT'What is the name of the log file ? ',L$\IFL$=''THEN5260
5270 INPUT'Which drive is it on ? ',D$\IFD$=''THEN 5270
5275 IF LEN(D$)>1THEN5270
5310 L1$=L$+','+D$ \ REM LOG FILE
5320 L2$='*'+L$+','+D$
5325 IF FILE(L1$)=3 THEN 5327 ELSEPRINT'LOG FILE IS NOT ON DISC'\GOTO5045
5327 IF FILE(L2$)=3 THEN 5330 ELSEPRINT'LOG POINTER FILE ERROR'\GOTO 5045
5330 OPEN#2,L2$
5340 READ#2,N
5350 CLOSE#2
5360 RETURN
5500 PRINT#A\PRINT#A,'STATION LOG  ',C9$,TAB(32),'PAGE',P2,
5510 PRINT#A,TAB(48),'FILE ',L1$\P1=P1-4\GOTO 5180
5550 X1$='STNDATA,'
5555 FOR I=1 TO 4\ I$=STR$(I) \ X$=X1$+I$(2,2)
5560 IF FILE(X$)=3 THEN EXIT 5590
5565 NEXT
5570 PRINT'STNDATA file is not on system'\ X=1\RETURN
5590 OPEN#2,X$ \ READ#2,L$,C9$ \ CLOSE#2
5595 RETURN
5600 READ #0,&D1,&D2,&D3,T,F,C$,S,R,M$,P,&Q1,&Q2,X$
5610 RETURN
5620 WRITE#1,&D1,&D2,&D3,T,F,C$,S,R,M$,P,&Q1,&Q2,X$
5630 RETURN
5640 FOR I=1 TO 4
5650 IF FILE ('SYSTEM,'+STR$(I))=2 THEN EXIT 5690
5660 NEXT
5670 PRINT'put the SYSTEM disc in the computer, then hit 'RETURN''
5680 INPUT '',A$ \ GOTO 5640
5690 CHAIN 'SYSTEM,'+STR$(I)
```

SYSTEM program (lines 5640 to 5695). It uses the error-trapping features of Northstar BASIC.

The data is stored in a log file with the following variables:

D1	Day
D2	Month
D3	Year
T	Time
F	Frequency or band
C$	Call sign
S	Signal report sent
R	Signal report received
M$	Mode
P	Power input
Q1	QSL sent (flag)
Q2	QSL received (flag)
X$	Comments

D1, D2, D3, Q1 and Q2 are stored as binary variables to conserve disk space.

Other variables used in the package include:

A	Which device output is routed to. (Normally, 0, which routes the output to the console, is assumed; routes output to the printer on a Northstar Horizon).
M1$	The first 3 letters of the month of the year. The value of D2 is converted to an index into M1$; prints the abbreviation for the month, while storing the value as a binary variable.
T$	The time string of T following conversion.
P1	The number of lines sent to the print device.
P2	The number of pages printed.
Tn	Tab controls for formatting the log listing.
L$	The log file name input.
L1$	The full name including the drive definition.
L2$	The pointer file name (includes the * as the first character).
N	The number of entries in a log data file.
X1$	Defined as *STNDATA.
X$	X1$ including the drive definition.
L$	The name of the current log file.
C1$	The callsign of the user.

STNINFO. STNINFO updates the STNDATA file. This file contains information that can be used by the various Amateur Radio programs running in your station. This information includes but is not limited to your callsign (C$), your latitude (L1), your longitude (L2), and the current log-file (L$) to remind you which file you accessed last. All updates are performed manually.

The program shown in Fig. 2-8 prints out the current contents of the files and asks for further input. The user is guided through several options; changes may or may not be made.

The callsign stored in the STNDATA file is read by the LOGPRINT program so it can be printed on each page of a log book listing. It may also be used in Contest programs. The latitude and logitude information is available for use by antenna pointing or OSCAR tracking programs.

If the current log file is changed (line 330) and the new log file does not exist, the NEWLOG program is automatically chained to allow space to be allocated for the new file.

Extensive error-handling routines detect absent files, hard disk errors, or a control-C abort. Note that if an error occurs that has not been trapped, it is listed at the console for further troubleshooting (line 590).

NEWLOG. NEWLOG allocates disk space for a log data file. The program listed in Fig. 2-9 computes the size of the data file (line 80) and creates a file on the floppy disk. A pointer file is also created (lines 90 to 130).

```
10 PRINT CHR$(26),CHR$(12)\REM VER 790906
20 PRINT'Station Customising Program, Version 0.2'
30 ERRSET 495,E1,E2
40 PRINT
50 PRINT'(C) Copyright Snow Micro System Inc. 1979'
60 REM By Joe Kasser G3ZC7 June 1979
70 PRINT \ PRINT\ X$='STNDATA,'
80 FOR I=1 TO 4 \ I$=STR$(I) \ I$=I$(2,2)
90 IF FILE(X$+I$)=3 THEN EXIT 120
100 NEXT
110 PRINT'file STNDATA - station data file missing'\ GOTO 400
120 OPEN#0,X$+STR$(I)
130 READ#0,L$,C$,L1,L2
140 CLOSE#0
150 PRINT'Station configuration for    ',C$
160 PRINT
170 PRINT'Current Log file is          ',L$
180 PRINT
190 INPUT'Do you want to update any information ? ',A$\IFA$='' THEN190
200 IF A$(1,1)='Y' THEN 210 ELSE 400
210 PRINT
220 INPUT'Do you want to update the callsign ? ',A$\IFA$=''THEN220
230 IF A$(1,1)='N' THEN 250
240 INPUT'What is your Call Sign ? ',C$\IFA$=''THEN240
250 PRINT'Current latitude is ',L1,' Longitude is ',L2
260 INPUT'Do you want to update the Geographical Co-ordinates ?',A$
270 IF A$='' THEN 260
280 IF A$(1,1)='Y' THEN 290 ELSE 310
290 INPUT'Your Station Latitude ? ',L1
300 INPUT'Your Station Longitude ? ',L2
310 INPUT'Do you want to change the current log file ?',A$\IFA$=''THEN310
320 IF A$(1,1)='Y' THEN 330 ELSE 360
330 INPUT'What is the name of the new log file ? ',L$\IFL$=''THEN330
340 INPUT'is it current ? ',A$\IFA$=''THEN340
350 IF A$(1,1)='Y' THEN A=0 ELSE A=1
360 OPEN #0,X$+I$
370 WRITE#0,L$,C$,L1,L2
380 CLOSE#0
390 IF A=1 THEN 460
400 FOR I=1TO4 \ I$=STR$(I)
410 IF FILE('SYSTEM,'+I$)=2 THEN EXIT 450
420 NEXT
430 PRINT'Put a system disc in any drive, then hit 'RETURN''
440 INPUT'', A$ \ GOTO 400
450 CHAIN 'SYSTEM,'+I$
460 FOR I=1TO4 \ I$=STR$(I)
470 IF FILE('NEWLOG,'+I$)=2 THEN EXIT 490
475 NEXT
480 PRINT 'File NEWLOG is not in any drive, put it in, then hit 'RETURN''
485 INPUT'',A$\ GOTO 460
490 CHAIN 'NEWLOG,'+I$
495 ERRSET 495,E1,E2
500 IFE2=15 THEN 400
510 IFE1=470 THEN 475
520 IF E1=410 THEN 420
550 IFE1=90 THEN100
560 IFE2<>7THEN580 ELSE PRINT'FILE ERROR'\GOTO400
580 IFE2<>8THEN590 ELSE PRINT'HARD DISC ERROR'\GOTO400
590 PRINT'ERROR ',E2,'AT LINE ',E1\GOTO400
```

Fig. 2-8. STNINFO program.

The pointer file is an invisible file which contains the number of entries in the log data file. At this time, the value 0 is stored in the pointer file. The program also contains the usual error-handling code.

LOGDEL. LOGDEL deletes log data files from the floppy disk directory. The program listed in Fig. 2-10 deletes both the log pointer and the log data files. It should be used rather than the Northstar DOS DE to delete files. The program also contains the usual error-handing code.

LOGEDIT. LOGEDIT allows the contents of a log data file to be edited. The operation is performed by creating a new log data file, and then editing the contents of the specified data file into the new data file. The contents of the 'old' file are not touched, and in the event of an error in the edit process, are available for reuse. Since this program sizes the new log data file based on the number of entries in the file to be edited,

```
10 REM NEWLOG VER 790704
20 ERRSET 200,E1,E2
30 N=0
40 INPUT'Name of new log file ? ',C$\IFC$=''THEN40
50 INPUT'Which drive do you want it on ? ',D \ D$=STR$(D)
60 IF D<0 OR D>4 THEN 50
70 INPUT'About how many entries do you want in it ? ',B
80 B= INT((B/4)+(B*.15))+1
90 CREATE C$+','+D$,B
100 CREATE '*'+C$+','+D$,2
110 OPEN #1,'*'+C$+','+D$
120 WRITE#1,N
130 CLOSE#1
140 FOR I=1 TO4
150 IF FILE('SYSTEM,'+STR$(I))=2 THEN EXIT 190
160 NEXT
170 PRINT'put a system disc on any drive, hit the 'RETURN''
180 INPUT' ',A$ \ GOTO 140
190 CHAIN 'SYSTEM,'+STR$(I)
200 ERRSET 200,E1,E2
210 IF E2=15 THEN 140
220 IF E1<>150 THEN 250 ELSE IF I>4 THEN 170 ELSE 160
250 IF E2<>7 THEN 270 ELSE PRINT'FILE ERROR' \ GOTO 140
270 IFE2<>8 THEN 280 ELSE PRINT'HARD DISC ERROR'\GOTO 140
280 PRINT'ERROR ',E2,' AT LINE ',E1\GOTO 140
```

Fig. 2-9. NEWLOG program.

```
10 REM LOGDEL VER 790709
20 REM BY JOE KASSER G3ZCZ
30 REM COPYRIGHT SNOW MICRO SYSTEMS INC.
40 ERRSET 160,E1,E2
50 INPUT"Which log file ? ",L$\IF L$=""THEN 50
60 INPUT"Which drive is it on ? ",I\IF I<0ORI>4THEN60
70 I$=STR$(I)
80 L1$=L$+","+I$(2,2) \ L2$="*"+L$+","+I$(2,2)
90 DESTROY L1$ \ DESTROY L2$
100 FOR I=1 TO 4
110 IF FILE("SYSTEM,"+STR$(I))=2 THEN EXIT 150
120 NEXT
130 PRINT"Put a system disc in any drive, then hit 'RETURN'"
140 INPUT" ",A$\GOTO 100
150 CHAIN "SYSTEM,"+STR$(I)
160 ERRSET 160,E1,E2
170 IF E2=15 THEN 100
180 IF E1<>110 THEN 190 ELSE IF I>4 THEN 130 ELSE 120
190 IF E2<>7 THEN 200 ELSE PRINT"FILE ERROR"\GOTO 100
200 IF E2<>8 THEN 210 ELSE PRINT"HARD DISC ERROR"\GOTO100
210 PRINT"ERROR ",E2," AT LINE ",E1\GOTO 100
```

Fig. 2-10. LOGDEL program.

it may be used for 'compacting' the disk by freeing up unused disk space if a large log data file contains only a few entries.

The program shown in Fig. 2-11 makes use of the subroutine library. The program begins by setting up the variables and constants (line 20). The name of the existing log data file, the name to be assigned to it after editing, and the drive numbers are requested and tested (lines 40 to 120). Next, a length error is tested for (line 65). Northstar DOS allows files to have up to 8 characters in their names. Since each log data file also has an invisible log pointer file that begins with an asterisk, this package only allows file names to contain up to 7 characters. The number of entries in the existing file is read (line 130) and the new log file is created (lines 150 and 160). The two files are then opened (lines 170 and 180) and the editing process begins.

Entries can be specified by number or by callsign. Since the log data file is a sequential file, entries can only be edited in sequence. If one is forgotten, the whole editing job will have to be done again. The file-to-file editing process is forgiving of errors and allows comments to be added so that the resulting file may be larger than the original file.

The entry to be edited is identified and searched for (lines 190 to 270). Data is written to the new file and the end of operation situation is detected (lines 280 to 320). The user is asked with what is to be edited and how it is to be changed (lines 330 to 650). The input is checked to see if it is within the correct range. It is impossible to detect every user input error, but many gross errors can be detected. Some of these are as follows:

A time that is not in the range of 0000 to 2400 is obviously wrong (line 480). A day that is not between 1 and 31 is also in error (line 510). A month greater than 12 does not exist (line 520). This last test is important because the month is printed out as a string computing an index into the M1$ matrix. An entry greater than 12, if not caught here, would cause the month conversion subroutine to generate a "bad subscript" error and crash the program.

The subroutines starting at lines 5015 and 5020 display the edited log data entry (line 660). If it is correct (see prompt sequence in lines 670 and 680), it is entered into the new log data file. Whether the sequence continues or terminates is determined (lines 690 to 700). The sequence is terminated (lines 710 to 790) by making sure that the remainder of the data in the old log is written into the new log. There is an error message that's used if the edit mode was by callsign and the desired callsign does not exist (lines 780). Last are the error-handling routines (lines 800 to 900), and the subroutine library (line 5000).

The following additional temporary variables are used in this program:

- Q Edit mode flag: 1 = by number, 2 = by callsign
- Q3 Callsign searching state in *Edit by Callsign* mode

```
10 REM LOGEDIT VER 790821
20 Q3=0\N1=0\N4=0 \ GOSUB 5010
30 ERRSET 800,E1,E2
40 INPUT"Name of LOG file ? ",L1$\IF L1$=""THEN 40
50 INPUT"Which drive is it on ? ",D1 \ IF D1<1ORD1>4 THEN50
60 INPUT"Name of file when edited ? ",L2$\IF L2$=""THEN60
65 IF LEN(L2$)<8THEN70ELSE!"LENTH ERROR (max is 7)"\GOTO60
70 IF L1$<>L2$ THEN80 ELSE PRINT"FILE NAME ERROR"\GOTO60
80 INPUT"Which drive do you want it put on ? ",D2 \ IF D2<1ORD2>4 THEN80
90 D1$=STR$(D1) \ D2$=STR$(D2) \D1$=D1$(2,2)\D2$=D2$(2,2)
100 IF FILE(L2$+","+D$(2,2))=2 THEN 120 ELSE 110
110 IF FILE(L2$+","+D$(2,2))=3 THEN 120 ELSE 130
120 PRINT"FILE ",L2$," ALREADY EXISTS ON DRIVE ",D$\GOTO 60
130 OPEN#0,"*"+L1$+","+D1$ \ READ#0,N1 \ CLOSE#0
140 PRINT"There are ",N1," entries in ",L1$
150 CREATE "*"+L2$+","+D2$ , 2 \ REM POINTER FILF
160 CREATE L2$+","+D2$ ,INT(N1/4+N1/10)+1
170 OPEN#0,L1$+","+D1$ \ REM OPEN OLD LOG FILE
180 OPEN#1,L2$+","+D2$ \ REM OPEN NEW LOG FILE
190Q3=0
195 INPUT"Edit by entry number or call sign ( N or C ) ? ",A$\IF A$=""THEN 195
200 IF A$(1,1)="N" THEN Q=1
210 IF A$(1,1)="C" THEN Q=2
220 IF Q=0 THEN 190
230 IFQ>1THEN260ELSE INPUT"Number ? ",N3\IF N3<1THEN230
240 IF N3>N4 THEN 250 ELSE PRINT "SEQUENCE ERROR" \ GOTO 230
250 GOSUB 5025 \N4=N4+1\ IF N4=N3 THEN 320 ELSE 280
260 IF Q>2 THEN 190 ELSE INPUT"Call sign ? ",C1$
270 GOSUB 5025 \N4=N4+1\ IF C$=C1$ THEN 320 ELSE 280
280 REM WRITE TO DISC
290 GOSUB 5035 \ IF N4>=N1 THEN 730 \ REM CLOSEOUT
300 IFQ=1 THEN 250
310 GOTO 270
320 REM
330 I=N4 \ GOSUB 5015 \ GOSUB 5020 \Q3=1
340 INPUT"Change QSL information ? ",A$\IF A$="" THEN 340
350 IF A$(1,1)="N"THEN 420
360 INPUT"Was a card sent ? ",A$\IF A$="" THEN 360
370 IF A$(1,1)="Y" THEN Q1=1 ELSE Q1=0
380 INPUT"Was one received ? ",A$\IF A$(1,1)="" THEN 380
390 IF A$(1,1)="Y" THEN Q2=1 ELSE Q2=0
400 INPUT"ANY OTHER CHANGES ?",A$\IFA$(1,1)=""THEN 400
410 IF A$(1,1)="N" THEN 660
420 INPUT"Change the comments ?",A$\IF A$(1,1)="" THEN 420
430 IF A$(1,1)="Y" THEN INPUT"...? ",X$
440 INPUT"ANY OTHER CHANGES ? ",A$\IF A$=""THEN440
450 IF A$(1,1)="N"THEN 660
460 INPUT"Change time ? ",A$\IF A$="" THEN 460
470 IF A$(1,1)<>"Y" THEN 490
480 INPUT "New time ? ",T \ IF T<0 OR T>2400 THEN 480
490 INPUT"Change the date  ?",A$\IFA$=""THEN490
500 IFA$(1,1)<>"Y"THEN 540
510 INPUT"Day ? ",D1\IF D1<1ORD1>31THEN510
520 INPUT"Month ( 1-12 ) ? ",D2\IF D2<1ORD2>12 THEN 520
530 INPUT"Year (19xx) ? ",D3\IF D3<1900 THEN 530 ELSE D3=D3-1900
540 INPUT"Change the  FREQ/BAND information ? ",A$\IF A$=""THEN540
550 IF A$(1,1)<>"Y"THEN560 ELSE INPUT"Freq/band ? ",F
560 INPUT"Change the signal reports  ?",A$\IF A$=""THEN560
570 IFA$(1,1)<>"Y"THEN 590
580 INPUT1"Sent ? ",S\PRINTTAB(24),\INPUT"Received ? ",R
590 INPUT "Change the power ? ",A$\IF A$=""THEN590
600 IF A$(1,1)<>"Y"THEN610 ELSE INPUT"Power ? ",P
610 INPUT"Change the mode ?",A$\IF A$=""THEN610
620 IFA$(1,1)<>"Y"THEN630 ELSE INPUT"Mode ? ",M$\IFM$=""THEN620
630 INPUT"Correct the call sign ? ",A$\IF A$=""THEN 630
640 IF A$(1,1)<>"Y" THEN 660
650 INPUT"Call sign ? ",C$\IF C$=""THEN 650
660 GOSUB 5015 \ GOSUB 5020
670 INPUT"OK ?",A$\IF A$=""THEN670
680 IF A$(1,1)="N" THEN 340 ELSE GOSUB 5035
685 IFN4>=N1THEN730
690 INPUT"Another entry ?",A$\IF A$="" THEN 690
700 IF A$(1,1)="Y" THEN 190
710 REM CLOSE OUT LOGS
720 GOSUB 5025 \N4=N4+1 \ GOSUB 5035 \ IF N4<N1 THEN 720
730 CLOSE#0
740 CLOSE#1
750 OPEN#0,"*"+L2$+","+D2$
760 WRITE#0,N4
770 CLOSE#0
780 IF Q3=1 THEN 790 ELSE IF Q=2 THEN PRINTC1$+" WAS NOT IN THE LOG"
790 GOTO 5045 \ REM TERMINATE
800 ERRSET 800,E1,E2
810 IFE2<>15THEN820ELSEIFE1<130THEN900ELSE880
820 IF E1<>5560 THEN 830 ELSE IF I>4 THEN 5570 ELSE 5565
830 IF E1<>5650 THEN 840 ELSE IF I>4 THEN 5670 ELSE 5660
840 IF E2<>7 THEN 850 ELSE PRINT "FILE ERROR"\GOTO 900
850 IF E2<>8 THEN 860 ELSE PRINT "HARD DISC ERROR"\GOTO 900
860 REM
870 PRINT"ERROR ",E2," AT LINE ",E1
880 DESTROY"*"+L2$+","+D2$
890 DESTROY    L2$+","+D2$
900 GOTO 790
5000 REM SUBROUTINE PACKAGE VER 790821
```

Fig. 2-11. LOGEDIT program.

N3 Number of entry to be edited in number mode
N4 Number of entry currently being accessed.

LOGENTER. LOGENTER enters data into a log data file. The information that can be entered into a data file is as follows: date (day, month, year) time, band frequency, callsign, signal reports exchanged, mode, transmitter power level, QSL (sent and received) status, and comments.

The program shown in Fig. 2-12 first sets up the variables, requests the name of the log, locates the last entry in the log data file, and prints it at the console (lines 10 to 170).

The subroutines (called by lines 190 to 240) request for and format the data to be entered into the log data entry (lines 550 to 750). Once the data is set up, it is displayed (line 250). If the user responds that the data is valid (line 260) it is written to disk (line 280) using a subroutine (lines 760 to 810). If further entries are desired, the process continues; if not, the log is closed and the pointer file updated (lines 300 to 420).

Error-trapping and handling routines are present (lines 820 to 1250). One user error that is trapped (line 860) and processed (lines 1070 to 1200) is the one that occurs when the space allocated to a file is exceeded. The subroutine reads the whole file area to see exactly how much was put onto the disk and displays the last entry.

This program breaks the rules of using one line of code as a subroutine to perform a single operation, in that disk reading is performed by lines 110 and 1130 and writing is done by line 790 instead of using the subroutine entry at lines 5035. This is done so that the error-handling routines can be told when an error occurred in the entry process as well as which error it was.

```
10 REM LOGENTER VER 791008
20 ERRSET 820,E1,E2
40 REM "Copyright Snow Micro Systems Inc. 1979"
50 GOSUB 5010 \REM SET UP VARIABLES
60 GOSUB 5050 \ REM GET DATA NAME
70 PRINT \ N1=N \E9=1\ REM SET INIT NUMBER
80 IFN>80 THEN PRINT"OPENING LOG NOW"
90 OPEN #1,L1$
100 FOR I=1 TO N
110 READ #1,&D1,&D2,&D3,T,F,C$,S,R,M$,P,&Q1,&Q2,X$
120 NEXT \ REM POINTER SHOULD NOW BE SET UP TO WRITE
130 PRINT
140 IF N>0THEN160
150 PRINT\PRINT "BRAND NEW LOG BOOK"\ PRINT\ GOTO 180
160 PRINT"Last entry was :-"
165 I=I-1
170 GOSUB5015 \ GOSUB5020
180 PRINT\PRINT"New Entry Information  "\PRINT
190 GOSUB 550 \ REM DATE
200 GOSUB 640 \ REM MODE
210 GOSUB 600 \ REM  POWER
220 GOSUB 620 \ REM  FREQ/BAND
230 Q1=0 \ Q2=0 \ REM MAIN LOOP
240 GOSUB 660 \ REM REST OF DATA
250 A=0 \ I=N+1 \ PRINT \GOSUB 5015 \ GOSUB 5020
260 INPUT"OK ?",A$
270 IF A$(1,1)="Y" THEN 280 ELSE 440
280 GOSUB 760 \ REM WRITE TO DISC
290 INPUT "Another Entry ?  ",A$
300 IF A$(1,1)="Y" THEN 440
310 IF A$(1,1)="A"  THEN 410
320 CLOSE #1 \ REMWRITING IS OVER
330 OPEN #0,L2$
340 WRITE #0,N
350 CLOSE #0
360 GOTO 5045
410 IF N>0 THEN N=N-1
420 GOTO 320
430 CHAIN "SYSTEM,"+STR$(I)
440 INPUT "Has DATE, MODE, POWER or FREQ/BAND Changed ?  ",A$\IFA$=""THEN440
450 IF A$(1,1)="N" THEN 230 \ REM KEEP GOING
460 INPUT "Has FREQ/BAND changed ?  ",A$\IFA$=""THEN460
470 IF A$(1,1)="Y" THEN GOSUB 620
480 INPUT "Has POWER changed ?  ",A$\IFA$=""THEN480
490 IF A$(1,1)="Y" THEN GOSUB 600
500 INPUT "Has MODE changed ?  ",A$\IFA$=""THEN500
510 IF A$(1,1)="Y" THEN GOSUB 640
520 INPUT"Has DATE changed ?  ",A$\IFA$=""THEN520
530 IFA$(1)="Y" THEN GOSUB 550
540 GOTO 230
550 INPUT"Day (1-31) ?  ",D1\IFD1<1ORD1>31THEN550
560 INPUT"Month ( 1 - 12 ) ?  ",D2\IFD2<0ORD2>12THEN560
570 INPUT "Year (19xx) ?",D3 \ IF D3<1900 THEN 570
580 D3 = D3 - 1900
590 RETURN
```

Fig. 2-12. LOGENTER program.

```
600 INPUT "Power (watts) ?  ",P
610 RETURN
620 INPUT "Freq/Band ?  ",F
630 RETURN
640 INPUT "Mode ?  ",M$\IFM$=""THEN640
645 IF M$(1,1)<>" "THEN650ELSEM$=M$(2,LEN(M$))\GOTO645
650 RETURN
660  PRINT \INPUT "Time ?  ",T
670 INPUT "Call sign ?  ",C$\IFC$=""THEN670
675 IF C$(1,1)<>" "THEN680ELSEC$=C$(2,LEN(C$))\GOTO675
680 INPUT1 "Report Received ?  ",R \ PRINT TAB(32),
690 INPUT "Report sent ?  ",S
700 INPUT1"QSL card sent ?  ",A$ \IFA$=""THEN700
710 PRINTTAB(32),\ IF A$(1,1)="Y" THEN Q1=1
720 INPUT"QSL card received ?  ",A$ \IFA$=""THEN720
730 IF A$(1,1)="Y" THEN Q2=1
740 INPUT "Comments ?  ",X$
750 RETURN
760 REM PUT IT ON DISC
770 D1=INT(D1) \D2=INT(D2)\D3=INT(D3)\S=INT(S)
780 R=INT(R)\P=INT(P)
790 WRITE#1,&D1,&D2,&D3,T,F,C$,S,R,M$,P,&Q1,&Q2,X$
800 N=N+1
810 RETURN
820 ERRSET 820,E1,E2
830 IF E1<>5650 THEN 840 ELSE IF I>4 THEN 5670 ELSE 5660
840 IF E2=15 THEN 360 \ REM CONTROL C INHIBIT
860 IF E1=790 AND E2=3 THEN 1070
870 IF E1=1130 AND E2=3 THEN 1150
890 IF E2<>8 THEN  900 ELSE PRINT"HARD DISC ERROR"\ GOTO 360
900 REM
970 IF E2<>7 THEN 980 ELSE PRINT"FILE ERROR"\GOTO360
980 REM
1000 PRINT"ERROR ",E2,"AT LINE ",E1\GOTO360
1020 PRINT "log file ",L$," does not exist "\GOTO60
1070 N=9999999
1080 PRINT"Log file is full, recovering to last disc entry"
1090 CLOSE#1
1100 OPEN #1,L1$
1110 FOR I=1 TO N
1120 I1=D1\I2=D2\I3=D3\C1$=C$ \ REM SAVE OLD CALL
1130 READ #1,&D1,&D2,&D3,T,F,C$,S,R,M$,P,&Q1,&Q2,X$
1140 NEXT
1150 N=I-1
1160 PRINT
1170 PRINT "LAST ENTRY ON DISC IS QSO WITH ",C1$," ON ",I1,I2,I3
1180 PRINT"file is now full, use 'NEWLOG' to open a new one"
1190 PRINT
1200 GOTO 320
1210 IF E9=0 THEN 360 \REM ABORT
1220 IF N<>N1 THEN 1240
1230 GOTO 330
1240 IF N>0 THEN N=N-1
1250 GOTO 360
5000 REM SUBROUTINE PACKAGE VER 790817
```

LOGMERGE. LOGMERGE joins two log data files end to end. It allows small log files to be kept on a day-to-day basis so that access time can be minimized. Remember this is a sequential file arrangement, so the more entries there are the longer it takes the computer to find the last entry in the log data file. Once a small file has been filled, it can be appended to the archive file by using the LOGMERE command. The two logs being merged are untouched. The joining processes creates a new file containing the contents of the two files.

The program is shown in Fig. 2-13. It begins by requesting the names of the two files to be merged (joined) the name of the new file (lines 30 to 60). The names are checked for validity and the new file created (lines 70 to 110). All three files are then opened (lines 120 to 140). The contents of the first file are read on a record or entry by entry basis and written to the new file using a subroutine (lines 300 and 310).

Once the contents of the first log file have been transferred to the new file, the last entry is displayed at the console (line 180). The contents of the second file are then read and written into the third file (lines 190 to 210). The program checks to see that the total count is correct (line 220). The pointer file for the newly joined logs is then written and the files are closed out (lines 230 to 280). Upon completion of the task, the SYSTEM program is chained.

LOGPRINT. LOGPRINT allows the contents of a log data file to be displayed at the console or to be printed. The user is given the choice between three output options—the whole log, by prefix, callsign or date.

If the whole log is requested, everything is listed. If the prefix option is chosen, all calls beginning with that prefix are tested. A prefix of G would result a display of calls like G3ZCZ, GW3ZCZ, GM3ZCZ, GI3ZCZ, and GU3ZCZ. If the prefix is G3, only G3ZCZ and any other call beginning with G3 would appear. If a prefix such as 4X was requested, both 4X4IL and G3ZCZ/4X would be displayed.

```
10 REM LOGMERGE VER 790704
20 ERRSET 500,E1,E2
30 INPUT"first file ?  ",L1$\IFL1$="" THEN 30
40 INPUT"second file ? ",L2$\IFL2$="" THEN 40
50 INPUT"name of NEW LOG file  ? ",L3$\IFL3$=""THEN 50
55 IF LEN(L3$)<8THEN60ELSE!"LENGTH ERROR (max is 7)"\GOTO50
60 INPUT "which drive do you want it on ? ",D\IFD<1ORD>4THEN60
70 L$=L1$\GOSUB320\N1=N\D1$=STR$(I)
80 L$=L2$\GOSUB320\N2=N\D2$=STR$(I)
90 B=INT((N1+N2)/4+0.1*(N1+N2))+2
100 CREATE L3$+","+STR$(D),B
110 CREATE "*"+L3$+","+STR$(D),2
120 OPEN#0,L1$+","+D1$
130 OPEN#1,L2$+","+D2$
140 OPEN#2,L3$+","+STR$(D)
150 FORI=1TON1
160 READ #0,&D1,&D2,&D3,T,F,C$,S,R,M$,P,&Q1,&Q1,X$
170 GOSUB300 \ N3=N3+1\NEXT
180 PRINT"LAST ENTRY IN ",L1$," WAS ",C$
190 FOR I=1TON2
200 READ #1,&D1,&D2,&D3,T,F,C$,S,R,M$,P,&Q1,&Q1,X$
210 GOSUB300 \ N3=N3+1 \NEXT
220 IF N3=N1+N2 THEN 230 ELSE PRINT"ENTRY COUNT ERROR"\GOTO 400
230 OPEN#3,"*"+L3$+","+STR$(D)
240 WRITE#3,N3
250 CLOSE#3
260 CLOSE#2
270 CLOSE#1
280 CLOSE#0
290 GOTO400
300 WRITE#2,&D1,&D2,&D3,T,F,C$,S,R,M$,P,&Q1,&Q2,X$
310 RETURN
320 FORI=1TO4
330 IF FILE("*"+L$+","+STR$(I))=3THENEXIT360
340 NEXT
350 PRINT"LOG FILE ",L$,"NOT ON SYSTEM"\GOTO400
360 OPEN#0,"*"+L$+","+STR$(I)
370 READ#0,N
380 CLOSE#0
390 RETURN
400 FOR I=1 TO 4
410 IF FILE("SYSTEM,"+STR$(I))=2 THEN EXIT 450
420 NEXT
430 PRINT"Put a system disc in any drive, then hit 'RETURN'"
440 INPUT" ",A$\ GOTO 400
450 CHAIN "SYSTEM,"+STR$(I)
500 ERRSET 500,E1,E2
510 IF E2=15 THEN 400
515 IF E1=100 THEN PRINT"CANNOT CREATE NEW FILE ",L3$
520 IF E1<>410 THEN530 ELSE IF I>4 THEN430 ELSE 420
530 IFE1<>330 THEN 540 ELSE IF I>4 THEN 350 ELSE 340
540 IFE2<>7THEN550 ELSE PRINT"FILE ERROR"\GOTO 400
550 IFE2<>8 THEN560 ELSEPRINT"HARD DISC ERROR"\GOTO400
560 PRINT"ERROR ",E2," AT LINE ",E1\GOTO400
```

Fig. 2-13. LOGMERGE program.

A bonus feature is that any callsign stored in the comments field like this: <G3ZCZ> (i.e., between a < and a >) can also be found. This allows the user to log operators who are at someone else's station and still recall the contact by the operator's callsign.

This feature allows the log to be scanned for contacts:

- □ with a particular country or call area
- □ with one particular station
- □ with someone whose entire callsign you don't remember (For example, if you worked a G3Z station, but don't remember the date or the whole call, all entries beginning with G3Z can be displayed.)

If the date option is selected, log entries on one day or falling between two specified dates can be displayed at the console and then listed to the printer. The output allocation technique allows you to confirm the entries as they are displayed before they are printed, which can prevent wasting paper.

The program listed in Fig. 2-14 uses the subroutine library to pick up the contents of the STNDATA file. Then it sets up the

```
10 REM LOGPRINT VER 791114
20 ERRSET 630,E1,E2
30 REM BY JOE KASSER G3ZCZ
40 REM COPYRIGHT SNOW MICRO SYSTEMS INC. 1979
50 GOSUB 5010\X=0\GOSUB5005\IFX=1THEN590
60 GOSUB 5050 \ REM GET DATA FILE NAME
70 INPUT'Scan/print whole log, by prefix or date (W, P or D ) ? ',A$
80 IFA$=''THEN70
90 P1=66 \ P2=1 \ Q3=0
100IF A$(1,1)='W' THEN 110 ELSE 120
110 Q=3 \ GOTO 270
120 IF A$(1,1)='P' THEN 130 ELSE 150
130 INPUT'Which prefix ? ',A$\IFA$=''THEN130
140 Q=2 \ GOTO 270
150 IF A$(1,1)='D' THEN 160 ELSE 70
160 INPUT 'Start Date ?  ',D7
170 INPUT 'Start Month (1-12) ? ',D8
180 INPUT 'Start Year (19xx) ?  ',D9 \ IF D9<1900 THEN 180
190 Q=1
200 INPUT 'Do you only want one day ?  ',A$\IFA$=''THEN200
210 IF A$(1,1)='Y' THEN 260
220 INPUT'End day ?  ',D4\IFD4>31ORD4<1THEN220
230 INPUT'End Month (1-12)?  ',D5\IFD5>12ORD5<1THEN230
240 INPUT'End year (19xx) ?  ',D6 \ IF D6<1900 THEN 240
250 GOTO 270
260 D4 = D7 \ D5 = D8 \ D6 = D9
270 PRINT \ GOSUB 5040
280 D9=(D9-1900)*10000+D8*100+D7
290 D6=(D6-1900)*10000+D5*100+D4
300 IF N=0 THEN 610
310 GOSUB 5030
320 OPEN #1,L1$
330 FOR I=1 TO N
340 READ #1,&D1,&D2,&D3,T,F,C$,S,R,M$,P,&Q1,&Q2,X$
350 IF Q=1 THEN 440
360 IF Q=3 THEN 470
362 FORJ=1TOLEN(X$)\IFX$(J,J)<>'<'THEN363ELSEEXIT364
363 NEXT J \ GOTO 370
364 J3=J\ FOR J=J3+1 TO LEN(X$) \ IF X$(J,J)<>'>' THEN365ELSEEXIT366
365 NEXT J \ GOTO 370
366 C2$=X$(J3+1,J-1)
368 IF LEN(C2$)<LEN(A$)THEN370
369 IF C2$(1,LEN(A$))=A$ THEN470
370 IF LEN(C$)<LEN(A$) THEN 430
380 FORJ=1TO LEN(C$)\ IF C$(J,J)='/'THENEXIT400
390 NEXT\ GOTO420
400 C1$=C$(J+1,LEN(C$)) \ IFLEN(C1$)=1 THEN 420
410 IF LEN(C1$)<LEN(A$)THEN420
415 IFC1$(1,LEN(A$))=A$ THEN 470
420 IF C$(1,LEN(A$))=A$ THEN 470
430 IF Q=2 THEN 480
440 D0=D3*10000+D2*100+D1
450 IF D0<D9 THEN 480
460 IF D0>D6 THEN EXIT 490
470 GOSUB 5020 \Q3=1\REM PRINT LINE OF DATA
480 NEXT
490 CLOSE#1
500 IF Q=2 THEN 510 ELSE 520
510 IFQ3=1THEN560ELSEPRINTA$,' was not in the log'\PRINT
520 IF Q=1 THEN 530 ELSE 540
530 IF Q3=1 THEN 560 ELSE PRINT'No contacts on that date'\GOTO560
540 IF Q=3 THEN 550 ELSE 560
550 IF Q3=1THEN560ELSEPRINT'No contacts in the log book'
560 INPUT'Do you want log page ejected ? ',A$\IF A$='' THEN560
570 IF A$(1,1)='Y' THEN 580 ELSE 590
580 FOR I=P2 TO 62\PRINT#A\NEXT
590 INPUT'Again ? ',A$\IF A$=''THEN590
600 IFA$(1,1)='Y'THEN70ELSEIFA$(1,1)='N'THEN605ELSE590
605 GOTO 5045
610 PRINT'LOG BOOK IS EMPTY '
620 GOTO 605
630 ERRSET 630,F1,E2
640 IFE2=15THEN605
670 IF E1<>5560 THEN680 ELSE IF I>4 THEN 5570 ELSE 5565
680 IF E1<>5650 THEN 690 ELSE IF I>4 THEN 5670 ELSE 5660
690 IFE2<>8THEN 700 ELSE PRINT'HARD DISC ERROR'\GOTO605
700 IFE2<>7THEN 710 ELSE  PRINT'FILE ERROR'\GOTO605
710 PRINT'ERROR ',E2,' AT LINE ',E1\GOTO605
5000 REM SUBROUTINE PACKAGE VER 790908
```

Fig. 2-14. LOGPRINT program.

variables, and requests the user to enter the name of the log data file (lines 50 to 60). The type of display option is then requested and set up (lines 70 to 300). If an illegal option is chosen, the program requests a valid alternative.

The printer page formatting variables are P1, lines per page, and P2, number of pages (line 90). Q3 is used as a flag to show that something was printed. The whole log option (lines 100 to 110), the prefix option, (lines 130 to 140) and the day option (lines 150 to 300) are set up. Q is used as the option flag. Q = 1 in the Day option, Q = 2 in the Prefix option, and Q = 3 in the whole log option.

The start date is requested in the day option (lines 160 to 180). If only one day is requested, the end date is set to equal the start date (line 260). If not, the ending date is requested (lines 220 to 240). There is input data error checking to make sure that entries fall into valid ranges—between 1 and 12 for months, and between 1 and 31 for days. (The entry is not checked to see if it is an illegal date such as 31 September, or if 29 February is a valid date for that year, although code to do those tests can be included in the package as a subroutine. Since illegal dates will always produce the message: "no entries on those dates," they are brought to your attention eventually.) The dates are processed into a number for later comparison with the log data entries (lines 280 and 290).

The log data file (L1$) is scanned by the FOR/NEXT loop (lines 330 to 480). Each entry is read (line 340). The print option is then tested (lines 350 to 360).

If the whole log option is chosen (Q = 3), the program flow branches to line 470 which prints out the entry using the subroutine package. The print flag Q3 is also set to 1.

If the date option is chosen (Q = 1), the program branches to line 440, where the date of the entry is converted to a number (D0). That number is tested against the starting and ending date numbers (lines 450 to 460). If the number is less than the starting number, the program branches to the NEXT statement (line 480) to get the next entry. If the date of the entry is greater than the ending date, the log data scanning process terminates as the program branches to line 490 and closes the input file.

If Q is not equal to 1 or 3, a 2 (Prefix option) is assumed. The assumption is valid, since invalid requests would already have been caught (lines 80, 100, 120, and 150). The contents of the comments column is tested (lines 362 to 365) to see if a callsign is imbeded between carets (< and >). If no callsign is present, the program flow skips to line 370. If a callsign is present, it is checked (lines 366 to 369) to see if it matches the desired prefix (A$). If it does, the program branches to the display process (line 470). If it does not, the callsign is matched. If the call contains a slash, the suffix is then matched; and if a match occurs the entry is printed (lines 380 to 420). Before the actual comparison or matching test occurs, a string length test is performed to verify that the callsign to be tested has the same or more characters than the reference or desired prefix (A$). This is to avoid a string compare error when the actual test takes place (lines 369, 415 and 420).

An appropriate message is displayed (lines 500 to 550) if no entries were found in the log that match the desired ones. A form feed is requested and performed at the end of the printout (lines 560 to 580). A repeat performance is requested and set up (lines 590 to 605). This allows the data to be displayed first. Then it either can be printed, or a new set of entries can be scanned. As usual, error-trapping (lines 630 to 710) and handling and a subroutine library (lines 5000 and following) appear.

LOGRENAM. LOGRENAM renames the log data and log printer files. The program

```
10 REM LOGRENAM VER 790906
20 Q=0
30 ERRSET 9060,E1,E2
40 REM DIRECTORY FILE IS <*> AND MUST BE TYPE 3
50 REM INITIALI7E STRING SPACE
60 DIM A(15),O$(8),N$(8),T$(8)
100 REM GET DATA AND OPEN FILE
110 INPUT 'Which disc drive ? ',D\IFD<0ORD>4THEN110
120 D$=STR$(D)
130 INPUT 'Name of existing log file? ',O$\IFO$=''THEN130
150 INPUT 'New name for file ? ',N$\IFN$=''THEN150
160 IF LEN(N$)<8THEN170ELSE!'LENTH ERROR (max is 7)'\GOTO150
170 N1$=N$
180 IF FILE(O$+','+D$(2,2))=2 THEN 190 ELSE 200
190 PRINT'FILE ',O$,' IS A COMMAND ON THIS DISC'\GOTO 130
200 IF FILE(N$+','+D$(2,2))=2 THEN 210 ELSE 220
210 PRINT'FILE ',N$,' IS A COMMAND ON THIS DISC'\GOTO 150
220 IF FILE(N$+','+D$(2,2))=3 THEN 230 ELSE 240
230 PRINT'FILE ',N$,' ALREADY EXISTS'\GOTO 150
240 OPEN #0,'<*>,'+D$
250 REM START READING THRU DIRECTORY TO FIND OLD FILE
260 FOR I=0 TO 62
270 READ #0%I*16,&A(0),&A(1),&A(2),&A(3),&A(4),&A(5)
280 READ #0,&A(6),&A(7),&A(8),&A(9),&A(10),&A(11)
290 READ #0,&A(12),&A(13),&A(14),&A(15)
300 IF CHR$(A(0))=' ' THEN GOTO 340
310 T$=CHR$(A(0))+CHR$(A(1))+CHR$(A(2))+CHR$(A(3))+CHR$(A(4))
320 T$=T$+CHR$(A(5))+CHR$(A(6))+CHR$(A(7))
330 IF O$=T$(1,LEN(O$)) THEN EXIT 370
340 NEXT I
350 IF Q=0 THEN 360 ELSE IF O$(1,1)='*' THEN PRINT'POINTER FILE MISSING'
360 !'FILE ',O$,' NOT FOUND'\GOTO 490
370 REM CONVERT NEW FILE NAME TO ASCII
380 J=1\N$=N$+'     '
390 FOR K=0 TO 7
400 A(K)=ASC(N$(J,J))
410 J=J+1
420 NEXT K
430 REM NOW WRITE OUT NEW DIRECTORY ENTRY
440 WRITE #0%I*16,&A(0),&A(1),&A(2),&A(3),&A(4),&A(5),NOENDMARK
450 WRITE #0,&A(6),&A(7),&A(8),&A(9),&A(10),&A(11),NOENDMARK
460 WRITE #0,&A(12),&A(13),&A(14),&A(15),NOENDMARK
470 CLOSE #0
480 IF Q=1 THEN 490 ELSE Q=1\O$='*'+O$\N$='*'+N$(1,LEN(N1$))\GOTO 170
490 REM ENDING
9000 FOR I=1 TO 4
9010 IF FILE('SYSTEM,'+STR$(I))=2 THEN EXIT 9030
9020 NEXT \ GOTO 9040
9030 CHAIN 'SYSTEM,'+STR$(I)
9040 PRINT'put a system disc in any drive, then hit 'RETURN''
9050 INPUT' ',A$ \ GOTO 9000
9060ERRSET 9060,E1,E2
9070 IF E2=15THEN9000
9080 IFE1=9010THENIFI<1ORI>4THEN9040ELSE9020
9090 IFE2<>7THEN9100ELSEPRINT'FILE ERROR'\GOTO9000
9100 IFE2<>8THEN9110ELSEPRINT'HARD DISC ERROR'\GOTO9110
9110 PRINT'ERROR ',E2,' AT LINE ',E1\GOTO9000
```

Fig. 2-15. LOGRENAM program.

(listed in Fig. 2-15) begins by identifying the old and new names of the log file (lines 100 to 170). The disk is then scanned to see if the names are valid (lines 180 to 230). If the old name is not present on the disk or if it is a type 2 (BASIC) file, an error is flagged. If the new name already exists, an error is also flagged.

The contents of the disk directory are read and the entry for the old name located (lines 240 to 340). The existence of any errors are pointed out (lines 350 to 360).

The ASCII file name is converted to binary (lines 370 to 420) which is then written back to the disk (lines 430 to 470). The process is repeated for the pointer file (line 480). The usual SYSTEM chaining feature extracted from the subroutine library (lines 9000 to 9050) and the error-handling procedures (lines 9060 to 9110) appear. In MICROSOFT Basic, the entire rename function is performed by one line: NAME "PRESENTNAME.LOG" AS "NEWNAME.LOG".

LOGRESTR. LOGRESTR restores log data files that have crashed due either to power failures occurring while the file is open, to a damaged pointer file. The program listed in Fig. 2-16 begins by setting up the variables (line 50) and requesting the name of the log data file to be restored (line 60). The pointer file is then read and the number of entries that it thinks are in the data file is displayed (lines 70 to 80).

The program flow then branches to line 1070, which sets up an extremely large value for the loop index counter N. The FOR/NEXT loop (lines 1110 to 1140) reads sequential entries in the log file until bad data is reached. Valid entry data is saved by overwriting the previous entry. When bad data is read, a *data read* error occurs (line 860). The loop counter is decremented to set the value of N to the last valid entry (line 1150). The new value for N and the callsign associated with the entry are displayed (lines 1160 to 1190). The log pointer file is then updated (lines 320 to 350). Error-handling routines appear (lines 820 to 980). The

```
10REM LOGRESTR VER 790822
20 ERRSET 820,E1,E2
40 REM "Copyright Snow Micro Systems Inc. 1979"
50 GOSUB 5010 \REM SET UP VARIABLES
60 GOSUB 5050 \ REM GET DATA NAME
70 PRINT \ N1=N \E9=1\ REM SET INIT NUMBER
80 PRINT"There are ",N," entries in the pointer file"
180GOTO 1070
320 CLOSE #1 \ REMWRITING IS OVER
330 OPEN #0,L2$
340 WRITE #0,N
350 CLOSE #0
360 GOTO 5045
820 ERRSET 820,E1,E2
830 IF E1<>5650 THEN 840 ELSE IF I>4 THEN 5670 ELSE 5660
840 IF E2=15 THEN 360 \ REM CONTROL C INHIBIT
860 IF E1=1130 THEN 1150
890 IF E2<>8 THEN  900 ELSE PRINT"HARD DISC ERROR"\ GOTO 360
970 IF E2<>7 THEN 980 ELSE PRINT"FILE ERROR"\GOTO360
980 REM
1000 PRINT"ERROR ",E2,"AT LINE ",E1\GOTO360
1070 N=9999999
1090 CLOSE#1
1100 OPEN #1,L1$
1110 FOR I=1 TO N
1120 I1=D1\I2=D2\I3=D3\C1$=C$ \ REM SAVE OLD CALL
1130 READ #1,&D1,&D2,&D3,T,F,C$,S,R,M$,P,&Q1,&Q2,X$
1140 NEXT
1150 N=I-1
1160 PRINT
1170 PRINT "LAST ENTRY ON DISC IS NUMBER ",N,
1180 PRINT " WHICH  IS QSO WITH ",C1$," ON ",I1,I2,I3
1190 PRINT
1200 GOTO 320
5000 REM SUBROUTINE PACKAGE VER 790821
```

Fig. 2-16. LOGRESTR program.

read function is not performed in the subroutine library to facilitate the error-trapping.

QSLPRINT. QSLPRINT prints QSL cards from information stored in log data files. An option allows for an automatic LOGEDIT feature in which the QSL sent/received information is updated and an updated log file is generated.

This version of the program prints QSL cards in the form of labels which can be stuck onto the back of conventional QSL cards. Samples of the printout are shown in Fig. 2-17. The exact type of information that

```
To Radio Station 4Z4ZR
Confirming our 2X FM QSO
on  30 Aug 1981 at 1913 GMT
Your report is RST 59
Freq/Band  2 MHz/Meters
Input power  2 Watts .
Pse QSL
73  .........  G3ZCZ
Northstar Computer Logbook
(C) Snow Micro Systems Inc.
```

```
To Radio Station K3STM
Confirming our 2X SSB QSO
on  1 Sep 1981 at 2210 GMT
Your report is RST 59
Freq/Band  20 MHz/Meters
Input power  200 Watts .
Pse QSL
73  .........  G3ZC7
Northstar Computer Logbook
(C) Snow Micro Systems Inc.
```

```
To Radio Station G3IOR
Confirming our 2X SSB QSO
on  13 Sep 1981 at 1905 GMT
Your report is RST 59
Freq/Band  15 MHz/Meters
Input power  200 Watts .
Pse QSL
73  .........  G3ZC7
Northstar Computer Logbook
(C) Snow Micro Systems Inc.
```

Fig. 2-17. Sample QSL cards generated by the QSLPRINT program.

```
10REM QSLPRINT VERSION 790821
15REM (C) SNOW MICROSYSTEMS INC
20ERRSET640,E1,E2
25REM PROGRAMMED BY JOE KASSER G3ZC7
30DIMM1$(36)\M1$="JanFebMarAprMayJunJulAugSepOctNovDec"
40 INPUT"What is the name of the log file ? ",L1$\IFL1$=""THEN40
50 INPUT"Which drive is it on ? ",D\IFD<1ORD>4THEN50
60 D$=STR$(D)\D$=D$(2,2)
70INPUT"Shall I edit the QSL information into the log  ? ",A$\IFA$=""THEN70
80IFA$(1,1)="N"THEN120\IFA$(1,1)="Y"THEN90ELSE70
90INPUT"What is the name of the new log file ? ",L2$\IFL2$=""THEN90
95IFLEN(L2$)<8THEN100ELSE!"LENGTH ERROR (max is 7)"\GOTO90
100INPUT"Which drive do you want it on ? ",D2\IFD2<0ORD2>4THEN100
110E8=1\D2$=STR$(D2)\D2$=","+D2$(2,2)
120INPUT"Which output device (0-7) ? ",A\IFA<0ORA>7THEN120
130IFFILE(L1$+","+D$)=3THEN170
140PRINTL1$+" is not present on drive ",D
150PRINT"Put it in drive ",D," then hit ","'RETURN'"
160INPUT"",A$\GOTO130
170GOSUB5005
180INPUT"How many cards do you want printed ?",C\IFC=0THEN250
190DIMZ$(C*10+1),Z(C),Q(C)\FORI=0TOC\Q(C)=0\Z(I)=0\NEXT
200FORJ=0TOC-1
210INPUT"Call sign ? ",E$\IFE$=""THEN250
220IFE8=0THEN240ELSEINPUT"HAS HE QSL'D ? ",A$\IFA$=""THEN220
230IFA$(1,1)="Y"THENQ(J)=1ELSEQ(J)=0
240Z$(J*10+1,J*10+10)=E$\NEXT
250C=J\ OPEN#1,"*"+L1$+","+D$
260 OPEN#0,L1$+","+D$
270 READ#1,N\CLOSE#1
280IFE8=0THEN320
290CREATEL2$+D2$,INT(N/4+N/10)+1
300CREATE"*"+L2$+D2$,2
310OPEN#1,L2$+D2$
320 FORI=0TON-1\GOSUB5025
330FORJ=0TOC-1\C2$=Z$(J*10+1,J*10+10)
340IFC$=C2$(1,LEN(C$))THEN350ELSE500
350IFZ(J)=1THEN500
360PRINT#A\Z(J)=1
370PRINT#A,"To Radio Station ",C$
380 PRINT#A,"Confirming our 2X ",M$," QSO"\T$=FNT$(T)
390 PRINT#A,"on ",D1," ",M1$((D2-1)*3+1,(D2-1)*3+3),1900+D3," at ",T$," GMT"
400PRINT#A,"Your report is RST",S
410PRINT#A,"Freq/Band ",F," MHz/Meters"
420PRINT#A,"Input power ",P," Watts ."
430IFE8=0THEN440ELSEQ2=Q(J)
440IFQ2=1THENE$="Tnx"ELSEE$="Pse"
450PRINT#A,E$+" QSL"\Q1=1
460PRINT#A,"73   .........  "+C1$
470PRINT#A,"Northstar Computer Logbook"
480PRINT#A,"(C) Snow Micro Systems Inc."
490FORX=1TO5\PRINT#A\NEXTX
500NEXTJ
510IFE8=1THENGOSUB5035
520NEXTI
530FORI=0TOC-1
540IFZ(I)=1THEN570
550IFZ(C)=0THEN!"THE FOLLOWING CALLS WERE NOT FOUND"
560PRINTZ$(I*10+1,I*10+10)\Z(C)=1
570NEXT
580CLOSE#0
590IFE8=0THEN630ELSECLOSE#1
600OPEN#0,"*"+L2$+D2$
610WRITE#0,N
620CLOSE#0
630GOTO5045
640ERRSET640,E1,E2
645IFE2=15THEN745
650IFE1<>5560THEN660ELSEIFI>4THEN5570ELSE5565
660IFE1<>5650THEN670ELSEIFI>4THEN5670ELSE5660
670IFE2<>7THEN680ELSEPRINT"FILE ERROR"\GOTO800
680IFE2<>8THEN690ELSEPRINT"HARD DISC ERROR"\GOTO630
690REM
700REM
710REM
720REM
730REM
740PRINT"ERROR ",E2," AT LINE ",E1
745IFE1<290THEN630ELSEIFE8=0THEN630
750DESTROYL2$+D2$
760DESTROY"*"+L2$+D2$
770GOTO630
800IFE1<290ORE1>300THEN820ELSE!"INSUFFICIENT DISC SPACE FOR "+L2$\GOTO750
820GOTO630
```

Fig. 2-18. QSLPRINT program.

is actually printed can be customized by changing the relevant lines in the program.

The program listed in Fig. 2-18 begins with a commercial (lines 10 to 15). The Month String used to print the month that a contact took place is set up (line 30). The log file is identified (lines 40 to 60). The automatic edit mode is set up (lines 70 to 110), and the output device is selected (line 120).

The presence of the log file on the desired disk drive tested for and verified (lines 130 to 160). The STNDATA information is picked up (line 170). The number of QSL cards to be printed is set up (lines 180 to 190). The list of callsigns of people who QSL cards will be prepared for is set up (lines 210 to 240). The pointer file is read (lines 250 to 270).

If the automatic edit flag (E8) is set to 1, then the new log file is created (lines 290 to 300). The log data file is opened (line 310) and the FOR/NEXT loop (lines 320 to 520) perform the whole operation.

Each entry is read in sequence using the read routine in the subroutine library. Each callsign (Z$) is compared with the currently read in entry (lines 330 to 340). If the callsign is found and a QSL card has not been generated for that station by the program, the QSL printed flag (Z[J]) is set (line 360) and the card is printed (lines 370 to 480). The feed to the next label is performed by the loop in line 490.

The auto edit flag is tested for (line 510) and if it is set, the updated information is sent to the new data file. After the loop is finished, any callsigns that QSL cards were not printed for are displayed (lines 530 to 570). This allows the user to find out if a bad callsign was requested or if a desired QSL was not in the assigned log file.

The log files are closed (lines 580 to 620) and the program terminates (line 630) causing the branch to the system chain routine in the subroutine library. Error trapping and handling routines appear (lines 640 to 820), and the subroutine library begins as usual (line 5000).

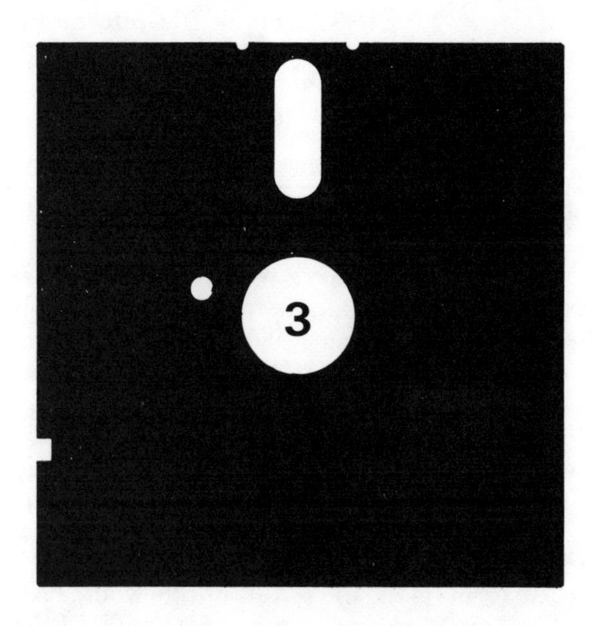

Awards

Thousands of awards are available to radio amateurs. These awards are usually offered for contacting certain stations in certain locations on certain frequencies. The two most famous awards are *Worked All Continents* (WAC) issued by the International Amateur Radio Union (IARU) and *Worked All States* (WAS) issued by the American Radio Relay League (ARRL).

A WAS RECORD-KEEPING PACKAGE

The WAS award is given for making contact with one amateur radio station in each of the 50 states in the USA. This chapter contains a program package that allows records of your WAS to be maintained.

Three programs or commands that conform to the operating system described in Chapter 2 are available. They are:

WASENTER	enters QSL data into a WAS data base
WASGEN	generates a WAS data base
WASPRINT	allows entries to be retrieved from a WAS data base

A WAS data base is a random-access file. There are 50 state entries in the file, and each entry can be up to 64 characters long. The 51st entry is the date that the file was last updated.

The contents of each entry are:

S	the state
C	the call area
W	the QSL status (worked status)
C$	the call sign of the station worked
B	the band/frequency
D	the date

M$ the mode of the contact (e.g., SSB or CW)

WASGEN. WASGEN is listed in Fig. 3-1. It formats a blank WAS data file. Up to 16 characters are allocated for the name of the state (line 50). The name of the new WAS data file is determined (lines 60 to 150). If that name already exists on the floppy disk, suitable error messages are generated and the user is requested to choose another name. In this case, the computer tells the user if he has picked the name of an existing data file (type 3) or the name of a command or program (type 2). Assuming that the name chosen is valid, the new file is created (line 150) opened (line 160), and the user input data is set (line 170) to zero (in the case of numerical data) or to a zero length string (in the case of string data). Then the FOR/NEXT loop (lines 180 to 210) formats the file by reading the state and call area information stored in the data statements (lines 240 to 360) and writes them, together with the blanks or zeros, to the floppy disk.

When the file has been formatted, the blank file information is written into the 51st entry location (line 215), and the program terminates using the now-familiar sequence (lines 9000 to 9050). Error-handling routines also appear (lines 9060 to 9110).

WASENTER. WASENTER is listed in Fig. 3-2. It begins with the usual comments and overhead statements (lines 10 to 40). The strings are set up (lines 50 to 60) and the WAS data file is identified (lines 70 to 110). An error message is displayed (line 110) if the desired data file is not present on the specified disk. An alternate method of error-handling would be to end line 110 with a statement that asked if the user wanted the FILES command run to display what files are on the selected disk. An even better alternative would be to automatically chain a version of FILES that remembered the selected disk, display the data files, and then chain back WASENTER to ask the user for the desired file name again.

```
10REM WASGEN VERSION 790822
20REM BY JOE KASSER G3ZC7
30REM COPYRIGHT SNOW MICRO SYSTEMS INC 1979
40 ERRSET 9060,F1,E2
50 DIM S$(16)
60 INPUT'What is the name of the log file ? ',L$\IFL$=''THEN60
70 INPUT'Which drive to you want it put on ?',D\IFD<1ORD>4THEN70
80 D$=STR$(D)\D$=','+D$(2,2)
90 IF FILE(L$+D$)=2THEN130
100 IF FILE(L$+D$)=3THEN120
110 GOTO 150
120 PRINT L$+' ALREADY EXISTS ',\GOTO140
130 PRINT L$+' IS A COMMAND NAME',
140 PRINT' ON THAT DISC'\GOTO 60
150 CREATE L$+D$,14
160 OPEN #1,L$+D$
170 C$=''\M$=C$\W=0\B=0\D=0
180 FOR I=1 TO 50
190 READ S$,C
200 WRITE#1 %I*64,S$,&C,&W,C$,B,D,M$
210 NEXT
215 WRITE#1 %51*64,'Blank File'
220 CLOSE #1
230 GOTO 9000
240 DATA'ALABAMA',4,'ALASKA',7,'ARIZONA',7,'ARKANSAS',5,
'CALIFORNIA',6
250 DATA'COLORADO',0,'CONNECTICUT',1,'DELAWARE',3,'FLORIDA',4,'GEORGIA',4
260 DATA'HAWAI',6,'IDAHO',7,'ILLINOIS',9,'INDIANA',9,'IOWA',0
270 DATA'KANSAS',0,'KENTUCKY',4,'LOUISIANA',5,'MAINE',1,'MARYLAND',3
280 DATA'MASSACHUSETTS',1,'MICHIGAN',8,'MINNESOTA',0,'MISSISSIPPI',5
290 DATA 'MISSOURI',0
300 DATA'MONTANA',7,'NEBRASKA',0,'NEVADA',7,'NEW HAMPSHIRE',1,'NEW
   JERSEY',2
310 DATA'NEW MEXICO',7,'NEW YORK',2,'N. CAROLINA',4,'N. DAKOTA',0,
   'OHIO',8
320 DATA'OKLAHOMA',5,'OREGON',7,'PENNSYLVANIA',3,'RHODE ISLAND',1
330 DATA'S. CAROLINA',4
340 DATA'S. DAKOTA',0,'TENNESSEE',4,'TEXAS',5,'UTAH',7,'VERMONT',1
350 DATA'VIRGINIA',4,'WASHINGTON',7,'WEST VIRGINIA',8,'WISCONSIN',9
360 DATA'WYOMING',7
9000 FOR I=1 TO 4
9010 IF FILE('SYSTEM,'+STR$(I))=2 THEN EXIT 9030
9020 NEXT \ GOTO 9040
9030 CHAIN 'SYSTEM,'+STR$(I)
9040 PRINT'put a system disc in any drive, then hit 'RETURN''
9050 INPUT' ',A$ \ GOTO 9000
9060ERRSET 9060,E1,E2
9070 IF E2=15THEN9000
9080 IFE1=9010THENIFI<1ORI>4THEN9040ELSE9020
9090 IFE2<>7THEN9100ELSEPRINT'FILE ERROR'\GOTO9000
9100 IFE2<>8THEN9110ELSEPRINT'HARD DISC ERROR'\GOTO9000
9110 PRINT'ERROR ',E2,' AT LINE ',E1\GOTO9000
```

Fig. 3-1. WASGEN program.

```
10REM WASENTER VERSION 790822
20REM BY JOE KASSER G3ZC7
30REM COPYRIGHT SNOW MICRO SYSTEMS INC. 1979
40ERRSET9060,E1,E2
50 DIM S$(16),M1$(36)
60 M1$='JanFebMarAprMayJunJulAugSepOctNovDec'
70INPUT'What is the name of the WAS data file ? ',L$\IFL$=''THEN70
80INPUT'Which drive is it on ? ',D\IFD(1ORD)4THEN80
90D$=STR$(D)\D$=','+D$(2,2)
100IF FILE(L$+D$)=3THEN120
110PRINT'NO IT ISN'T'\GOTO80
120 OPEN#0,L$+D$\OPEN#1,L$+D$
130READ #0 %51*64,S$
140 PRINT L$+' was last updated on '+S$
150 RESTORE
160 INPUT'Which state ? ',S1$\IFS1$=''THEN150
170 FORI=1TO50
180 READ S$
190IFLEN(S$)(LEN(S1$)THEN210
200 IF S1$=S$(1,LEN(S1$)) THEN EXIT 220
210NEXT \ PRINT'ENTRY ERROR' \ GOTO 150
220READ #0 %I*64,S$,&C,&W,C$,B,D,M$
230 GOSUB590
240INPUT'Do you want to update the entry ? ',A$\IFA$=''THEN430
250IFA$(1,1)='Y'THEN260ELSE430
260INPUT'QSL status or whole entry (Q or W) ? ',A$\IFA$=''THEN260
270IFA$(1,1)='Q'THEN350ELSEIFA$(1,1)='W'THEN280ELSE260
280INPUT'Call sign ? ',C$\IFC$=''THEN280
290INPUT'Day of QSO ? ',D1\IFD1(1ORD1)31THEN290
300INPUT'Month (1-12) ? ',D2\IFD2(1ORD2)12THEN300
310INPUT'Year (19xx) ? ',D3\IFD3(1900THEN310
320 D=D1+D2*100+(D3-1900)*10000
330INPUT'Band ? ',B
340INPUT'Mode ? ',M$\IFM$=''THEN340
350INPUT'QSL Status (Q, W, S) ? ',A$\IFA$=''THEN350
360IFA$(1,1)='W'THEN W=1
370IFA$(1,1)='Q'THEN W=2
380IFA$(1,1)='S'THEN W=3
390GOSUB590
400INPUT'OK ? ',A$\IFA$=''THEN400
410IFA$(1,1)='Y'THEN420ELSEIFA$(1,1)='N'THEN260ELSE400
420 WRITE #1 %I*64,S$,&C,&W,C$,B,D,M$
430INPUT'An other entry ? ',A$\IFA$=''THEN450
440IFA$(1,1)='Y'THEN150
450INPUT'What is today's date ? ',S$
460 WRITE #1 %51*64,S$
470 CLOSE#0\CLOSE#1
480GOTO9000
490DATA'ALABAMA','ALASKA','ARIZONA','ARKANSAS','CALIFORNIA'
500DATA'COLORADO','CONNECTICUT','DELAWARE','FLORIDA','GEORGIA'
510DATA'HAWAI','IDAHO','ILLINOIS','INDIANA','IOWA'
520DATA'KANSAS','KENTUCKY','LOUISIANA','MAINE','MARYLAND'
530DATA'MASSACHUSETTS','MICHIGAN','MINNESOTA','MISSISSIPPI','MISSOURI'
540DATA'MONTANA','NEBRASKA','NEVADA','NEW HAMPSHIRE','NEW JERSEY'
550DATA'NEW MEXICO','NEW YORK','N. CAROLINA','N. DAKOTA','OHIO'
560DATA'OKLAHOMA','OREGON','PENNSYLVANIA','RHODE ISLAND','S. CAROLINA'
570DATA'S. DAKOTA','TENNESSEE','TEXAS','UTAH','VERMONT'
580DATA'VIRGINIA','WASHINGTON','WEST VIRGINIA','WISCONSIN','WYOMING'
590IF W=0THENE$=''
600IF W=1THENE$='WORKED'
610IF W=2THENE$='QSL'D'
620IF W=3THENE$='SENT'
630 PRINT#A, S$,TAB(18),E$,\IFW)0THEN650
640PRINT#A\GOTO680
650PRINT#A,TAB(25),C$,TAB(36),\D3=INT(D/10000)\D2=INT((D-D3*10000)
   /100)
660D1=INT((D-D3*10000)-(D2*100))
670PRINT#A, %2I,D1,%,TAB(39),M1$((D2-1)*3+1,(D2-1)*3+3),1900+D3,
   TAB(50),M$
680RETURN
9000 FOR I=1 TO 4
9010 IF FILE('SYSTEM,'+STR$(I))=2 THEN EXIT 9030
9020 NEXT \ GOTO 9040
9030 CHAIN 'SYSTEM,'+STR$(I)
9040 PRINT'put a system disc in any drive, then hit 'RETURN''
9050 INPUT' ',A$ \ GOTO 9000
9060ERRSET 9060,E1,E2
9070 IF E2=15THEN9000
9080 IFE1=9010THENIFI(1ORI)4THEN9040ELSE9020
9090 IFE2()7THEN9100ELSEPRINT'FILE ERROR'\GOTO9000
9100 IFE2()8THEN9110ELSEPRINT'HARD DISC ERROR'\GOTO9000
9110 PRINT'ERROR ',E2,' AT LINE ',E1\GOTO9000
```

Fig. 3-2. WASENTER program.

The WAS data file is opened twice (line 120). This allows both reading from and writing to the same file. In this instance new data is written over old data, so errors in data entry may result in erroneous data being written to the floppy disk. The last updating entry is read (line 130) and then displayed (line 140).

The data pointer is restored to the beginning of the data in line 440 (line 150). The desired state is requested (line 160). The FOR/NEXT loop (lines 170 to 210) reads the name of each state stored in the data statements (lines 490 to 580) and matches them to the desired state (requested in line 160). The matching statement (line 200) only compares the number of characters in the name of the state that is equal to the number of characters in the desired name. Thus, if A is entered as the desired state, ALABAMA passes the test since only the A in ALABAMA would be tested. Similarly, ALASKA is requested by typing AL or ALA, ALABAMA still passes the test and shows up on the screen. As you can see, it's important to enter sufficient characters to distinguish mode. Only when there is an error or mismatch is the whole FOR . . . NEXT loop is the state you want from all others. N. and S. have been used as the abbreviation for North and South whereas the West in WEST VIRGINIA is stored in full. (WE is sufficient to distinguish the name of that state.)

A quick test is done on the length of the strings (line 190) and the matching test (line 200) is skipped if the desired state name is shorter than the name currently read in from the data table. This avoids a string length compare error if a match is attempted (line 200) between IDAHO (which contains 5 characters) and ALABAMA (which contains 7 characters).

If a match is not found, an error message is displayed (line 210). If a match is found, the program loop exists from the loop and continues (line 220); the normal exit performed. When the program flow branches to line 220, the value of I is equal to the number of tests that have been performed. For example, if the desired state is IOWA, the value of I is equal to 15 because IOWA is the 15th entry in the data table. The WAS data file on the floppy disk is read to find the information associated with the desired state (line 220). Since each entry can contain up to 64 bytes, the position of the desired data can be calculated by multiplying I by 64).

The entry display subroutine (which begins at line 590) is called (line 230). Once the entry has been displayed, you are asked if a change is to be made (line 240). If the first character of the user response is not the letter Y, a negative is assumed and the program flow skips to line 430.

If the entry is to be updated, you're again requested to choose between updating the whole entry or just the QSL information (line 260). Only allow a Q or a W is allowed as a response (line 270). The instructions are repeated if any other response is given.

The enter function is most often used to enter details of a contact and then to update the QSL information when a card is received. The order in which information is entered can be set up so that the QSL update section is the last operation. This way, if the QSL information is the only part of the entry that is to be updated, it is easy to make the program skip the other updating lines and branch straight to line 350, the beginning of the QSL information updating routine.

The callsign, date band, and mode information are asked for and stored (lines 280 to 340). The date is stored as a single number. The QSL information is also stored as a number. It can be:

- 1 (W) Station was worked, but no QSL has been sent or received
- 2 (Q) QSL has been received
- 3 (S) QSL has been sent, but not received

The user enters the letter (line 350) which is converted to the appropriate number (lines 360 to 380). If another letter is entered (in response to the prompt of line 350), the value of W is not changed.

The entry display subroutine which starts at line 590 is called (line 390), and the updated information is displayed. The confirmation that the new data is valid is requested and processed (lines 400 to 410) There's a string length test (line 400) and a positive test for a Y or an N response (line 410). If an unknown letter is entered (line 400), the question is asked again. If an N or a NO was entered, the program branches back to the start of the updating process (line 260). If a YES or a Y is entered the new data is written to disk (line 420). You're then asked if another entry is to be written or examined (line 430). If it is, the program branches back to restore the data pointer (line 150) and the whole sequence recommences. If not, the update day is itself updated (lines 450 to 460), the disk files are closed (line 470) and the termination routine starting at line 9000 begins (line 480).

The names of the states stored in the data statements (lines 490 to 580) are in alphabetical order. As long as the order is the same in both programs (WASGEN and WASENTER), the exact order is immaterial. Alphabetical order was chosen to make the listing more readable. Alternatively, the states can be sorted by call area.

The entry display subroutine begins (line 590). The value of W is converted to E$ (lines 590 to 620). If W = 0, then E$ is a blank string or a zero length string. The name of the state (S$) is printed (line 630) as is its contact status (E$). If no contact has been entered, E$ is a zero length string, and there is no point in trying to display any other data; hence, the test for W > 0 (line 630). If W = 0, a carriage return/line feed is sent to the console and the subroutine is exited (line 640).

If a contact is present in the entry for the state, the callsign (C$) is displayed, followed by the date (line 650).

The date as stored on the disk is changed to the day-month-year format and displayed (lines 650 to 670) before returning from the subroutine (line 680). The TAB (—) feature is used to format the display on the screen.

Note the #A in the PRINT statements (lines 630 to 670). A is not defined anywhere in the program and is thus assumed to be 0, representing the console. A routine asking which device the output should be sent to can be inserted into this program, setting up the value of A. However, that routine was not added because the purpose of this routine is to enter/edit the WAS data file from the console.

The terminating and error-handling routines should be familiar to you by now (lines 9000 to 9110).

Only 64 bytes are allocated to each entry, and no test is performed to ensure that the entry is less than 64 bytes in length. With up to 16 bytes for the name of the state and up to 10 bytes each for the callsign and mode information, the total number of bytes in each entry will not exceed 64.

64 was chosen to simplify program debugging. It is easy enough to perform a direct read to memory from the disk file using the Disc Operating System (DOS) debug features, when trying to see what is in the file. Using the value of 64 the entries will line up on a 64 × 16 memory-mapped video display neatly. The extra overhead of the unused space is not that significant.

WASPRINT. WASPRINT causes the contents of a WAS data file to be printed at a selected output device. The states are always listed in alphabetical order. Alphabetizing is simple here because the WASGEN command set up the data file so as to store the entries for the states in alphabetical order.

The program is listed in Fig. 3-3. It begins with the usual overhead (lines 10 to 30). This overhead is important because it shows who wrote the program and when it was last updated. Next, it sets up the length of the state name string (S$), defines the name of the month string (M1$), identifies the name of the WAS file, and ensures that it actually is present on the specified floppy disk (line 40 to 110, identical to line 40 to 110 of WASENTER). The value of various variables are set to zero (line 120). W1 is the number of states in the callarea worked, W2 is the number of stations QSL'd, and W3 is the number to whom cards have been sent but who have not replied. These correspond to the 1, 2, and 3 values for the worked variable (W) in each of the state entries. L1 is the number of states that have been printed out. The variables are set to zero (line 120) even though BASIC assumes that they are zero unless otherwise instructed because if a repeat is requested later, or if another call area is requested, the variables must be reset. Since different computer languages and different dialects

```
10REM WASPRINT VERSION 790822
20REM COPYRIGHT SNOW MICRO SYSTEMS 1979
30REM BY JOE KASSER G3ZCZ
40 ERRSET 9060,E1,E2
50 DIM S$(16),M1$(36)
60 M1$='JanFebMarAprMayJunJulAugSepOctNovDec'
70INPUT'What is the name of the WAS data file ? ',L$\IFL$=''
  THEN70
80INPUT'Which drive is it on ? ',D\IFD<1ORD>4THEN80
90D$=STR$(D)\D$=','+D$(2,2)
100IF FILE(L$+D$)=3THEN120
110PRINT'NO IT ISN'T'\GOTO80
120W1=0\W2=W1\W3=W1\L1=W1
130INPUT'Which call area (0-9) ? ',C1
140INPUT'Which output device (0-7) ? ',A\IFA<0ORA>7THEN140
150OPEN#0,L$+D$
160READ #0 %51*64,S$
170PRINT
180 PRINT#A,L$+ ' WAS Status as of ',S$,\IFC1>9THEN200
190PRINT#A,' for the',STR$(C1),' call area',
200 PRINT#A
210PRINT#A
220 PRINT#A,'STATE',TAB(18),'STATUS',TAB(26),'CALL',TAB(42),'DATE',
230 PRINT#A,TAB(50),'MODE'
240FORJ=1TO63\PRINT#A,'-',\NEXT\PRINT#A
250 FOR I=1 TO 50
260READ #0 %I*64,S$,&C,&W,C$,B,D,M$
270IF W=0THENE$=''
280IF W=1THENE$='WORKED'
290IF W=3THENE$='SENT'
300IF W=2THENE$='QSL'D'
310IFC=C1THEN320ELSEIFC1>9THEN320ELSE410
320 PRINT#A, S$,TAB(18),E$,\IFW>0THEN340
330 PRINT#A\GOTO400
340PRINT#A,TAB(25),C$,TAB(36),\D3=INT(D/10000)\D2=INT((D-D3*10000)/100)
350D1=INT((D-D3*10000)-(D2*100)) \ ON W GOTO 360,370,380
360 W1=W1+1 \ GOTO 390
370 W2=W2+1 \ GOTO 390
380 W3=W3+1
390PRINT#A, %2I,D1,%,TAB(39),M1$((D2-1)*3+1,(D2-1)*3+3),1900+D3,
   TAB(50),M$
400L1=L1+1
410 NEXT
420 CLOSE #0
430PRINT
440 PRINT#A, W2,' QSL'd,',W1,' worked and ',W3,' cards in the mail'
450IFL1=W2THENPRINT'You've got them all'
460IFC1>9THEN480
470INPUT'An other one ? ',A$\IFA$=''THEN500ELSEIFA$(1,1)='Y'THEN120E
   LSE500
480IFA=0THEN500
490FORI=60 TO 66 \ PRINT#A\ NEXT
500REM ENDING
9000 FOR I=1 TO 4
9010 IF FILE('SYSTEM,'+STR$(I))=2 THEN EXIT 9030
9020 NEXT \ GOTO 9040
9030 CHAIN 'SYSTEM,'+STR$(I)
9040 PRINT'Put a system disc in any drive, then hit 'RETURN''
9050 INPUT' ',A$ \ GOTO 9000
9060ERRSET 9060,E1,E2
9070 IF E2=15THEN9000
9080 IFE1=9010THENIFI<1ORI>4THEN9040ELSE9020
9090 IFE2<>7THEN9100ELSEPRINT'FILE ERROR'\GOTO9000
9100 IFE2<>8THEN9110ELSEPRINT'HARD DISC ERROR'\GOTO9000
9110 PRINT'ERROR ',E2,' AT LINE ',E1\GOTO9000
```

Fig. 3-3. WASPRINT program.

then opened (line 150) and the date of the of these languages make different assumptions as to the value of variables that have not been set up in the program, it is good practice to clear all variables that are not implicitly set up by the program before processing them. In other words, don't assume that variables whose values are unspecified have a value of zero unless you know for sure that the dialect of the compiler/interpreter that you are using assumes the same.

The call area to be displayed is determined (line 130). A response greater than 9 tells the program to display all 50 entries.

Which device the output is sent to is determined (line 140). The WAS data file is last update is read (line 160). A heading for the listing is printed (lines 180 to 210). If a particular call area has been specified, the heading includes that information (line 190). The column headings are tabulated (lines 220 to 230) and underscored (line 240).

The FOR/NEXT loop (lines 250 to 410) fetches, examines and displays the contents of each state entry. The entry is red (line 260) and the value of E$ is set up based on the value of W in the same manner as in the WASENTER program (lines 270 to 300).

The call area (C) that the current entry is in is tested (line 310). If it is not in the wanted call area, and if only a specified call area is to be listed, the program skips to line 410 which causes the next iteration of the loop to begin. If line 310 is inserted between lines 260 and line 270, the execution of the program is speeded up, since if the entry is not to be displayed there is no point in processing the value of W to create E$.

The name of the state and the QSL

A

WASOSCAR WAS Status as of 31 DEC 1977

STATE	STATUS	CALL	DATE	MODE
ALABAMA				
ALASKA				
ARIZONA				
ARKANSAS				
CALIFORNIA	QSL'D	W6CG	24 May 1976	CW
COLORADO				
CONNECTICUT	QSL'D	K1HTV	16 Jun 1976	SSB
DELAWARE				
FLORIDA	QSL'D	K4KQ	31 May 1976	CW
GEORGIA				
HAWAI				
IDAHO				
ILLINOIS	QSL'D	W9QQG	28 Dec 1975	CW
INDIANA				
IOWA	QSL'D	W0II	12 Jan 1976	CW
KANSAS	SENT	W0CY	23 Oct 1975	CW
KENTUCKY	QSL'D	W4MOP	16 Jun 1976	SSB
LOUISIANA				
MAINE				
MARYLAND	QSL'D	WA3LND	29 Sep 1974	CW
MASSACHUSETTS	QSL'D	W1CRL	13 Mar 1977	SSB
MICHIGAN	QSL'D	WB8BGY	29 Sep 1974	CW
MINNESOTA	QSL'D	W0PHD	17 Oct 1975	SSB
MISSISSIPPI				
MISSOURI	QSL'D	W0SL	17 Mar 1977	SSB
MONTANA				
NEBRASKA				
NEVADA				
NEW HAMPSHIRE	QSL'D	W1JSM	29 Dec 1975	CW
NEW JERSEY	QSL'D	K2QBW	2 Jan 1976	CW
NEW MEXICO				
NEW YORK	QSL'D	W2GN	7 Oct 1975	CW
N. CAROLINA				
N. DAKOTA	QSL'D	W0EOZ	1 Feb 1976	CW
OHIO	WORKED	K8NU	25 Mar 1977	SSB
OKLAHOMA	QSL'D	WA5ETV	4 Jan 1976	SSB
OREGON				
PENNSYLVANIA	QSL'D	AC3BWU	4 Jan 1976	SSB
RHODE ISLAND	SENT	WA1RFT	25 Dec 1975	CW
S. CAROLINA				
S. DAKOTA	QSL'D	W0IT	1 Nov 1976	CW
TENNESSEE				
TEXAS	QSL'D	W5VY	17 Mar 1977	SSB
UTAH				
VERMONT				
VIRGINIA	QSL'D	W4ART	6 Oct 1975	CW
WASHINGTON				
WEST VIRGINIA				
WISCONSIN	QSL'D	W9OII	17 Oct 1975	CW
WYOMING				

21 QSL'd, 1 worked and 2 cards in the mail

WASOSCAR WAS Status as of 31 DEC 1977 for the 0 call area

STATE	STATUS	CALL	DATE	MODE
COLORADO				
IOWA	QSL'D	W0II	12 Jan 1976	CW
KANSAS	SENT	W0CY	23 Oct 1975	CW
MINNESOTA	QSL'D	W0PHD	17 Oct 1975	SSB
MISSOURI	QSL'D	W0SL	17 Mar 1977	SSB
NEBRASKA				
N. DAKOTA	QSL'D	W0EOZ	1 Feb 1976	CW
S. DAKOTA	QSL'D	W0IT	1 Nov 1976	CW

5 QSL'd, 0 worked and 1 cards in the mail

B

WASDATA WAS Status as of 15 AUG 1979

STATE	STATUS	CALL	DATE	MODE
ALABAMA				
ALASKA				
ARIZONA				
ARKANSAS	QSL'D	W5SW	22 Dec 1972	SSB
CALIFORNIA	QSL'D	WB6HAR	8 Aug 1973	SSB
COLORADO	WORKED	WB0MIV	21 Nov 1976	SSB
CONNECTICUT				
DELAWARE				
FLORIDA	QSL'D	WB4FSV	12 Aug 1973	SSB
GEORGIA	QSL'D	WA4YWR	11 Jan 1979	SSB
HAWAII	QSL'D	KH6IJ	3 Mar 1974	SSB
IDAHO	QSL'D	KJ7BSA	1 Aug 1973	SSB
ILLINOIS	QSL'D	KW9ITU	12 May 1974	SSB
INDIANA				
IOWA				
KANSAS	QSL'D	WB0FIS	6 Sep 1972	SSB
KENTUCKY				
LOUISIANA				
MAINE	QSL'D	W1VF	22 Dec 1972	SSB
MARYLAND	QSL'D	VK1RY/W3	1 Nov 1972	SSB
MASSACHUSETTS	QSL'D	K1FJR	17 Dec 1972	SSB
MICHIGAN	QSL'D	WC8ITU	14 May 1974	SSB
MINNESOTA				
MISSISSIPPI				
MISSOURI				
MONTANA	WORKED	WA7KKN	21 Nov 1976	SSB
NEBRASKA	QSL'D	KT0NEB	7 Sep 1972	SSB
NEVADA				
NEW HAMPSHIRE				
NEW JERSEY				
NEW MEXICO				
NEW YORK	QSL'D	WA2JTX	11 Jan 1979	SSB
N. CAROLINA	QSL'D	KR4ITU	16 May 1974	SSB
N. DAKOTA				
OHIO				
OKLAHOMA				
OREGON				
PENNSYLVANIA				
RHODE ISLAND				
S. CAROLINA	QSL'D	K4II/4	24 Jun 1973	SSB

Fig. 3-4. (A) WASOSCAR WAS status as of 31 Dec. 1977. (B) WASDATA WAS status as of 15 Aug. 1979.

```
S. DAKOTA
TENNESSEE
TEXAS           QSL'D  KH5ITU    12 May 1974   SSB
UTAH
VERMONT
VIRGINIA        QSL'D  KE4ITU    12 May 1974   SSB
WASHINGTON
WEST VIRGINIA
WISCONSIN
WYOMING
 18 QSL'd, 2 worked and  0 cards in the mail
```

status are printed out (line 320). If the value of W shows that a contact has not been made, the line is terminated at this time (line 330). If not, the callsign is printed, the date is computed, and the values of W1, W2, or W3 are updated, depending on the value of W (lines 340 to 380). Then the formatted date information and the mode that was used to make the contact are printed (line 390). The state displayed counter (L1) is incremented (line 400) and the next iteration of the loop takes place (line 410).

When all 50 states have been processed, the WAS data file is closed (line 420) and a summary is printed (lines 430 to 440). If all the states in the specified listing have been QSL'd, the computer (prompted by line 450) displays a congratulatory message. The user can request a rerun of the program only if a total WAS listing for all 50 states was not requested (lines 460 to 470).

If the output was not routed to the console (checked by line 480) six line feeds are sent (line 490).

The same ending and error-handling appears as seen in the WASGEN and WASENTER programs (lines 9000 to 9110).

A DXCC RECORD KEEPING PACKAGE

One of the criteria used to measure the performance of an amateur radio station is the number of countries with which contacts have been made and confirmed (QSL'd). Many DX chasers dream of rising to the top of the DX Century Club (DXCC) honor roll. The minimum number of confirmed countries required to join is 100. This means that a substantial amount of record keeping is needed to keep track of the status of various countries. Has a country been worked? If so, has a QSL been sent? Was the QSL sent direct, or via a QSL manager, or via the bureau? Has a QSL received? The computer can make this record-keeping chore very easy.

A package of programs designed to keep track of the DXCC status of an amateur radio station is described below. Written in MICROSOFT Basic, it enables the operator to enter the data, store it on a floppy disk, and print summary listings sorted in alpha numerical order. Options exist for selected printouts such as only those stations QSL'd (useful for submitting a claim to the ARRL) or those contacts which are still unconfirmed.

The WAS package described earlier in this chapter set up blank entries for all the states. These were filled in as the states were worked. This package requires the user to enter the prefix at the same time as the rest of the data. The computer then totals the number of prefixes, assuming that the user has allocated one prefix to a country. The same package of programs can thus be used to keep track of the different prefixes worked, an even more complex data processing situation. The listings generated by the package can be kept at the operating position and referenced at any time. Handwritten changes can be made as new countries are worked, and the computer updated as needed. The lists are particularly useful during contests when a decision has to be made as to how much time to spend in a pileup for a rare station. Obviously, more time should be spent if that country has never been worked, or if a QSL has been difficult to obtain.

The programs in this section have been written in the MICROSOFT dialect of BASIC as opposed to the NORTHSTAR dialect used in the WAS package. The differences between the dialects will become

```
10 REM DXCCGEN BY JOE KASSER 1982
20 ON ERROR GOTO 620
30 INPUT "What is the name of the DXCC record file ";F$
40 IF LEN(F$) = 0 THEN 30
50 INPUT "Is this a new file ";N$
60 IF LEN(N$) = 0 THEN 50
70 N = 0
80 IF LEFT$(N$,1) = "Y" THEN N = 1
90 IF LEFT$(N$,1) = "N" THEN N = 2
100 IF N = 0 THEN 50
110 OPEN "O",#2,F$ + ".$$$"
120 IF N = 1 THEN 180
130 OPEN "I",#1,F$ + ".DX"
140 IF EOF(1) THEN 200
150 LINE INPUT#1,A$
160 PRINT#2,A$
170 GOTO 140
180 OPEN "O",#3,F$ +".BAK" : CLOSE#3
190 GOTO 210
200 CLOSE#1
210 INPUT "Which country prefix ";P$
220 IF LEN(P$) = 0 THEN 210
230 IF LEFT$(P$,1) = "*" THEN 550 : REM exit
240 INPUT "Which Band ";B$
250 IF LEN(B$) = 0 THEN 240
260 INPUT "Which Mode ";M$
270 IF LEN(M$) = 0 THEN 260
280 INPUT "Call Sign ";C$
290 IF LEN(C$) = 0 THEN 280
300 INPUT "Date YY/MM/DD ";D$
310 IF LEN(D$) = 0 THEN 300
320 INPUT "Time ";T
330 Q = 0 : INPUT "QSL Status (Q W D B) " ;Q$
340 IF LEN(Q$) = 0 THEN 330
350 IF LEFT$(Q$,1) = "Q" THEN Q = 1
360 IF LEFT$(Q$,1) = "W" THEN Q = 2
370 IF LEFT$(Q$,1) = "D" THEN Q = 3
380 IF LEFT$(Q$,1) = "B" THEN Q = 4
390 IF Q = 0 THEN  330
400 PRINT P$ TAB(6) B$ TAB(10) M$ TAB(14);
410 PRINT C$ TAB(25) D$ TAB(34) T ;
420 PRINT TAB(40) ;
430 ON Q-1 GOTO 450,460,470
440 PRINT "QSL'D" : GOTO 480
450 PRINT "WORKED" : GOTO 480
460 PRINT "MAILED DIRECT" : GOTO 480
470 PRINT "SENT VIA BUREAU" : GOTO 480
480 INPUT "Are the data correct ";A$
490 IF LEN(A$) = 0 THEN 480
500 IF LEFT$(A$,1) = "N" THEN 210
510 IF LEFT$(A$,1) = "Y" THEN 520 ELSE 480
520 REM NOW WRITE TO DISK
530 PRINT #2,P$;",";C$;",";B$;",";M$;",";D$;",";T;",";Q
540 GOTO 210 : REM GET NEXT
550 REM exit routine
560 CLOSE#2
570 IF N = 1 THEN 600
580 KILL F$ + ".BAK"
590 NAME F$ + ".DX"  AS F$ + ".BAK"
600 NAME F$ + ".$$$" AS F$ + ".DX"
610 GOTO 640
620 IF ERR = 53 AND ERL = 580 THEN RESUME 600
630 RESUME 640 : REM ALL OTHER ERRORS ARE YOUR FAULT
640 END
```

Fig. 3-5. The DXCCGEN program.

apparent during the discussion of each of the three programs in the package.

THE DXCCGEN PROGRAM

The DXCCGEN program listed in Fig. 3-5 is used to put new entries into a data file. The program begins with an overhead statement (line 10). The error-trapping routine lets the program bypass any reference to a non-existent backup (.BAK) file on the disk (line 20). The user is asked to identify the name of the record file (line 30). This program (in MICROSOFT BASIC running under CP/M) assumes a filetype of ".DX." The user response is checked to make sure an entry was made (line 40).

The status of the record file is determined (line 50). The value of the file state flag (N) is set up (lines 70 to 90). If the user response is neither a "yes" nor a "no," the question is repeated (line 100).

A temporary file (filetype .$$$) which will be used to hold the data being entered by the user is created and opened (line 110). When all the data has been entered, the old file (if any) is converted to backup status (.BAK) and the temporary file becomes the permanent one by being renamed to the .DX filetype. This is MICROSOFT BASIC's way of adding data to sequential files. If the data file is new, the program flow skips forward (line 120) to line 180, bypassing the statements that read in data from an existing file.

Data is moved from the existing file (.DX) to the new one (.$$$) (lines 130 to 170). The addition of data to a sequential file in MICROSOFT BASIC works like this: A new file is created and opened as an output file with a temporary filetype (line 110). The existing file is then opened as an input file (line 130). A loop that reads a line of data from the existing file (line 150) and writes it back to the new file (line 160). The end of file condition is detected by the EOF statement (line 140) that is inserted prior to the INPUT line of the loop.

A dummy backup file is created on the disk (line 180) if the record file is a new one to avoid the possibility of an error later when the filetype names are updated at the end of the program.

The old data file is closed (line 200). The data entry sequence begins (line 210). The user enters the prefix as a response to the question asked in line 210. If the RETURN key is accidentally hit before the first character of the reply string is entered, it's seen as an error (line 220). If the user wishes to terminate the entry sequence, the "*" character can be entered (line 230). Band information (line 240), mode data (line 260), and callsign (line 280) are requested. The date of the contact is also requested (lines 300 to 310). A sample format is given in the prompt question: (YY/MM/DD). The 19th of March 1982 would be entered as 82/03/19. Using this format, the dates can be scanned by year and month relatively easily. (See Fig. 3-8 and 3-10). The time of the contact is entered (line 320) as a number. This package puts the time of each contact into the record file so that contacts can be cross-referenced to log contacts later. Many contacts are made during contests, and finding the contact in the log even knowing the date may be difficult.

The QSL status data is requested and accepted (lines 330 to 340). Four different QSL states are allowed: QSL'd, worked and a card sent direct, worked and a card sent via the bureau, and worked with no card sent (lines 350 to 380). If the test fails due to an illegal entry, the program flow to branches back to line 330 to let the user enter a valid status (line 390).

The data that has been entered is displayed (lines 390 to 460), and the user must then confirm that the data is correct before it is written to disk (line 470). If the user sees an error, the program flow branches back to line 210 to repeat the entry sequence. If the data is good, it is written to the disk (line 530), and the program branches back to line 210 to ask the user for the next set of data (line 540).

The closeout sequence begins (line 550). The output file is closed (line 560). The program flows to line 600 if the file is being written for the first time (line 570). If it is not, the old backup file is deleted from the disk (line 580), and the filetype of the existing file is changed from .DX to .BAK (line 590). If the .BAK file does not exist on the disk, the error is trapped by the error-trapping routine. In any case, the program continues, naming the temporary file (.$$$) as the new .DX file (line 590). The program flow then skips over the error-catching routine to the final line (line 630).

The error-trapping routine begins (line 610) and, in this case, is only used to trap the error that arises if the .BAK file is not present on the disk.

The data is stored in the disk file in ASCII format with a comma between each string so that it can be read back by another BASIC program later. A sample file is shown in Fig. 3-6. This file is in ASCII, and can be edited by any text editor or word processing program. This eliminates the need for a DXCCEDIT program such as was used in the WAS package written in Northstar BASIC.

```
4U,4U1ITU,15,SSB,81/10/31, 2028 , 4
CT2,CT2CR,15,SSB,81/08/30, 1939 , 4
DL,DL2NAI,10,SSB,81/09/01, 920 , 1
DM,Y46XF,20,SSB,81/09/12, 1720 , 1
F,F8ZS,15,SSB,81/09/20, 1910 , 1
FS7,FG7TD/FS7,20,SSB,81/08/31, 2030 , 4
G,G3IOR,15,SSB,81/09/13, 1905 , 1
GJ,GJ3VLX,20,SSB,81/08/30, 1819 , 4
I,I5TDJ,20,SSB,81/08/30, 1443 , 2
J2,J28DL,10,SSB,81/11/12, 1939 , 2
JA,JA3BOA,15,SSB,81/09/16, 1515 , 4
OK,OK3JW,15,SSB,81/09/12, 1711 , 1
PA,PA3BNT,10,SSB,81/09/04, 1321 , 1
SM,SM0AVK,20,SSB,81/08/30, 1519 , 1
SV0,SV0BL,20,SSB,81/09/07, 1340 , 2
W,KA1HMW,10,SSB,82/01/31, 1515 , 1
YN,HT3JQ,20,SSB,81/11/12, 448 , 1
ZL,ZL3HH,20,SSB,81/08/31, 510 , 4
LZ,LZ1XI,20,SSB,81/08/30, 1440 , 2
YU,YU4EBL,15,SSB,81/08/31, 1013 , 2
EA,EA7CEL,10,SSB,81/08/31, 1230 , 2
YO,YO3AJN,10,SSB,81/09/01, 1354 , 2
HB,HB9XH,15,SSB,81/09/02, 1859 , 2
UQ,UQ2GHT,10,SSB,81/09/04, 1402 , 2
HA,HA4XH,20,SSB,81/09/07, 1748 , 2
OE,OE6HZG,15,SSB,81/09/12, 1707 , 2
GM,GM5AIW,15,SSB,81/09/12, 1731 , 2
SP,SP6IXF,15,SSB,81/09/12, 1736 , 2
4Z,4Z4ZB,20,SSB,81/10/06, 420 , 2
8P,8P6QL,15,SSB,81/10/24, 1949 , 2
LU,LU9FFA,10,SSB,81/10/31, 1802 , 4
UH8,RH8HCV,10,SSB,81/11/06, 1213 , 2
J7,J73PD,15,SSB,81/11/11, 1826 , 2
5N,OE5BS/5N2,10,SSB,81/11/13, 1259 , 4
UL,UL7NAJ,10,SSB,82/01/01, 714 , 2
UC,UC2AHF,10,SSB,82/01/01, 714 , 2
VP8,VP8ANT,20,SSB,82/01/02, 1954 , 2
FK,FK8DH,15,SSB,82/01/03, 522 , 4
M1,M1J,15,SSB,82/01/03, 752 , 4
9H,9H1R,10,SSB,82/01/03, 948 , 4
GD,GD2HCX,10,SSB,82/01/03, 1316 , 4
PY,PY5EG,10,SSB,82/01/03, 1417 , 2
Z2,Z21EI,10,SSB,82/01/03, 1618 , 4
TA,TA1CT,20,SSB,82/01/09, 1826 , 2
UB,RB5MWK,10,SSB,82/01/08, 1257 , 2
3B8,3B8CA,10,SSB,81/01/10, 1413 , 2
VK,VK7GE,20,SSB,82/01/13, 646 , 4
CX,CX4AB,15,SSB,82/01/16, 1738 , 4
UF,UF6FFF,20,SSB,82/01/19, 537 , 2
SV8,SV8KX,20,SSB,82/01/19, 541 , 2
3A,3A2CP,15,SSB,82/02/25, 1807 , 2
FM,FM7CD,10,SSB,82/02/28, 1254 , 2
LX,PA3AIR/LX,20,SSB,81/11/01, 1715 , 4
UA3,UA3DQE,10,SSB,81/09/01, 811 , 2
UA2,UK2BBB,15,SSB,81/09/12, 1746 , 2
UA9,UA9MAZ,10,SSB,81/11/06, 1222 , 2
IS,IS0IIT,20,SSB,82/01/15, 734 , 2
VE,VE3MFT,10,SSB,82/03/07, 1319 , 2
```

Fig. 3-6. A sample file.

THE DXCCREAD PROGRAM

The data stored on the disk can be selectively read out so that listings of counties worked or QSL'd can be printed in a formatted manner. The program used to perform this function is the DXCCREAD program listed in Fig. 3-7.

The overhead statement (line 10) identifies the program. The user is asked to identify the DXCC record file to be scanned (line 20), and the response is checked for error (line 30). The user is asked to enter the date for inclusion in the printout (line 40), and the date is checked (line 50).

The user is asked what kind of listing should be generated (line 60) and the reply is processed (lines 70 to 100). If the first letter of the response is "N," the print control flag (Q9) is set to the value 5 (to list everything), and the program flows on to line 190 bypassing the statements that determine the selection criteria. If the reply begins with the letter "Y," the program flow continues (line 110), and if the response is neither a "Y" nor an "N," the program branches back to line 60 to phrase the question another time (line 100).

Each option in the menu is allocated a number and the user chooses a number

```
10 REM DXCCREAD VER 820330
20 INPUT "Which DXCC record file " ; F$
30 IF LEN(F$) = 0 THEN 20
40 INPUT "What is the date today "; D1$
50 IF LEN(D1$) = 0 THEN 40
60 INPUT "Dou you wnat a selection only " ; Z$
70 IF LEN(Z$) = 0 THEN 60
80 IF LEFT$(Z$,1) = "N" THEN Q9 = 5 : GOTO 190
90 IF LEFT$(Z$,1) = "Y" THEN 110
100 GOTO 60
110 PRINT "The folowing options are available"
120 PRINT "Stations QSL'd                    1"
130 PRINT "Stations worked but no card sent 2"
140 PRINT "Cards in the mail direct          3"
150 PRINT "Cards sent via the bureau         4"
160 PRINT "All contacts in the list          5"
170 INPUT "Enter number " ; Q9
180 IF Q9 < 1 OR Q9 > 5 THEN 110
190 INPUT "CONSOLE OR PRINTER " ; A$
200 IF LEN(A$) = 0 THEN 190
210 IF LEFT$(A$,1) = "P" THEN P6 = 1
220 IF P6 = 0 THEN 590
230 INPUT "Wait between pages (Y/N) " ; Z$
240 IF LEN(Z$) = 0 THEN 230
250 IF LEFT$(Z$,1) = "Y" THEN P5 = 1 ELSE P5 = 0
260 GOTO 590
270 REM SUBROUTINES BEGIN HERE
280 IF L9 =< 0 THEN GOSUB 470
290 L9 = L9 - 1 : REM LINE COUNT
300 T$ = STR$(T) : T$ = "0000" + MID$(T$,2) : T$ = RIGHT$(T$,4)
310 I = I + 1
320 IF P6 = 1 THEN 360
330 PRINT I;TAB(5) P$;TAB(10);C$;TAB(20);D$;TAB(30);
340 PRINT T$;TAB(37);B$;TAB(41);M$;TAB(47);
350 GOTO 380
360 LPRINT I;TAB(5) P$;TAB(10);C$;TAB(20);D$;TAB(30);
370 LPRINT T$;TAB(37);B$;TAB(41);M$;TAB(47);
380 IF Q = 1 THEN Q$ = "QSL'D"
390 IF Q = 2 THEN Q$ = "WORKED"
400 IF Q = 3 THEN Q$ = "CARD SENT DIRECT"
410 IF Q = 4 THEN Q$ = "CARD SENT VIA BUREAU"
420 IF P6 = 1 THEN 450
430 PRINT Q$
440 GOTO 460
450 LPRINT Q$
460 RETURN
470 REM HEADING ROUTINE
480 P9 = P9 + 1 : REM PAGE NUMBER
490 IF P6 = 1 THEN 510
500 GOTO 570 : REM DONT PAGE IT
510 IF P9 > 1 THEN LPRINT CHR$(26),CHR$(12)
520 IF P5 = 1 THEN INPUT "CHANGE PAPER AND TYPE P TO CONTINUE " ; Z$
530 LPRINT F$,D1$,"PAGE ";P9
540 LPRINT
550 LPRINT "   PX   CALL       DATE    TIME  BAND MODE   QSL STATUS"
560 LPRINT"-------------------------------------------------------------"
570 L9 = 50 : REM 50 IS HALF OF DXCC
580 RETURN
590 OPEN"I",#2,F$ + ".DX"
600 IF P6 = 1 THEN LPRINT CHR$(26)
610 IF EOF(2) THEN 690
620 INPUT#2,P$,C$,B$,M$,D$,T,Q
630 IF Q = 1 THEN Q1 = Q1 + 1
640 IF Q = 2 THEN Q2 = Q2 + 1
650 IF Q = 3 THEN Q3 = Q3 + 1
660 IF Q = 4 THEN Q4 = Q4 + 1
670 IF Q9 = Q OR Q9 = 5 THEN GOSUB 280
680 GOTO 610
690 CLOSE #2
700 IF Q9 = 5 THEN 710 ELSE 880
710 IF P6 = 1 THEN 800
720 PRINT
730 PRINT "SUMMARY"
740 PRINT "TOTAL = " , , Q1 + Q2 + Q3 + Q4
750 PRINT "QSL'S RECEIVED = ",Q1
760 PRINT "QSL'S SENT DIRECT = ",Q3
770 PRINT "QSL'S SENT VIA BUREAU = ", Q4
780 PRINT "WORKED BUT NO CARD SENT = ", Q2
790 GOTO 880
800 GOSUB 470
810 LPRINT
820 LPRINT "SUMMARY"
830 LPRINT "TOTAL = " , , Q1 + Q2 + Q3 + Q4
840 LPRINT "QSL'S RECEIVED = ",Q1
850 LPRINT "QSL'S SENT DIRECT = ",Q3
860 LPRINT "QSL'S SENT VIA BUREAU = ", Q4
870 LPRINT "WORKED BUT NO CARD SENT = ", Q2
880 I = 0
890 L9 = 0 : P9 = 0 : P5 = 0 : P6 = 0
900 Q1 = 0 : Q2 = 0 : Q3 = 0 : Q4 = 0
910 INPUT "Again " ; Z$
920 IF LEN(Z$) = 0 THEN 910
930 IF LEFT$(Z$,1) = "Y" THEN 60
```

Fig. 3-7. The DXCREAD program.

corresponding to the desired option (line 170). The response is tested (line 180) to see that it is a valid number. The user must decide if the listing is to be displayed on the screen (console) or printed at the printer (lines 180 to 190). If the printer option was chosen (line 210), the printout flag (P6) is set. If not, the program branches to line 590, bypassing the printer customisation area. The program assumes that all responses to the question other than those beginning with the letter "P" mean that the list should be displayed. The printer can pause between pages so that separate pages can be used instead of a continuous length of paper (lines 230 to 250). The print pause flag (P5) is set (line 250), and the program flow bypasses the subroutines and continues at line690 (line 260).

The printing subroutine begins (line 280) by testing the lines-printed counter (L9). If it is zero, the page heading subroutine (starting at line 470) is invoked. If not, the line count is decremented (line 290). The time of the contact data, stored on the disk as a number, is converted to a string and formatted to include leading zeros (line 300). The number of prefixes printed (I) is then incremented (line 310). The program branches to line 330 or line 360 depending where the listing is to be sent (console or printer). The MICROSOFT dialect of BASIC does not allow a PRINT# statement, and separate lines containing the PRINT or LPRINT statements have to be incorporated in the program. The contents of lines 330 to 340 and lines 360 to 370 are identical in every respect except which device the output is sent. The LPRINT statements are bypassed if the console route is being followed (line 350). The QSL status byte (Q) is converted to a string (lines 380 to 410) before the QSL status is sent to the relevant output device (lines 430 to 450). The subroutine terminates (line 460).

The subroutine beginning line 470 prints the page heading. The page counter (P9) is incremented (line 480). The heading is skipped if the output is sent to the screen (lines 490 to 500). If line 490 had been written as

```
490  IF P6 = 0 THEN 570
```

then line 500 would not have been necessary. If the line to be printed is the last line of the page (or the first line ever of a new listing) a form feed character is sent (line 510). The CHR$(26) and CHR$(12) characters are inserted in the program to cover almost all output devices. The printer waits between pages if that mode was desired (line 520). The same time, the user is told to change the paper and notify the computer when that has been done, using the dummy variable Z$.

The name of the file, the date, and the page number of the listing is printed (line 530). A space is printed (line 540) and then the column headings (line 550), which are underlined (line 560).

The line count is set to 50 (line 570). This means that DXCC requires at least two pages.

The main program flow continues (line 590), opening the record file as an input file. The user entered the file name back in line 20, and it is assumed to be a ".DX" filetype. The printer buffer is cleared if the output is to be routed to the printer (line 600). The data is read from the file (lines 610 to 620), and the end of the file is detected (i.e., all the entries have been scanned) and the data is read (line 620).

The QSL status in each category is totalled (lines 630 to 660). The QSL status is tested to determine if it is to be printed (line 670). If it is, the subroutine starting at line 280 is invoked. The loop continues the program branches back to line 610 (line 680).

When the end of the file condition is recognised (line 610), the loop terminates with a branch and the file is closed (line 690). If any option other than the entire list is chosen, the program flow branches to line 800 (line 700); otherwise, the summary is printed (lines 710 to 870). The two groups of print statements are used to route the output to either the screen or the printer, depending on the state of the output device flag (P6).

All the variables are reset to zero (lines 880 to 900) and the user is asked if the program is to be run again (line 910). The response to the question is tested (lines 920 to 930) and if the first letter of the response is a "Y" the program branches back to line 60. If not, the program terminates. The branch-back feature allows the file to be scanned at the screen and printed later.

A sample of the output of the DXCC-READ program is shown in Fig. 3-8. The data is formatted and tagged with the name of the file and the date of the printout. A summary of all the QSL states is printed at the end. The only problem with the listing is that the prefixes are not in alphabetical order, a situation that can be corrected with

```
G3ZCZ/4X      82/03/03      PAGE  1

   PX  CALL          DATE     TIME  BAND MODE   QSL STATUS
--------------------------------------------------------------
 1 4U  4U1ITU      81/10/31  2028   15  SSB   CARD SENT VIA BUREAU
 2 CT2 CT2CR       81/08/30  1939   15  SSB   CARD SENT VIA BUREAU
 3 DL  DL2NAI      81/09/01  0920   10  SSB   QSL'D
 4 DM  Y46XF       81/09/12  1720   20  SSB   QSL'D
 5 F   F8ZS        81/09/20  1910   15  SSB   QSL'D
 6 FS7 FG7TD/FS7   81/08/31  2030   20  SSB   CARD SENT VIA BUREAU
 7 G   G3IOR       81/09/13  1905   15  SSB   QSL'D
 8 GJ  GJ3VLX      81/08/30  1819   20  SSB   CARD SENT VIA BUREAU
 9 I   I5TDJ       81/08/30  1443   20  SSB   WORKED
10 J2  J28DL       81/11/12  1939   10  SSB   WORKED
11 JA  JA3BOA      81/09/16  1515   15  SSB   CARD SENT VIA BUREAU
12 OK  OK3JW       81/09/12  1711   15  SSB   QSL'D
13 PA  PA3BNT      81/09/04  1321   10  SSB   QSL'D
14 SM  SM0AVK      81/08/30  1519   20  SSB   QSL'D
15 SV0 SV0BL       81/09/07  1340   20  SSB   WORKED
16 W   KA1HMW      82/01/31  1515   10  SSB   QSL'D
17 YN  HT3JQ       81/11/12  0448   20  SSB   QSL'D
18 ZL  ZL3HH       81/08/31  0510   20  SSB   CARD SENT VIA BUREAU
19 LZ  LZ1XI       81/08/30  1440   20  SSB   WORKED
20 YU  YU4EBL      81/08/31  1013   15  SSB   WORKED
21 EA  EA7CEL      81/08/31  1230   10  SSB   WORKED
22 YO  YO3AJN      81/09/01  1354   10  SSB   WORKED
23 HB  HB9XH       81/09/02  1859   15  SSB   WORKED
24 UQ  UQ2GHT      81/09/04  1402   10  SSB   WORKED
25 HA  HA4XH       81/09/07  1748   20  SSB   WORKED
26 OE  OE6HZG      81/09/12  1707   15  SSB   WORKED
27 GM  GM5AIW      81/09/12  1731   15  SSB   WORKED
28 SP  SP6IXF      81/09/12  1736   15  SSB   WORKED
29 4Z  4Z4ZB       81/10/06  0420   20  SSB   WORKED
30 8P  8P6QL       81/10/24  1949   15  SSB   WORKED
31 LU  LU9FFA      81/10/31  1802   10  SSB   CARD SENT VIA BUREAU
32 UH8 RH8HCV      81/11/06  1213   10  SSB   WORKED
33 J7  J73PD       81/11/11  1826   15  SSB   WORKED
34 5N  OE5BS/5N2   81/11/13  1259   10  SSB   CARD SENT VIA BUREAU
35 UL  UL7NAJ      82/01/01  0714   10  SSB   WORKED
36 UC  UC2AHF      82/01/01  0714   10  SSB   WORKED
37 VP8 VP8ANT      82/01/02  1954   20  SSB   WORKED
38 FK  FK8DH       82/01/03  0522   15  SSB   CARD SENT VIA BUREAU
39 M1  M1J         82/01/03  0752   15  SSB   CARD SENT VIA BUREAU
40 9H  9H1R        82/01/03  0948   10  SSB   CARD SENT VIA BUREAU
41 GD  GD2HCX      82/01/03  1316   10  SSB   CARD SENT VIA BUREAU
42 PY  PY5EG       82/01/03  1417   10  SSB   WORKED
43 Z2  Z21EI       82/01/03  1618   10  SSB   CARD SENT VIA BUREAU
44 TA  TA1CT       82/01/09  1826   20  SSB   WORKED
45 UB  RB5MWK      82/01/08  1257   10  SSB   WORKED
46 3B8 3B8CA       81/01/10  1413   10  SSB   WORKED
47 VK  VK7GE       82/01/13  0646   20  SSB   CARD SENT VIA BUREAU
48 CX  CX4AB       82/01/16  1738   15  SSB   CARD SENT VIA BUREAU
49 UF  UF6FFF      82/01/19  0537   20  SSB   WORKED
50 SV8 SV8KX       82/01/19  0541   20  SSB   WORKED
51 3A  3A2CP       82/02/25  1807   15  SSB   WORKED
52 FM  FM7CD       82/02/28  1254   10  SSB   WORKED
53 LX  PA3AIR/LX   81/11/01  1715   20  SSB   CARD SENT VIA BUREAU
54 UA3 UA3DQE      81/09/01  0811   10  SSB   WORKED
55 UA2 UK2BBB      81/09/12  1746   15  SSB   WORKED
56 UA9 UA9MAZ      81/11/06  1222   10  SSB   WORKED
57 IS  IS0IIT      82/01/15  0734   20  SSB   WORKED
58 VE  VE3MFT      82/03/07  1319   10  SSB   WORKED
SUMMARY
TOTAL =                        58
QSL'S RECEIVED =               9
QSL'S SENT DIRECT =            0
QSL'S SENT VIA BUREAU =        16
WORKED BUT NO CARD SENT =      33
```

Fig. 3-8. A sample of the output of the DXCCREAD program.

```
10 REM DXCCSORT Version 820313
20 ON ERROR GOTO 440
30 M = 350
40 DIM P$(M),C$(M),B$(M),M$(M),T(M),Q(M),D$(M)
50 INPUT "Which DXCC record file " ; F$
60 IF LEN(F$) = 0 THEN 50
70 OPEN"I",#2,F$ + ".DX"
80 OPEN "O",#1,F$ + ".$$$"
90 IF EOF(2) THEN 150
100 INPUT#2,P$,C$,B$,M$,D$,T,Q
110 I = I + 1
120 P$(I) = P$ : C$(I) = C$ : B$(I) = B$ : M$(I) = M$
130 D$(I) = D$ : T(I) = T : Q(I) = Q
140 GOTO 90
150 CLOSE #2
160 PRINT "THERE ARE";I;" ENTRIES IN THE FILE"
170 PRINT "LIST IS LOADED, SORT BEGINING"
180 N = 1
190 B = 0
200 FOR J = 1 TO I - N
210 IF P$(J) <= P$(J+1) THEN 320
220 X$ = P$(J+1) : P$(J+1) = P$(J) : P$(J) = X$
230 X$ = C$(J+1) : C$(J+1) = C$(J) : C$(J) = X$
240 X$ = B$(J+1) : B$(J+1) = B$(J) : B$(J) = X$
250 X$ = M$(J+1) : M$(J+1) = M$(J) : M$(J) = X$
260 X$ = D$(J+1) : D$(J+1) = D$(J) : D$(J) = X$
270 X = T(J+1) : T(J+1) = T(J) : T(J) = X
280 X = Q(J+1) : Q(J+1) = Q(J) : Q(J) = X
290 B = 1
300 N = N + 1
310 PRINT P$(J),P$(J+1)
320 NEXT J
330 IF B = 1 THEN 180
340 PRINT "LIST IS SORTED"
350 FOR J = 1 TO I
360 PRINT P$(J),C$(J)
370 PRINT#1, P$(J);",";C$(J);",";B$(J);",";M$(J);",";D$(J);",";T(J);",";Q(J)
380 NEXT
390 CLOSE#1
400 KILL F$ + ".BAK"
410 NAME F$ + ".DX"  AS F$ + ".BAK"
420 NAME F$ + ".$$$" AS F$ + ".DX"
430 GOTO 490
440 ON ERROR GOTO 440
450 IF ERR = 53 AND ERL = 400 THEN RESUME 410
460 IF ERL = 70 THEN RESUME 50
470 IF ERL = 390 AND ERR = 61 THEN PRINT "DISK IS FULL"
480 RESUME 490
490 END
```

Fig. 3-9. The DXCCSORT program.

A

```
G3ZCZ/4X      82/03/16      PAGE  1
   PX  CALL          DATE     TIME  BAND MODE   QSL STATUS
--------------------------------------------------------------------
 1 3A  3A2CP       82/02/25  1807   15  SSB   WORKED
 2 3B8 3B8CA       81/01/10  1413   10  SSB   WORKED
 3 4U  4U1ITU      81/10/31  2028   15  SSB   CARD SENT VIA BUREAU
 4 4Z  4Z4ZB       81/10/06  0420   20  SSB   WORKED
 5 5N  OE5BS/5N2   81/11/13  1259   10  SSB   CARD SENT VIA BUREAU
 6 8P  8P6QL       81/10/24  1949   15  SSB   WORKED
 7 9H  9H1R        82/01/03  0948   10  SSB   CARD SENT VIA BUREAU
 8 CT2 CT2CR       81/08/30  1939   15  SSB   CARD SENT VIA BUREAU
 9 CX  CX4AB       82/01/16  1738   15  SSB   CARD SENT VIA BUREAU
10 DL  DL2NAI      81/09/01  0920   10  SSB   QSL'D
11 DM  Y46XF       81/09/12  1720   20  SSB   QSL'D
12 EA  EA7CEL      81/08/31  1230   10  SSB   WORKED
13 F   F8ZS        81/09/20  1910   15  SSB   QSL'D
14 FK  FK8DH       82/01/03  0522   15  SSB   CARD SENT VIA BUREAU
15 FM  FM7CD       82/02/28  1254   10  SSB   WORKED
16 FS7 FG7TD/FS7   81/08/31  2030   20  SSB   CARD SENT VIA BUREAU
17 G   G3IOR       81/09/13  1905   15  SSB   QSL'D
18 GD  GD2HCX      82/01/03  1316   10  SSB   CARD SENT VIA BUREAU
19 GJ  GJ3VLX      81/08/30  1819   20  SSB   CARD SENT VIA BUREAU
20 GM  GM5AIW      81/09/12  1731   15  SSB   WORKED
21 HA  HA4XH       81/09/07  1748   20  SSB   WORKED
22 HB  HB9XH       81/09/02  1859   15  SSB   WORKED
23 I   I5TDJ       81/08/30  1443   20  SSB   WORKED
24 IS  IS0IIT      82/01/15  0734   20  SSB   WORKED
25 J2  J28DL       81/11/12  1939   10  SSB   WORKED
26 J7  J73PD       81/11/11  1826   15  SSB   WORKED
27 JA  JA3BOA      81/09/16  1515   15  SSB   CARD SENT VIA BUREAU
28 LU  LU9FFA      81/10/31  1802   10  SSB   CARD SENT VIA BUREAU
29 LX  PA3AIR/LX   81/11/01  1715   20  SSB   CARD SENT VIA BUREAU
30 LZ  LZ1XI       81/08/30  1440   20  SSB   WORKED
31 M1  M1J         82/01/03  0752   15  SSB   CARD SENT VIA BUREAU
32 OE  OE6HZG      81/09/12  1707   15  SSB   WORKED
33 OK  OK3JW       81/09/12  1711   15  SSB   QSL'D
34 PA  PA3BNT      81/09/04  1321   10  SSB   QSL'D
35 PY  PY5EG       82/01/03  1417   10  SSB   WORKED
36 SM  SM0AVK      81/08/30  1519   20  SSB   QSL'D
37 SP  SP6IXF      81/09/12  1736   15  SSB   WORKED
38 SV0 SV0BL       81/09/07  1340   20  SSB   WORKED
39 SV8 SV8KX       82/01/19  0541   20  SSB   WORKED
40 TA  TA1CT       82/01/09  1826   20  SSB   WORKED
41 UA2 UK2BBB      81/09/12  1746   15  SSB   WORKED
42 UA3 UA3DQE      81/09/01  0811   10  SSB   WORKED
43 UA9 UA9MAZ      81/11/06  1222   10  SSB   WORKED
44 UB  RB5MWK      82/01/08  1257   10  SSB   WORKED
45 UC  UC2AHF      82/01/01  0714   10  SSB   WORKED
46 UF  UF6FFF      82/01/19  0537   20  SSB   WORKED
47 UH8 RH8HCV      81/11/06  1213   10  SSB   WORKED
48 UL  UL7NAJ      82/01/01  0714   10  SSB   WORKED
49 UQ  UQ2GHT      81/09/04  1402   10  SSB   WORKED
50 VE  VE3MFT      82/03/07  1319   10  SSB   WORKED
51 VK  VK7GE       82/01/13  0646   20  SSB   CARD SENT VIA BUREAU
52 VP8 VP8ANT      82/01/02  1954   20  SSB   WORKED
53 W   KA1HMW      82/01/31  1515   10  SSB   QSL'D
54 YN  HT3JQ       81/11/12  0448   20  SSB   QSL'D
55 YO  YO3AJN      81/09/01  1354   10  SSB   WORKED
56 YU  YU4EBL      81/08/31  1013   15  SSB   WORKED
57 Z2  Z21EI       82/01/03  1618   10  SSB   CARD SENT VIA BUREAU
58 ZL  ZL3HH       81/08/31  0510   20  SSB   CARD SENT VIA BUREAU
SUMMARY
TOTAL =                        58
QSL'S RECEIVED =               9
QSL'S SENT DIRECT =            0
QSL'S SENT VIA BUREAU =        16
WORKED BUT NO CARD SENT =      33
```

B

```
G3ZCZ/4X      82/03/16      PAGE  1

   PX  CALL          DATE     TIME  BAND MODE   QSL STATUS
--------------------------------------------------------------------
 1 DL  DL2NAI      81/09/01  0920   10  SSB   QSL'D
 2 DM  Y46XF       81/09/12  1720   20  SSB   QSL'D
 3 F   F8ZS        81/09/20  1910   15  SSB   QSL'D
 4 G   G3IOR       81/09/13  1905   15  SSB   QSL'D
 5 OK  OK3JW       81/09/12  1711   15  SSB   QSL'D
 6 PA  PA3BNT      81/09/04  1321   10  SSB   QSL'D
 7 SM  SM0AVK      81/08/30  1519   20  SSB   QSL'D
 8 W   KA1HMW      82/01/31  1515   10  SSB   QSL'D
 9 YN  HT3JQ       81/11/12  0448   20  SSB   QSL'D
G3ZCZ/4X      82/03/16      PAGE  1

   PX  CALL          DATE     TIME  BAND MODE   QSL STATUS
--------------------------------------------------------------------
 1 4U  4U1ITU      81/10/31  2028   15  SSB   CARD SENT VIA BUREAU
 2 5N  OE5BS/5N2   81/11/13  1259   10  SSB   CARD SENT VIA BUREAU
 3 9H  9H1R        82/01/03  0948   10  SSB   CARD SENT VIA BUREAU
 4 CT2 CT2CR       81/08/30  1939   15  SSB   CARD SENT VIA BUREAU
 5 CX  CX4AB       82/01/16  1738   15  SSB   CARD SENT VIA BUREAU
 6 FK  FK8DH       82/01/03  0522   15  SSB   CARD SENT VIA BUREAU
 7 FS7 FG7TD/FS7   81/08/31  2030   20  SSB   CARD SENT VIA BUREAU
 8 GD  GD2HCX      82/01/03  1316   10  SSB   CARD SENT VIA BUREAU
 9 GJ  GJ3VLX      81/08/30  1819   20  SSB   CARD SENT VIA BUREAU
10 JA  JA3BOA      81/09/16  1515   15  SSB   CARD SENT VIA BUREAU
11 LU  LU9FFA      81/10/31  1802   10  SSB   CARD SENT VIA BUREAU
12 LX  PA3AIR/LX   81/11/01  1715   20  SSB   CARD SENT VIA BUREAU
13 M1  M1J         82/01/03  0752   15  SSB   CARD SENT VIA BUREAU
14 VK  VK7GE       82/01/13  0646   20  SSB   CARD SENT VIA BUREAU
15 Z2  Z21EI       82/01/03  1618   10  SSB   CARD SENT VIA BUREAU
16 ZL  ZL3HH       81/08/31  0510   20  SSB   CARD SENT VIA BUREAU
```

Fig. 3-10. (A) The whole listing from a sorted DXCC data file. (B) the first printout is the QSL'd entries, and the second is the listing of countries for which cards have been sent via the bureau.

a sorting program.

THE DXCCSORT PROGRAM

A simple sorting program called DXCCSORT is listed in Fig. 3-9. The program begins with the usual identifying statement (line 10), the error-trapping vector (line 20), and the maximum number of entries to be sorted (line 30). For large lists, this number (M) must be changed, or, better yet, a faster sorting technique must be used. The matrices used to hold the data being sorted are dimensioned (line 40). The user is asked to enter the name of the DX record file to be sorted (line 50). Error-trapping assumes a user pressing only RETURN has made an error (line 60). The DX record file and the temporary file that will contain the sorted data are opened (lines 70 to 80). The entries are read sequentially (line 90 detects the end of file condition and line 100 reads in an actual data line). The lines of data counter (I) are incremented (line 110) and the data is loaded into the various matrices used to store it (lines 120 to 130). The program flow to branches back to line 90 to test and read the next data entry (line 140).

When the last entry has been read (detected by line 90), the program flow is diverted to line 150 which closes the input file. The user is notified that the file has been read and that the sort is beginning (lines 160 to 170). A modified bubble sort is performed (lines 180 to 330) in which two successive entries in the prefix array are tested (line 210). If the second one has a lower alpha numerical value than the first, the positions of the two entries in all the matrices are changed using the dummy variables X$ and X (lines 220 to 280). A bubble flag (B) is also set to signify that something bubbled. The two callsigns are printed to reassure the user that the computer is still alive and is doing something. The variables N and B are used to cut short the sorting process when all the prefixes have been sorted. A message appears when the sort is complete (line 340).

The sorted data is written to the disk file (lines 350 to 380) and displayed. The file is then closed (line 390). Even though the data was only written to the disk in the loop of lines 350 to 380, the actual file was opened in line 80, allowing any errors to be detected as soon as possible, not after an hour or two of sorting data.

The backup file is deleted from the disk (line 400). The old unsorted .DX file becomes the new backup file (line 410), and the temporary file becomes the new .DX file (line 420). The program terminates with a branch to line 490 (line 430).

The error-trapping code begins (line 440), resetting the error-trapping vector. If a backup file doesn't exist, it's compensated for (line 450). If the user requests a non-existant DXCC record file, the program requests a new name (line 460). If a "Disk Full" error is detected (line 470) the RESUME statement (line 480) may allow the user to manually delete a file or two from the disk and save the sorted data by directing the program to start again at line 350.

Samples of a sorted DXCC data file are shown in Fig. 3-10, which contains the sorted version of the file illustrated in Fig. 3-8. Figure 3-10A contains the whole listing, and Fig. 3-10B contains two selected printouts. The first printout is the QSL'd entries, and the second is the listing of countries for which cards have been sent via the bureau.

OTHER AWARDS

The DXCC record-keeping package described above can also be used to keep track of WPX records, US Counties QRA squares, or any other award. For example, the DOK code can be entered instead of the prefix for the DOK award issued by the DARC.

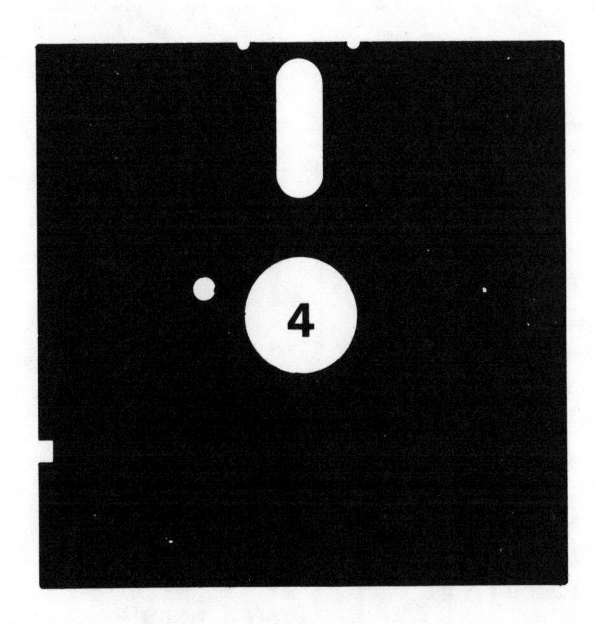

Contests

The objective of most amateur radio contests, besides affording enjoyment to those who take part in it, is to contact as many other stations as possible within a limited period of time and exchange some information. Duplicate contacts are not usually allowed, depending on the particular contest. Of all the techniques used to minimize duplicate contacts, use of a computer is the only one that can guarantee zero duplicates. This chapter presents a contest package written in Northstar BASIC that follows on from the BASIC logging package described in Chapter 2. The earlier NEWLOG command has to be used to allocate disc space for the contest log. The entries in the log are in the same format as in the LOG ENTER command. This means that LOGEDIT, QSLPRINT, and the other commands will work on log data files generated by either of the contest programs.

Two such contest programs, as well as the support programs, are listed in this chapter. One contest is a general purpose contest program; the other, SWEEPSTK, is optimized for the ARRL Sweepstakes contest. In the Sweepstakes contest, duplicates do not score points on any band, although in other contests they may. The contest program, however, does not distinguish between bands when testing for duplicates; thus, if the rules allow stations to be worked for points on different bands, separate log data files should be used for each band.

CONTEST

The general contest program is listed in Fig. 4-1. It's a modified version of LOG-ENTER which stores the log entry data in the disk data file and also keeps a checklist in memory. The commands that perform these operations are shown in Fig. 4-2.

```
10 PRINT"G3ZCZ General Contest Program Version 1.2"\REM VER 820329
20 INPUT"Roughly how many QSO's do you expect to make ? ",M9
25 ERRSET 2000,E1,E2
30 IFM9<100 THEN PRINT"OPTIMIST !"
40 DIMW$(M9*10),C4$(17)
50 C4$="QRCXFLBMDWT*"
110 GOSUB 5010\X=0\GOSUB5005\IFX=1THEN5045ELSEGOSUB5050
120 OPEN #1,L1$
145 IFN=0THEN220
150PRINT"Loading check list"
160 FOR I1=1TON
170 READ #1,&D1,&D2,&D3,T,F,C$,S,R,M$,P,&Q1,&Q2,X$
175 IFC$<>"/*"THEN177ELSEN4=N4-1\IFN4<0THENN4=0\GOTO200
177 PRINT C$
180 GOSUB840\IFI=0THEN190ELSEGOSUB960
190 REM
195 N4=N4+1
200 NEXTI1\C8$=C$
210 I=I1-1\GOSUB1020\N=I
220 GOSUB1040
230N=N+1
240 GOSUB 1000
250 IFN>M9-5THENPRINT"CHECK LIST IS ALMOST FULL"
251 INPUT1"-? ",A$\IFA$=""THEN250
255 IFLEN(A$)>1THEN250
260 FOR J2=1TOLEN(C4$)\IFA$(1,1)=C4$(J2,J2)THENEXIT280
270 NEXT \!CHR$(7)\ GOTO250
280 PRINT TAB(5),\GOSUB 980\GOTO250
290 INPUT"REPORT RECEIVED ? ",X$
300 RETURN
310 INPUT"REPORT SENT ? ",S
315 IF S>599 THEN310
320 RETURN
350 INPUT"Day (1-31) ?  ",D1\IFD1<1ORD1>31THEN350
360 INPUT"Month ( 1 - 12 ) ?  ",D2\IFD2<0ORD2>12THEN360
370 INPUT "Year (19xx) ? ",D3 \ IF D3<1900 THEN 370
380 D3 = D3 - 1900
390 RETURN
400 INPUT "Power (watts) ?  ",P
410 RETURN
420 INPUT "Freq/Band ?  ",F\IF INT(F)>999THEN420
430 RETURN
440 INPUT "Mode ?  ",M$\IFM$=""THEN440
450 IF M$(1,1)<>" "THEN460ELSEM$=M$(2,LEN(M$))\GOTO450
460 RETURN
470 INPUT"Call sign ?  ",A$\IFLEN(A$)=0THEN510
480 IFA$(1,1)<>" "THEN490ELSEA$=A$(2,LEN(A$))\GOTO480
490 C$=A$\GOSUB840
500 IFI=1THEN!"OK"ELSE!"WORKED",CHR$(7)
510 RETURN
520 INPUT "Time ?  ",T\IFT>2400THEN520
530 RETURN
660 IF D2>0THEN670ELSE!"Enter the date first please !"\RETURN
670 REM
675 I=N \ PRINT
685 GOSUB 5020 \ RETURN
690 RETURN
695 INPUT"ARE YOU SURE ? ",A$\IF A$=""THEN695ELSE IF A$(1,1)<>"Y"
    THENRETURN
698 GOSUB1040\C$="/*"\N4=N4-1\IFN4<0THENN4=0
700 D1=INT(D1) \D2=INT(D2)\D3=INT(D3)\S=INT(S)
710 R=INT(R)\P=INT(P)
720 S8=0\GOSUB840\IFI=0THEN730ELSEGOSUB960
730 GOSUB5035
735 N=N+1\IF C$="/*"THEN760
738 N4=N4+1
740 C8$=C$
750 GOSUB 1040
760 GOTO 1000
770 IFN>0THENN=N-1
790 CLOSE #1
800 OPEN#0,L2$
810 WRITE #0,N
820 CLOSE #0
830 GOTO 5045\REM EXIT
840 IF C$="/*"THEN940
850 J=0 \ FOR I=1TOLEN(C$)
860 J=J+ASC(C$(I,I))-47\NEXT
870 J=J*10
880 IF J<M9 THEN 890 ELSE J=J-M9\GOTO880
890 D$=W$(J*10+1,J*10+10)
900 IF D$(1,1)=" "THEN 950
910 IF D$(1,LEN(C$))=C$THEN940
920 J=J+1\IFJ>=M9THENJ=J-M9
930 GOTO 890
940 I=0\RETURN
950 I=1\RETURN
960 W$(J*10+1,J*10+10)=C$\RETURN
980ONJ2GOTO660,290,470,310,695,700,420,440,350,400,520,770
1000 !CHR$(26),CHR$(12)\!\!"LAST QSO = ",C8$
1005 PRINT "THERE ARE ",N-1," ENTRIES IN THE LOG"
1006 PRINT"NEXT QSO = ",N4+1
1010 RETURN
1020GOSUB5015\GOSUB5020
1030INPUT"HIT RETURN WHEN READY",A$\RETURN
1040C$="?"\Z$=C$\Z2$=Z$\Z3$=Z$\Z4$=Z$\R=0\S=0\Y1=0\X$=""
1045 S=59
1050RETURN
2000 ERRSET 2000,E1,E2
2005 IFE1=480 THEN 470
2015 IF E1=5560 THEN 5570
2020 IF E1=5650 THEN 5670
2030 IF E2<>8 THEN 2040ELSE!"HARD DISC ERROR"\GOTO 2100
2040 IF E2<>7 THEN 2050ELSE!"FILE ERROR"\GOTO 2100
2050 IF E2=15 THEN 790
2090 !"ERROR ",E2," AT LINE ",E1 \ GOTO 790
2100 GOTO 5045
5000 REM SUBROUTINE PACKAGE VER 791008
```

Fig. 4-1. CONTEST program.

The listing begins with the usual REM statement that identifies the program and the revision date. The user is asked to estimate the number of contacts (M9) that will be made (line 20). The error trapping vectors are set up (line 25). The value of M9 is

J2	Letter	Line Number	Command
1	Q	660	QSL or confirm log entry
2	R	290	Received report data for log file
3	C	470	Call sign
4	X	310	Report sent (transmitted) for log file
5	F	695	Fudge last logged entry
6	L	700	Log entry to log file and checklist
7	B	420	Band/frequency information for log file
8	M	440	Mode information for log file
9	D	350	Date information for log file
10	W	400	Power (watts) information for log file
11	T	520	Time information for log file
12	*	770	Terminate program, close log file and return to the system program

Fig. 4-2. General contest commands.

contacts is less than 100. The array (W$) checked, and a suitable sarcastic comment is displayed if the estimated number of that will be used to store the checklist is set up (line 40). Ten letters are allowed in each callsign. The Command Matrix (C4$) is also dimensioned in this line because Northstar BASIC only allocates 10 letters to an ASCII string unless otherwise directed. C4$ (defined in line 50) contains the first letter of each command usable in the contest.

The remaining variables and constants are then set up, the log file to be used is identified, and the STNDATA file is accessed. If the STNDATA file is not on any of the disk drives, X is untouched (i.e., 0) and the program terminates. Otherwise, the log file is opened (line 120). The number of entries in the log data file (N) is tested (line 145). The value was obtained in the subroutine beginning on line 5005 (called by line 110). If N is zero, the log file is assumed to be empty and the program branches to line 220.

The checklist-updating sequence (line 150) is performed by the FOR . . . NEXT loop (lines 160 to 200). Each log entry is read (line 170). If the callsign associated with that entry is not a /*, the entry is assumed valid; if it is, the NEXT statement (line 200) is invoked. The callsign is examined and its position in the checklist matrix computed (line 840).

If the callsign has not been worked before, it is inserted into the checklist (line 960). The number of contacts (N4) is then incremented, and the loop goes onto process the next entry. After processing all the calls in the log, the callsign of the last valid contact (C8$) is set up (line 200). The last complete entry is displayed (line 210). The value of N is set up, the received data is cleared, and the contest loop begins.

The main contest loop begins (line 240) by calling the subroutine at line 1000 which displays the current QSO status. If the checklist is almost full, a warning message is printed (line 250); otherwise, a prompt character is displayed (line 251) which the user must respond to. The user response is one ASCII character (in the set contained in C4$). If it is less than or greater than 1 character, it is rejected and the prompt is repeated. A FOR/NEXT loop is used to scan each character in C4$ and test to see if it is the same as the character that was entered (line 260). If a match is found, the loop exits to line 280. If a match is not found, the loop exits to line 270. A bell character is sent, producing an audio tone to alert the operator of an error, and the program branches back to line 250 to request another instruction.

The cursor advances and the subroutine in line 980 which performs the selected command is invoked (line 280). The loop then continues (line 250).

The subroutine that sets up the received report (lines 290 to 320) is set up as an ASCII string; either 59-1000, 59KW, or 59 15 can be entered. This scheme accommodates power, zone, or sequential number exchanges.

Data is set up for the RST report sent or received (lines 310 to 320). Anything larger than 599 is obviously an error, and is caught by error-trapping (line 315).

Day, month, and year information is set up (lines 350 to 390). Note the usual tests for correct order of magnitude of the data.

The transmitter power, frequency/band, and mode (CW, SSB, etc) information for the log entry are set up (lines 400 to 460). A short routine deletes leading space characters in the mode data (line 450).

The callsign of the station being worked or about to be worked is entered and processed (lines 470 to 510). The callsign is entered (line 470) and any leading blank characters are deleted (line 480). The callsign's position in the checklist is calculated, and it's checked to see if it is already in it's proper place (lines 490 to 510). If the station has already been worked, the bell character is sent in addition to a visual message. (At 3 a.m., after many hours of contest operating, a beep is often necessary to get the attention of a sleepy operator.)

The time is entered by the program (lines 520 to 530). If a hardware clock is available inside the computer, these lines can be changed to read the clock circuit.

The log entry is displayed so the user can confirm that the data is valid before writing it to the disk log data file (lines 660 to 690).

The fudge flag (/*) is set up in case a bad entry was previously written to the log (lines 695 to 698). The log data can be corrected after the contest using LOGFUDGE. The program falls through to the routine that writes to the log data file (beginning at line 700). All the numeric data except the frequency or band information must be integers (lines 700 to 710). The callsign is written into the checklist (line 720). The entry is then entered into the log (lines 730 to 735). If the entry is not a fudge (/*) entry, the program updates the QSO counter (N4) and sets up the last callsign string (C8$). If the entry is a fudge entry, only the number of entries (N) is updated. The contact data is cleared (line 750) and the program flow branches to line 1000 to print out the next QSO prompt before returning to the beginning of the main loop.

The log data file is closed (lines 770 to 820) and the log pointer file is updated. N is decremented (line 770) because it was incremented (line 735). The program then branches to line 5045 which invokes the chain SYSTEM routine in the log package subroutine library.

A routine computes the position of each callsign in the checklist and also if that callsign has been worked before (lines 840 to 950). There are two basic techniques for storing callsigns in the checklist. The first technique is to store the callsigns sequentially starting from the beginning. This is a straightforward programming method, but as the checklist is filled, the search time gets longer.

The second technique is to use the callsign to compute an offset position from the start of the table, and then store the callsign there. This means that some kind of "hashing" routine is performed on the letters in the callsign and a number is computed. This number then becomes the callsign's position in the checklist.

This program uses a compromise. A simple hashing technique is used to find a starting slot in the table, but it is filled sequentially, effectively splitting the checklist into a number of small checklists. This has some of the advantages and disadvantages of both techniques, while still giving reasonable search times for each callsign.

The fudge entry condition is detected and the remainder of the subroutine is skipped (line 840).

The callsign is converted to a number (lines 850 to 870) using the following algorithm. Each character is converted to a number and then summed with the previous characters. The result is multiplied by 10. As an example, consider the callsign WA3LOS. The (decimal) ASCII code for W is 87, which, when 47 is subtracted from it, becomes 40. Thus W + A + 3 + L + O + S becomes 47 + 18 + 4 + 29 + 32 + 36 or 166. This number (J) multiplied by 10 (line 870) to suggest a slot WA3LOS might fit into. Notice that WA3OSL or WA3LSO will return the same value of J. Then multiply by 10 (line 870) allocates 10 spaces to callsigns having a numerical equivalent to 166 before reserving space for the callsigns have a numerical equivalent to 167. Overflow of the string area is tested for (line 880). For example, if space had been allocated for 1000 callsigns (M9 = 1000) then WA3LOS, which would point to the 1670th, position would point outside the area allocated to the checklist. This potential bomb is taken care of by converting the pointer to point to 1670–1000 or the 670th position (line 880). This technique allows any size checklist to be used and still expedites the search. Of course, the smaller the area assigned to the checklist, the greater the probability of a collision. The intermediate callsign variable (D$) is set up (line 890) equal to the contents of the position in the checklist pointed to by the callsign (i,e., the 670th slot).

D$ is tested to see if the slot is empty (line 900). If it is, the program flow branches forward to line 950. If it is not, the contents of the slot in the checklist (D$) is compared to the current callsign (C$) (line 910). If they are the same, then a previous QSO has taken place and the program flow branches to line 940. If the value of D$ doesn't equal that of C$ (that is, another call, such as WA3LSO, was in the slot), the value of J is incremented and the next position is tested. Tests continue sequentially until either a blank slot is discovered or a match occurs as illustrated in Fig. 4-3. In this manner, even if the table has folded back in itself, a blank slot or a match, will eventually be discovered. Of course, as the table gets full, the search time is correspondingly increased because the number of empty slots gets smaller. Since even an 48K system can allocate space for about 2000 calls (about 20K bytes) this situation should not arise.

The QSL'd flag (I) is set and the program flow returns from the subroutine (line 940), while the QSL'd flag is cleared (line 950). A one-line subroutine stores the current callsign into the checklist at a previously-computed position (line 960).

The subroutine beginning at line 1000 first sends two commonly-used screen clear characters to the terminal. They will clear the screen on an ADM-3 and on most memory-mapped video displays. The callsign of the last logged QSO is then displayed. The number of entries in the log and the number allocated to the next QSO are shown (lines 1005 to 1006). The two numbers

Example of some calls stored in the list in memory

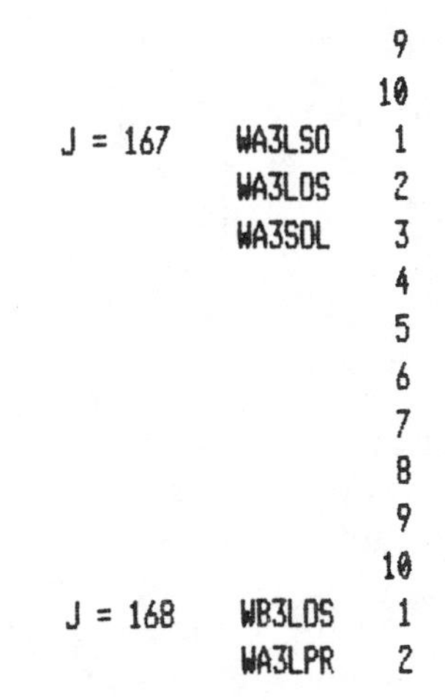

Fig. 4-3. Storage of calls in checklist.

may be different because "/*" entries are not assigned QSO numbers. A subroutine (lines 1020 to 1030) displays the last contact as read in from the log in full when bringing up the program. The user is asked for a response, which the program waits for before continuing (line 1030).

Another subroutine (lines 1040 to 1050) clears the QSO data. Since most reports are 59 or 599 in a contest, the report is automatically set to 59 (line 1045). It can be overwritten by the X command if a different report is given. If the contest is a CW contest, line 1045 should be changed to read

```
1045  S = 599
```

The error-trapping routines begin (line 2000). A control C character (E2 = 15) closes the log files and returns to the SYSTEM Program. It will not decrement N before closing the file; hence, reading the log again results in an error. LOGRESTR must be run before working on a log file that was closed when the contest program was aborted. Line 2050 could have been changed to branch to 770 instead of 790, but then there would have been no way to flag the log file as having had an abort or error sequence happen.

The subroutine library begins as usual (line 5000).

THE SWEEPSTAKES CONTEST

The SWEEPSTAKES program is a version of the CONTEST program, optimized for the ARRL Sweepstakes Contest. The program listed in Fig. 4-4 even has the same line numbers where identical functions are performed.

The basic difference between the Sweepstakes Contest and any other contest is that the Sweepstakes Contest is a traffic-handling contest. The contest exchange simulates a standard format message transfer, and the multipliers are the number of ARRL sections worked. The contest is discussed in detail in Section 8.

Since lines of code in this program that have been used in the CONTEST program were discussed in that section, this section discusses only the differences between the two programs and the additional software used in the SWEEPSTAKES program.

The Section Matrix String (S$) and Section string (S( )) are defined (line 40). Since each section name is stored as a two byte string in S$, S$ has to be dimensioned to twice the number of sections. The section string matrix is set up (lines 60 to 90). The technique of letting S$ = S$+ ". . . . ." used in lines 70 to 90 allow the long string to be set up by statements on different lines.

Examples of the section abbreviations (line 60) are:

CT = Connecticut
EM = East Mass
ME = Maine
NH = New Hampshire
RI = Rhode Island, etc.

The Command String Matrix (C4$—line 50) is expanded due to the additional commands optimized for this contest. These commands are summarized in Fig. 4-5.

Several routines accept Sweepstakes contest information (lines 540 to 650). The number, Z$ (lines 540 to 550), is the regular QSO number. The Precedence, Z3$ (lines 560 to 570), is the power level and can be either A for low power (less than 250 watts) or B for high power stations. The check, Z2$ (lines 580 to 590), is the last two digits of the year that the contester received his first amateur radio license. If an operator first received a license in 1968, the check becomes 68. The section is entered (line 600). If the input string contains no characters or more than two characters it is rejected. A subroutine beginning at line 1140 is called (line 620). If the section is valid (i.e., the two letters match a real section and it has not been worked yet) a message to that effect is displayed. The section can be identified by checking to see if a callsign has been

```
5  REM VER 820329
10 PRINT"G3ZCZ Sweepstakes Contest Program Version 2.0"
15 ERRSET 2000,E1,E2
20 INPUT"Roughly how many QSO's do you expect to make ? ",M9
30 IFM9<100 THEN PRINT"OPTIMIST !"
40 DIMW$(M9*10),C4$(17),S$(75*2),S(75)
50 C4$="QSRPNCKXFLBMDIWT*"
60  S$="CTEMMENHRIVTWMMAENNYNNSNWNQBDEEPMDWPON"
70  S$=S$+"ALGAKYNCNFSCSFTNVAWIMNAKLAMSNMNTOKSTCZSK"
80  S$=S$+"EBLAORSBSCSDSFSJSVPAABAZIDMTNVORUTWAAKWYBCMIOHWVNWILINWS"
90  S$=S$+"COIAKSMNMONBNDSD"
100 DATA 7,13,18,29,38,49,59,63,66,74
110 GOSUB 5010\X=0\GOSUB5005\IFX=1THEN5045ELSEGOSUB5050
120 OPEN #1,L1$
145 IFN=0THEN220
150PRINT"Loading check list"
160 FOR I1=1TON
170 READ #1,&D1,&D2,&D3,T,F,C$,S,R,M$,P,&Q1,&Q2,X$
175 IFC$<>"/*"THEN180ELSEN4=N4-1\IFN4<0THENN4=0\GOTO200
180 GOSUB840\IFI=0THEN190ELSEGOSUB960\GOSUB1060
190 GOSUB1080\GOSUB970
195 N4=N4+1
196 PRINT I1,TAB(6),C$,TAB(18),Z4$
200 NEXTI1\C8$=C$
210 I=I1-1\GOSUB1020\N=I
220 GOSUB1040
230N=N+1
240 GOSUB 1000\GOSUB1260
250 IF N>M9-5 THEN PRINT"CHECK LIST IS ALMOST FULL"
251 INPUT1"-?",A$\IFA$=""THEN250
255 IFLEN(A$)>1THEN250
260 FOR J2=1TOLEN(C4$)\IFA$(1,1)=C4$(J2,J2)THENEXIT280
270 NEXT \PRINTCHR$(7)\ GOTO250
280 PRINT TAB(5),\GOSUB 980\GOTO250
290 INPUT"REPORT RECEIVED ? ",R
300 RETURN
310 INPUT"REPORT SENT ? ",S
315 IFS>599THEN315
320 RETURN
350 INPUT"Day (1-31) ?  ",D1\IFD1<1ORD1>31THEN350
360 INPUT"Month ( 1 - 12 ) ?  ",D2\IFD2<0ORD2>12THEN360
370 INPUT "Year (19xx) ?",D3 \ IF D3<1900 THEN 370
380 D3 = D3 - 1900
390 RETURN
400 INPUT "Power (watts) ?  ",P
410 RETURN
420 INPUT "Freq/Band ?  ",F\IFF>999THEN420
430 RETURN
440 INPUT "Mode ?  ",M$\IFM$=""THEN440
450 IF M$(1,1)<>" "THEN460ELSEM$=M$(2,LEN(M$))\GOTO450
460 RETURN
470 INPUT"Call sign ?  ",A$\IFLEN(A$)=0THEN510
480 IFA$(1,1)<>" "THEN490ELSEA$=A$(2,LEN(A$))\GOTO480
490 C$=A$\GOSUB840
500 IFI=1THENPRINT"OK"ELSEPRINT"WORKED",CHR$(7)
510 RETURN
520 INPUT "Time ?  ",T\IFT>2400 THEN520
530 RETURN
540 INPUT"Number ? ",A$\IFLEN(A$)>0THENZ$=A$
```

Fig. 4-4. SWEEPSTAKES program.

```
550 RETURN
560 INPUT"Prec ? ",A$\IFLEN(A$)>0THENZ3$=A$
570 RETURN
580 INPUT"Check ? ",A$\IFLEN(A$)>0THENZ2$=A$
590 RETURN
600 INPUT"Section ? ",A$\IFLEN(A$)=0THEN650
610 IFLEN(A$)<2THEN600ELSEZ4$=A$
620 GOSUB1140\IFI=1THENIFS(I3)=0THENPRINT" New One"
630 IF I<>2THEN640ELSEPRINT"ENTER CALL AREA FIRST"\GOTO650
640 RETURN
650 RETURN
660 X$=Z$+Z3$+","+Z2$+","+Z4$
670 T$=FNT$(T)\PRINTT$,TAB(10),"# ",Z$," ",Z3$,TAB(20),C$,TAB(30),
680 PRINT Z2$,TAB(35),Z4$
690 RETURN
695INPUT"ARE YOU SURE ? ",A$\IFA$=""THEN695ELSEIFA$(1,1)="Y"THEN
   698ELSERETURN
698 GOSUB1040\C$="/*"\IFS8=1THENS9=S9-2\N4=N4-1\IFN4<0THENN4=0
700 D1=INT(D1) \D2=INT(D2)\D3=INT(D3)\S=INT(S)
705 X$=Z$+Z3$+","+Z2$+","+Z4$
710 R=INT(R)\P=INT(P)
720 S8=0\GOSUB840\IFI=0THEN730ELSEGOSUB960\GOSUB1060
730 WRITE#1,&D1,&D2,&D3,T,F,C$,S,R,M$,P,&Q1,&Q2,X$
735 N=N+1\IF C$="/*"THEN760
738 N4=N4+1
740 GOSUB1080\GOSUB970\C8$=C$
750 GOSUB 1040
760 GOTO 1000
770 IFN>0THENN=N-1
790 CLOSE #1
800 OPEN#0,L2$
810 WRITE #0,N
820 CLOSE #0
830 GOTO 5045
840 IF C$="/*"THEN940
850 J=0 \ FOR I=1TOLEN(C$)
860 J=J+ASC(C$(I,I))-47\NEXT
870 J=J*10
880 IF J<M9 THEN 890 ELSE J=J-M9\GOTO880
890 D$=W$(J*10+1,J*10+10)
900 IF D$(1,1)=" "THEN 950
910 IF D$(1,LEN(C$))=C$THEN940
920 J=J+1\IFJ>=M9THENJ=J-M9
930 GOTO 890
940 I=0\RETURN
950 I=1\RETURN
960 W$(J*10+1,J*10+10)=C$\RETURN
970 S(I3)=S(I3)+1\RETURN
980ONJ2GOTO660,600,290,560,540,470,580,310,695,700,420,440,350,1255
   ,400,520,770
1000 PRINT CHR$(26),CHR$(12)\PRINT\PRINT"THERE ARE ",N-1," ENTRIES IN
     THE LOG"
1005 PRINT "LAST QSO WAS ..",C8$
1006 PRINT"NEXT QSO = ",N4+1
1010 RETURN
1020GOSUB5015\GOSUB5020
1030INPUT"HIT RETURN WHEN READY",A$\RETURN
1040C$="?"\Z$=C$\Z2$=Z$\Z3$=Z$\Z4$=Z$\R=0\S=0\Y1=0\X$=""
1045 S=59
1050RETURN
```

```
1060FORI=1TOLEN(X$)\IFX$(I,I)="?"THENEXIT1050
1070NEXT\S9=S9+2\S8=S8+1\GOTO1050
1080FORI=1TOLEN(X$)\IFX$(I,I)=","THENEXIT1100
1090NEXT\GOTO1115
1100FORX4=I+1TOLEN(X$)\IFX$(X4,X4)=","THENEXIT1120
1110NEXT
1115PRINT"COMMENTS ARE NOT CONTEST DATA"\RETURN
1120Z4$=X$(X4+1,LEN(X$))
1140 IFZ4$(1,1)="?"THEN1250
1150 IFC$="?"THEN1250
1155 IF C$="/*"THEN1250
1160 FOR I3=LEN(C$) TO 1 STEP -1
1170 IFASC(C$(I3,I3)))=65THEN1230
1180 IFASC(C$(I3,I3))<=47THEN1230
1190Y=VAL(C$(I3,I3))\IFY=0THENY=10
1200 X=0\FOR I=1 TO Y \X1=X\ READ X\NEXT\RESTORE
1210 FORI3=X1TOX\IF S$(I3*2+1,I3*2+2)=Z4$(1,2)THENEXIT1240
1220 NEXT\GOTO1250
1230 NEXT\I3=0\I=0\RETURN
1240 I=1\RETURN
1250 I3=0\I=2\RETURN
1255 GOSUB 1000
1260 PRINT"DAY ",D1,D2,D3,TAB(16),"BAND ",F,TAB(32),"MODE ",M$\
     PRINT
1280 S7=0\Y=0\FOR I=1 TO 10
1290 READ X \IFI<9THEN1300
1295 X=X+1\IFI=10THENY=Y-1
1300 PRINTI,TAB(6),
1310 FOR J=Y TO X-1\IFS(J)=0 THEN 1320 ELSEPRINT"-- ",\S7=S7+1\GOT
     O1330
1320 PRINT S$(J*2+1,J*2+2)," ",
1330 NEXTJ\Y=X+1\PRINT\NEXTI
1340 RESTORE
1350 PRINT" VE",TAB(6),
1360 FORI=1 TO 8\READ X\IFS(X)=0THEN1370ELSEPRINT"-- ",\S7=S7+1\GO
     TO1380
1370 PRINTS$(X*2+1,X*2+2)," ",
1380 NEXT\PRINT\RESTORE\PRINT
1385 PRINT"POINTS = ",S9,TAB(16),"SECTIONS ",S7,TAB(32),"SCORE",S9
     *S7
1390 GOTO1005
2000 ERRSET 2000,E1,E2
2005 IF E1=480 THEN 470
2010 IF E1=5560 THEN 5570
2020 IF E1=5650 THEN 5670
2030 IF E2<>8 THEN 2040 ELSEPRINT"HARD DISC ERROR"\ GOTO2100
2040 IF E2<>7 THEN 2050 ELSE PRINT"FILE ERROR"\ GOTO 2100
2050 IF E2=15 THEN 790
2090 PRINT"ERROR ",E2," AT LINE ",E1\ GOTO790
2100 GOTO 5045
5000 REM SUBROUTINE PACKAGE VER 790822
```

Fig. 4-4. SWEEPTAKES program. (continued)

previously entered (line 630). This is required because section abbreviations such as LA and OR can each apply to two sections. LA can be Los Angeles (W6) and Louisiana (W5) while OR can be Orange (W6) and Oregon (W7).

All the data (lines 520 to 600) was first assigned to the temporary variable A$ and later, if found to be valid, was transferred to its permanent variable by the use of the Zn$ = A$. This technique minimizes unwanted changes in the data. If, for example, the section was requested instead of the precedence, entering a carriage return will not change any section data already entered.

The log/QSO entry is displayed before it is logged (lines 660 to 690). The QSO data is built into the comment field (X$) for the log entry (line 660), while only the data required for the sweepstakes contest is displayed (lines 670 to 680).

Before writing the data to the log (line 705), the program checks to make sure that the X$ reflects the latest input data. If the data revealed by the Q command is wrong, it can be changed and logged correctly without another iteration of the Q command.

The sections worked totals are now updated (line 720), and the subroutines starting at lines 1080 and 970 are culled, which ensure that X$ contains contest data and flags the section as being worked (line 740).

The QSO data is tested to ensure that all the required data is present (lines 1060 to 1070). A QSO does not get any points unless it is complete. S8 is a data complete flag, and S9 is the total points made at any time. These are updated as the contacts are logged so that the score and multipliers can be displayed at any time. The contest data in X$ is scanned to ensure that it is formatted correctly (lines 1080 to 1115). Line 1120 separates the section data (Z4$). If it is unknown, or if the callsign is unknown or a fudge entry (/*), the program branches to line 1250 (lines 1140 to 1155). The FOR/NEXT loop (lines 1160 to 1230) scans the callsign from the suffix side searching for the call area number. Thus, WA3LOS returns the third call area, and G3ZCZ/W4 returns the fourth call area. The ASCII character in the 'number'

J2	Letter	Line Number	Command
1	Q	660	QSL or confirm log entry
2	S+	600	Enter section information
3	R	290	Received report for log file
4	P+	560	Enter precedence information
5	N+	540	Enter number information
6	C	470	Enter callsign
7	K+	580	Enter check information
8	X	310	Report transmitted for log file
9	F	695	Fudge last logged entry
10	L	700	Log entry and update checklist
11	B	420	Change band information for log file
12	M	440	Mode information for log file
13	D	350	Enter date information for log file
14	I+	1255	Information and status display
15	W	400	Enter power (watts) information for log file
16	T	520	Enter time for log file
17	*	770	Terminate program, close log file, and return to the system program

+Sweepstakes only

Fig. 4-5. Sweepstakes contest commands.

position is converted to a decimal value (line 1190), and a zero becomes a 10. The loop (line 1200) reads the data stored in line 100. This determines the call area delimiters in the section table. There are thus 6 sections in the W1 call area, for example. The sections in the call area in which the callsign (C$) is located are then tested (lines 1210 to 1220) to see if one of them is the same as the section data entered by the operator [Z4$]. If so, the routine exits normally. If not, an error flag is set when the loop completes. The search/error flags are set as follows (lines 1230 to 1250):

I3 = section number if section is valid
I3 = zero, if test fails.
I = 0 if callsign is bad
I = 1 if section is valid
I = 2 if data is not completed (i.e., ? or /*)

The subroutine which begins at line 1255 performs the I or Information Display Command. After using the line 1000 subroutine to display the QSO status, the current operating conditions (date, band and mode) are displayed (line 1260). Then a summary of call areas and sections are displayed (lines 1280 to 1380). Any section that has not been worked has its two-letter abbreviation displayed. Any section that has been worked is not displayed; '--' is displayed instead. The variable S7 is used to keep a running count of the sections worked. A summary of the total score is shown (line 1385). The routine flow then branches back and repeats the next QSO display function (line 1005). Two typical information status displays are shown in Fig. 4-6.

LOGFUDGE

LOGFUDGE is used to clean up a contest log that may have fudged entries. It scans the log data file and deletes any entries that are followed by a fudge flag (/*).

The program listing is shown in Fig. 4-7. The program sets up the standard variables, identify the log data file, and request the name of and create the new log file (lines 10 to 115). Since the old file is untouched, it may be examined later by the LOGPRINT program.

The new log data file is opened (line 120). The first entry in the old log data file is

```
1   CT EM ME NH RI VT WM
2   EN NY NN SN WN
3   DE EP MD WP
4   AL GA KY NC NF SC SF TN VA WI
5   AK LA MS NM NT OK ST CZ
6   EB LA OR SB SC SD SF SJ SV PA
7   AZ ID MT NV OR UT WA AK WY
8   MI OH WV
9   IL IN WS
10  CO IA KS MN MO NB ND SD
VE  MA QB ON MN SK AB BC NW

1   -- EM ME NH RI VT WM
2   EN NY NN SN WN
3   DE EP MD WP
4   AL -- KY NC NF SC SF TN VA WI
5   AK LA MS -- -- -- ST CZ
6   EB LA OR SB SC SD SF SJ SV PA
7   AZ ID MT NV OR UT WA AK WY
8   -- -- WV
9   -- IN WS
10  CO IA KS MN MO NB ND SD
VE  MA QB ON MN SK AB BC NW

POINTS= 16    SECTIONS  8     SCORE 16
LAST QSO WAS ..WA3VXE
NEXT QSO = 9
-?
```

Fig. 4-6. Contest status and information display.

then read (line 130). The data just read is stored in temporary variables (lines 140 to 150). Then a test is made (line 160) to see if all the entries in the old log data file have been read. If they have, the program flow branches to line 180. If not, the next log entry is read (line 170). If this entry contains a "/*" the program flow branches to line 220. If not, the data in the temporary variables are written out to the new log data file, and the new log data file entry counter (N4) incremented (lines 190 to 200).

As long as there is still data to be processed, the program then slips back to line 140 where it moves the data read in from the last entry (which wasn't a /*) to the temporary storage variables (line 210). If the program has detected a /* in line 180 and has reached line 220, it will branch back to line 130 as long as there is data to be processed. At line 130 it will read the next entry and continue the processing sequence. In this manner, both the /* entry and the line or entry preceding it are not written out to the new log data file.

The program then closes the log files and updates the new log pointer file (lines 240 to 280). The number of entries in the new log file is displayed (line 290) and the program flow then continues (lines 9000 to 9050) which contain the chain SYSTEM routine.

Error-handling routines are included (lines 9060 to 9110).

```
10 REM LOGFUDGE VER 800130
20 ERRSET 9060,E1,E2
30 DIM X1$(64),X2$(64)\I=0
40 Q3=0\N1=0\N4=6
50 INPUT'Name of LOG file ? ',L1$\IF L1$=''THEN 50
60 INPUT'Which drive is it on ? ',D1 \ IF D1<1ORD1>4 THEN60
70 INPUT'NAME OF NEW LOG FILE ? ',L2$\IFL2$=''THEN70
75 INPUT'WHICH DRIVE IS IT TO BE PUT ON ? ',D2\IFD2<1ORD2>4THEN75
80 D1$=STR$(D1) \ D2$=STR$(D2) \D1$=D1$(2,2)\D2$=D2$(2,2)
90 OPEN#0,'*'+L1$+','+D1$ \ READ#0,N1 \ CLOSE#0
100 PRINT'There are ',N1,' entries in ',L1$
110 OPEN#0,L1$+','+D1$ \ REM OPEN OLD LOG FILE
115CREATE '*'+L2$+','+D2$,2 \ CREATE L2$+','+D2$,INT((N1/4)+(N1*.
   15))+1
120 OPEN#1,L2$+','+D2$ \ REM OPEN NEW LOG FILE
130  READ #0,&D1,&D2,&D3,T,F,C$,S,R,M$,P,&Q1,&Q2,X$
135 I=I+1
140 D4=D1\D5=D2\D6=D3\T1=T\F1=F\S1=S\R1=R\P1=P
150 Q3=Q1\Q4=Q2\C1$=C$\M3$=M$\X1$=X$
160 IF I>=N1THEN180
170  READ #0,&D1,&D2,&D3,T,F,C$,S,R,M$,P,&Q1,&Q2,X$
180 I=I+1\IF C$='/*'THEN220\REM SKIP
190  WRITE#1,&D4,&D5,&D6,T1,F1,C1$,S1,R1,M3$,P1,&Q3,&Q4,X1$
200 N4=N4+1
210 IF I>N1 THEN 240 ELSE 140
220 IF I>N1 THEN 240 ELSE 130
240 CLOSE#0
250 CLOSE#1
260 OPEN#0,'*'+L2$+','+D2$
270 WRITE#0,N4
280 CLOSE#0
290 !'There are now ',N4,' entries in the log file'
9000 FOR I=1 TO 4
9010 IF FILE('SYSTEM,'+STR$(I))=2 THEN EXIT 9030
9020 NEXT \ GOTO 9040
9030 CHAIN 'SYSTEM,'+STR$(I)
9040 PRINT'put a system disc in any drive, then hit 'RETURN''
9050 INPUT' ',A$ \ GOTO 9000
9060ERRSET 9060,E1,E2
9070 IF E2=15THEN9000
9080 IFE1=9010THENIFI<1ORI>4THEN9040ELSE9020
9090 IFE2<>7THEN9100ELSEPRINT'FILE ERROR'\GOTO9000
9100 IFE2<>8THEN9110ELSEPRINT'HARD DISC ERROR'\GOTO9000
9110 PRINT'ERROR ',E2,' AT LINE ',E1\GOTO9000
```

Fig. 4-7. LOGFUDGE program.

CKLSTGEN

CKLSTGEN is used to generate a checklist from a log file. The log file may have been created by LOGENTER, LOGEDIT, or one of the contest programs.

The program is listed in Fig. 4-8. It begins in the usual way by setting up the variables and constants (line 30). The log file is then identified (lines 40 to 50). The user is asked if the checklist is to be saved on disk (line 55). If it is, the program flow continues (line 60); if not, F1 is set and the program flows skips ahead to line 130. The program then requests the name to be allocated to the checklist data file and ensures that the number of characters in it is legal and that it was not given the same name as the log file (line 60 to 80). F1 is cleared (line 85). The name and disk of the checklist file are then converted to an ASCII string (line 90) and the file name tested (lines 100 to 110) to see if it is already present on the disk. If it is, an error message is displayed (line 120) and the program branches back to line 60 to give the user a chance to give another name to the checklist.

The log pointer file is opened. The number of entries in the log data file is fetched (line 130), and displayed (line 140). If the disk "write into" flag F1 is cleared, space is created on the disk for the checklist file (line 160). The log data file is opened (line 170). Space in memory for the checklist is allocated (line 180). If there is not enough RAM in the computer, an error occurs here.

The FOR/NEXT loop (lines 240 to 270) reads each entry in the log data file (line 250), displays each callsign at the console (line 255), and stores the callsign in the checklist (line 265). In this program the callsigns are stored sequentially in the order in which they were worked. There is no point in hashing the calls into the list because they are going to be sorted later anyway. Once all the calls are loaded into memory a message confirming that that stage has been reached is displayed (line 280).

```
10 REM CKLSTGEN VER 820305
20 ERRSET 590,E1,E2
30 Q3=0\N1=0\N4=0 \ GOSUB 5010
40 INPUT'Name of LOG file ? ',L1$\IF L1$=''THEN 40
50 INPUT'Which drive is it on ? ',D1 \ IF D1<1ORD1>4 THEN50
55  D1$=STR$(D1) \ D1$=D1$(2,2)
60 INPUT'Do you want the check list saved on disc ? ',A$\IFA$=''THEN60
70 IF A$(1,1)='Y'THEN80 ELSE F1=1 \ GOTO 170
80 INPUT'What is the name of the check list file ? ',L2$\IF L2$=''THEN80
90 IF LEN(L2$)<8THEN100ELSE!'LENGTH ERROR (max is 7)'\GOTO80
100 IF L1$<>L2$ THEN110 ELSE PRINT'FILE NAME ERROR'\GOTO80
110 INPUT'Which drive do you want it put on ? ',D2 \ IF D2<1ORD2>4 THEN110
120 F1=0
130 D2$=STR$(D2) \ D2$=D2$(2,2)
140 IF FILE(L2$+','+D2$)=2 THEN 160 ELSE 150
150 IF FILE(L2$+','+D2$)=3 THEN 160 ELSE 170
160 PRINT'FILE ',L2$,' ALREADY EXISTS ON DRIVE ',D2$\GOTO 80
170 OPEN#0,'*'+L1$+','+D1$ \ READ#0,N1 \ CLOSE#0
180 PRINT'There are ',N1,' entries in ',L1$
190 IF F1=1 THEN210
200 CREATE L2$+','+D2$ , INT(N1*12/256)+1
210 OPEN#0,L1$+','+D1$ \ REM OPEN OLD LOG FILE
220 DIM W$(N1*10),B1(N1)
230 FOR X=0 TO N1-1
240 GOSUB 5025 \REM READ ENTRY
250 PRINT X+1,TAB(5),C$
260 W$(X*10+1,X*10+10) = C$
270 NEXT X
280 PRINT 'CHECK LIST LOADED'
290 GOSUB 430
300 FOR I=0 TO N1-1
310 D$=W$(I*10+1,I*10+10)
320 PRINT D$
330 NEXT I
340 IF F1=1 THEN 410
350 OPEN #2,L2$+','+D2$
360 FOR I=0 TO N1-1
370 WRITE #2,W$(I*10+1,I*10+10)
380 NEXT
390 WRITE #2,'-*-'
400 CLOSE#2
410 GOTO 5045
420 GOTO 5550 \ REM GET STNDATA
430 PRINT'STARTING TO SORT'
440 N9=1
450 C=0\FORI=N1-1TO N9 STEP -1
460 IFW$((I-1)*10+1,(I-1)*10+10)<W$(I*10+1,I*10+10)THEN500
470 A$=W$((I-1)*10+1,(I-1)*10+10)
480 W$((I-1)*10+1,(I-1)*10+10)=W$(I*10+1,I*10+10)
490 W$(I*10+1,I*10+10)=A$ \ C=1 \B2=B1(I-1)\B1(I-1)=B1(I)\B1(I)=B2
500 NEXT \IFC=0THEN510 ELSE N9=N9+1\!N9,\IF N9/20=INT(N9/20)THEN!\GOTO450
510 PRINT'SORT OVER'
520 RETURN
530 FOR I = 1 TO 4
540 IF FILE('SYSTEM,'+STR$(I))=2 THEN EXIT 560
550 NEXT \ GOTO 570
560 CHAIN 'SYSTEM,'+STR$(I)
570 PRINT'put a system disc in any drive, then hit 'RETURN''
580 INPUT' ',A$ \ GOTO 530
590ERRSET 590,E1,E2
600 IF E2=15THEN530
610 IFE1=540 THENIFI<1ORI>4THEN570ELSE550
620 IFE2<>7THEN630ELSEPRINT'FILE ERROR'\GOTO530
630 IFE2<>8THEN640ELSEPRINT'HARD DISC ERROR'\GOTO530
640 PRINT'ERROR ',E2,' AT LINE ',E1\GOTO530
5000 REM SUBROUTINE PACKAGE VER 790817
```

Fig. 4-8. CKLSTGEN program.

A sorting routine is called (line 290) to arrange the callsigns in the checklist in alphanumerical order. The FOR/NEXT loop (lines 500 to 560) displays each callsign and, if the disk write flag (F1) is cleared (line 570), the sorted list is written to the checklist file (lines 580 to 630). The "-*-" in line 620 is used as the end of file flag. The program branches to the chain SYSTEM routine in the subroutine library (line 640).

The sort routine (lines 1000 to 1080) performs a bubble sort algorithm. The bubble sort may not be the fastest sorting routine available but it is the simplest to program. Since the routine has been treated as a subroutine even though it is only used once, it may be replaced by any other sorting algorithm of your choice. However, that since it does not require any memory other than the list being sorted, it may be the optimal method to use when sorting large lists in systems with small memories.

Status messages are displayed on the screen (lines 1010 and 1070). The print statement (line 6060) displays the iteration of the sort to indicate that the computer is still at work.

The error-handling routines also appear (lines 2060 to 2110).

```
10 REM CKLSTRD VER 791207
20 ERRSET 9060,F1,E2
30 INPUT"Name of CHECK LIST file ? ",L1$\IF L1$=""THEN 30
40 INPUT"Which drive is it on ? ",D1 \ IF D1<1ORD1>4 THEN40
50 D1$=STR$(D1) \ D1$=D1$(2,2)
60 INPUT"Which output device (0-7) ? ",Z\IFZ>7 OR Z<0 THEN60
70 OPEN #1,L1$+","+D1$
80 I=0
90 PRINT#Z,"Check list ......",TAB(20),L1$
100 FOR J=0 TO 5
110 READ #1,C$
120 IF LEN (C$)<3 THEN EXIT230
130 IF C$(1,3)="-*-"THEN EXIT 230
140 IF C$=C1$THEN 150 ELSE 160
150 PRINT#Z, TAB(J*10),"******", \ GOTO 180
160 PRINT#Z, TAB(J*10),C$,
170 C1$=C$
180 NEXT
190 PRINT#Z
200 I=I+1 \ IF I<60 THEN 100
210 REM FORM FEED
220 FOR I=1 TO 5 \ PRINT #Z \ NEXT \ GOTO 80
230 PRINT#Z
240 CLOSE#1
9000 FOR I=1 TO 4
9010 IF FILE("SYSTEM,"+STR$(I))=2 THEN EXIT 9030
9020 NEXT \ GOTO 9040
9030 CHAIN "SYSTEM,"+STR$(I)
9040 PRINT"put a system disc in any drive, then hit 'RETURN'"
9050 INPUT" ",A$ \ GOTO 9000
9060ERRSET 9060,E1,E2
9070 IF E2=15THEN9000
9080 IFE1=9010THENIFI<1ORI>4THEN9040ELSE9020
9090 IFE2<>7THEN9100ELSEPRINT"FILE ERROR"\GOTO9000
9100 IFE2<>8THEN9110ELSEPRINT"HARD DISC ERROR"\GOTO9000
9110 PRINT"ERROR ",E2," AT LINE ",E1\GOTO9000
```

Fig. 4-9. CKLSTRD program.

CKLSTRD

CKLSTRD displays the contents of a checklist file on a selected output device. The program listing is given in Fig. 4-9. It begins with the usual identification and error trapping (lines 10 to 20). The desired checklist file is determined (lines 30 to 50), and the output device is set (line 60).

The data file is opened (line 70), the line counter (I) is set to zero (line 80), and a header for the list is set (line 90). Five callsigns are read from the file and printed horizontally or in one line in the format shown in Fig. 4-10 (lines 100 to 180). A callsign is read from the checklist file (line 110). If it has less than 3 characters, something is wrong, so the program skips to the termination sequence (line 230).

If the callsign just read in was the end of list characters (-*-) the program branches to line 230 because all the calls have been processed (line 130). The callsign just read in (C$) is compared to the callsign read in previously (C1$) (line 140). If they are the same, a flag (******) is output and the program flow skips to line 180). This flags duplicate callsigns. Remember that when the list was sorted, duplicate entries were not deleted and ended up adjacent to each other. If the calls are different, the callsign is output (line 160) and the callsign cur-

```
Check list ......   CK1981
4N2DX     4X4GI     ******    ******    4X4IK     4X4IL
4X4IO     4X4JW     4X4US     4X6AS     4X6CQ     ******
4X6DD     4Z4AQ/P   4Z4JS     4Z4MK     4Z4US     ******
******    4Z4ZA     4Z4ZB     ******    ******    ******
******    CT2CQ     CT2CR     DA2DC     DF0DX     DF5CW
DF9ZP/P   DH1EAG    DJ1ZU     DJ9KH     DJ9MT     DK3GI
DK9ZP     DL0RC/P   DL1YAL    DL2NAI    DL5RBG    DL6WT
EA2ABJ    EA2YD     EA7CEL    F0GMM     F8ZS      FG7TD/FS7
G2CKQ     G3IOR     G4BYK     G4CVZ     GI4FUM/4X GJ3VLX
GM5AIW    HA4XH     HA5KKC/7  HA5KKG    HA7KPL    HA7TM
HB9XH     HG5A      I0JU      I0UXK     I5RCR     I5TDJ
IX1BGJ    IZ5ARI    JA3BOA    K3JW      ******    K3STM
******    ******    ******    LZ1XI     LZ2TG     N4DII/4X
OE3JSA    OE6HZG    OK1AGN    OK2BSA    OK3CFP    OK3JW
PA0QX     PA3AJG    PA3BNT    PA6KEI    SK0EJ     SM0AVK
SM6JNT    SN0WPC    ******    SP2PDI    SP6IXF    SP8ECV
SP8GTS    SV0BL     UA3AOV    UA3DQE    UA3FT     UA4CCF
UA4HFG    UA9WGJ    UK2BBB    UK2BCR    UK3XAM    UK5MAF
******    UK5MCU    UQ2GHT    W2VLX     W2VOX     W4YLU
WA2LQQ    WA2UNO/MM3WA4WTG    Y38UL     Y44ZI/P   Y46XF
YB2DI     YO3AJN    YO6AWR/P  YU1KQ     YU2CKH    YU3DBA
YU4EAW    YU4EBL    ZC4ESB    ZL3HH     ZR6AX/4X
```

Fig. 4-10. Sample checklist printout.

rently occupying C$ is transferred to C1$ (line 170). The carriage return-line feed sequence is sent (line 190), and the line counter (I) is incremented (line 200). If less than 60 lines have been output, the program branches back to line 100 to process the next five callsigns.

Form feeds are sent (lines 210 to 220). The line counter is reset to zero (line 80), the new page is then titled (line 90), and the printout continues.

The termination sequence closes the file (lines 230 to 240) and chains the SYSTEM program using the routines in lines 9000 to 9050.

Error-trapping appears (lines 9060 to 9110).

MULTIBANDING THE CONTENTS PACKAGE

The contest package presented in this chapter does not concern itself with which band a station is worked on. It always flags a duplicate contact irrespective of the fact that the previous contact(s) may have been on a different band. In a contest such as the ARRL Sweepstakes this is fine, because stations may only be contacted once for points, and contacts on an other band are also considered as duplicates. The general purpose contest program listed in Fig. 4-1 can be used for all contests if separate logs are kept for different bands. This works out all right for multi-operator contest stations, but a single operator cannot change bands quickly if the logs have to be reloaded each time. The advantage of the program is that it uses significantly less memory than a program which stores the band information as well as the callsign in the contact check list.

A modified version of the general purpose contest program that allows for multiple band operation is listed in Fig. 4-11. It is a patched version of the original, and the modifications are as follows.

A frequency matrix (F1) is added to the dimension statement (line 40). Matrices that contain the allowable bands and a weighted binary representation of each of them are set up (lines 60 and 65). Since this program is designed for HF contests the allowable bands are 10, 15, 20, 40, and 80 Meters. These values are stored in the B1 matrix (line 65). The B matrix (line 60) contains the weighted binary equivalents 1, 4, 16, 64, and 256. The use of a weighted binary technique to store the band information allows a duplicate contact on the same band to be recorded as such and not as a contact on an other band. The subroutine in lines 962 and 964 enters the frequency/band information into the F1 matrix when logging a contact is culled (line 190). The user input band information is checked (line 422) to verify that a chosen band is one that matches the values stored in the band matrix (B1). If an illegal band is entered, the program flow branches back to rephrase the question. A better technique would have been to display the allowable bands, like this:

```
10 PRINT'G3ZCZ General Contest Program Version 1.2'\REM VER 820329
20 INPUT'Roughly how many QSO's do you expect to make ? ',M9
25 ERRSET 2000,E1,E2
30 IFM9<100 THEN PRINT'OPTIMIST !'
40 DIMW$(M9*10),C4$(17),F1(M9)
50 C4$='QRCXFLBMDWT*S'
60 DIMB(5)\B(1)=1\B(2)=4\B(3)=16\B(4)=64\B(5)=256
65 DIMB1(5)\B1(1)=10\B1(2)=15\B1(3)=20\B1(4)=40\B1(5)=80
110 GOSUB 5010\X=0\GOSUB5005\IFX=1THEN5045ELSEGOSUB5050
120 OPEN #1,L1$
145 IFN=0THEN220
150PRINT'Loading check list'
160 FOR I1=1TON
170 READ #1,&D1,&D2,&D3,T,F,C$,S,R,M$,P,&Q1,&Q2,X$
175 IFC$ <> '/*'THEN177ELSEN4=N4-1\IFN4<0THENN4=0\GOTO200
177 PRINT C$
180 GOSUB840\IFI=0THEN190ELSEGOSUB960
190 GOSUB962
195 N4=N4+1
200 NEXTI1\C8$=C$
210 I=I1-1\GOSUB1020\N=I
220 GOSUB1040
230N=N+1
240 GOSUB 1000
250 IFN>M9-5THENPRINT'CHECK LIST IS ALMOST FULL'
251 INPUT1'-? ',A$\IFA$=''THEN250
255 IFLEN(A$)>1THEN250
260 FOR J2=1TOLEN(C4$)\IFA$(1,1)=C4$(J2,J2)THENEXIT280
270 NEXT \PRINTCHR$(7)\ GOTO250
280 PRINT TAB(5),\GOSUB 980\GOTO250
290 INPUT'REPORT RECEIVED ? ',X$
300 RETURN
310 INPUT'REPORT SENT ? ',S
315 IF S>599 THEN310
320 RETURN
350 INPUT'Day (1-31) ?  ',D1\IFD1<1ORD1>31THEN350
360 INPUT'Month ( 1 - 12 ) ?  ',D2\IFD2<0ORD2>12THEN360
370 INPUT 'Year (19xx) ? ',D3 \ IF D3<1900 THEN 370
380 D3 = D3 - 1900
390 RETURN
400 INPUT 'Power (watts) ?  ',P
410 RETURN
420 INPUT 'Freq/Band ?  ',F\IF INT(F)>999THEN420
422FORQ=1TO5\IFB1(Q)=FTHENEXIT430\NEXT\GOTO420
430 RETURN
440 INPUT 'Mode ?  ',M$\IFM$=''THEN440
450 IF M$(1,1)<>' 'THEN460ELSEM$=M$(2,LEN(M$))\GOTO450
460 RETURN
470 INPUT'Call sign ?  ',A$\IFLEN(A$)=0THEN510
480 IFA$(1,1)<>' 'THEN490ELSEA$=A$(2,LEN(A$))\GOTO480
490 C$=A$\GOSUB840
500 IFI=1THENPRINT'OK'ELSEPRINT'WORKED',CHR$(7),' on ',
502 IFI=1THEN510\F2=F1(J)
503 FORQ=5TO1STEP-1
504 IFF2<B(Q)THEN508
506 PRINTB1(Q),' ',\F2=F2-B(Q)\IFB1(Q)=FTHENPRINT'DUPLICATE ',
508 NEXT\IFF2>1THEN503
510 PRINT\RETURN
520 INPUT 'Time ?  ',T\IFT>2400THEN520
530 RETURN
660 IF D2>0THEN670ELSEPRINT'Enter the date first please PRINT'\RETURN
670 REM
675 I=N \ PRINT
685 GOSUB 5020 \ RETURN
690 RETURN
695 INPUT'ARE YOU SURE ? ',A$\IF A$=''THEN695ELSE IF A$(1,1)<>'Y'THENRETURN
698 GOSUB1040\C$='/*'\N4=N4-1\IFN4<0THENN4=0
700 D1=INT(D1) \D2=INT(D2)\D3=INT(D3)\S=INT(S)
710 R=INT(R)\P=INT(P)
720 S8=0\GOSUB840\GOSUB962\IFI=0THEN730ELSEGOSUB960
730 GOSUB5035
735 N=N+1\IF C$='/*'THEN760
738 N4=N4+1
740 C8$=C$
750 GOSUB 1040
760 GOTO 1000
770 IFN>0THENN=N-1
790 CLOSE #1
800 OPEN#0,L2$
810 WRITE #0,N
820 CLOSE #0
830 GOTO 5045\REM EXIT
840 IF C$='/*'THEN940
850 J=0 \ FOR I=1TOLEN(C$)
860 J=J+ASC(C$(I,I))-47\NEXT
870 J=J*10
880 IF J<M9 THEN 890 ELSE J=J-M9\GOTO880
890 D$=W$(J*10+1,J*10+10)
900 IF D$(1,1)=' 'THEN 950
910 IF D$(1,LEN(C$))=C$THEN940
920 J=J+1\IFJ>=M9THENJ=J-M9
930 GOTO 890
940 I=0\RETURN
950 I=1\RETURN
960 W$(J*10+1,J*10+10)=C$\RETURN
962 FORQ=1TO5\IFF=B1(Q)THENF1(J)=F1(J)+B(Q)
964 NEXT\RETURN
980ONJ2GOTO660,290,470,310,695,700,420,440,350,400,520,770,1000
1000 PRINTCHR$(26),CHR$(12)\PRINT'NEXT QSO= ',N4+1\PRINT'LAST QSO = ',C8$,' TIME= ',T
1002 PRINT'Band/Freq =  ',F\PRINT'Log file is ',L$
1005 PRINT 'THERE ARE ',N-1,' ENTRIES IN THE LOG'
1010 RETURN
1020GOSUB5015\GOSUB5020
1030INPUT'HIT RETURN WHEN READY',A$\RETURN
1040C$='?'\Z$=C$\Z2$=Z$\Z3$=Z$\Z4$=Z$\R=0\S=0\Y1=0\X$=''
1045 S=59
1050RETURN
2006 ERRSET 2000,E1,E2
2005 IFE1=480 THEN 470
2015 IF E1=5560 THEN 5570
2020 IF E1=5650 THEN 5670
2030 IF E2<>8 THEN 2040ELSEPRINT'HARD DISC ERROR'\GOTO 2100
2040 IF E2<>7 THEN 2050ELSEPRINT'FILE ERROR'\GOTO 2100
2050 IF E2=15 THEN 790
2055 IF E1=251THEN251
2090 PRINT'ERROR ',E2,' AT LINE ',E1 \ GOTO 790
2100 GOTO 5045
5000 REM SUBROUTINE PACKAGE VER 791008
```

Fig. 4-11. Multiband version of CONTEST.

FOR I = 1 TO 5 \ PRINT B1(I), " ", \
NEXT \ PRINT \ GOTO 420

instead of using GOTO 420. However, that would take more memory.

The band the previous contact(s) took place on is displayed (lines 500 to 510). The B matrix is scanned to see if the corresponding weighted binary value for the band (F2) is equal to or greater than the value in the F1 matrix. If it is, a contact was made and that fact is displayed. If the contact was made on the same band as is in current use, that fact is also shown when the message "DUPLICATE" is displayed. The scanning order is 80, 40, 20, 15, and 10.

When modifying the program to incorporate multi-banding several other modifications were put in as well. An "S" com-

```
10 REM MULTIBANDED CKLSTGEN VER 820329
20 ERRSET 590,E1,E2
30 Q3=0\N1=0\N4=0 \ GOSUB 5010
40 INPUT"Name of LOG file ? ",L1$\IF L1$=""THEN 40
50 INPUT"Which drive is it on ? ",D1 \ IF D1<1ORD1>4 THEN50
55  D1$=STR$(D1) \ D1$=D1$(2,2)
60 INPUT"Do you want the check list saved on disc ? ",A$\IFA$=""THEN60
70 IF A$(1,1)="Y"THEN80 ELSE F1=1 \ GOTO 170
80 INPUT"What is the name of the check list file ? ",L2$\IF L2$=""THEN80
90 IF LEN(L2$)<8THEN100ELSEPRINT"LENGTH ERROR (max is 7)"\GOTO80
100 IF L1$<>L2$ THEN110 ELSE PRINT"FILE NAME ERROR"\GOTO80
110 INPUT"Which drive do you want it put on ? ",D2 \ IF D2<1ORD2>4
   THEN110
120 F1=0
130 D2$=STR$(D2) \ D2$=D2$(2,2)
140 IF FILE(L2$+","+D2$)=2 THEN 160 ELSE 150
150 IF FILE(L2$+","+D2$)=3 THEN 160 ELSE 170
160 PRINT"FILE ",L2$," ALREADY EXISTS ON DRIVE ",D2$\GOTO 80
170 OPEN#0,"*"+L1$+","+D1$ \ READ#0,N1 \ CLOSE#0
180 PRINT"There are ",N1," entries in ",L1$
190 IF F1=1 THEN210
200 CREATE L2$+","+D2$ , INT(N1*6/50)+1
210 OPEN#0,L1$+","+D1$ \ REM OPEN OLD LOG FILE
220 DIM W$(N1*10),B1(N1),F3(N1)
230 FOR X=0 TO N1-1
240 GOSUB 5025 \REM READ ENTRY
250 PRINT X+1,TAB(5),C$
260 W$(X*10+1,X*10+10) = C$
262 F3(X)=F
270 NEXT X
280 PRINT "CHECK LIST LOADED"
290 GOSUB 430
292 OPEN #2,L2$+","+D2$
295 FOR J=1TO5 \ READ F5 \ GOSUB299\ NEXT \WRITE#2,"TOTAL"\GOTO360
296 DATA10,15,20,40,80
299 Q5=0\ PRINT\PRINT "LIST FOR BAND/FREQ= ",F5\WRITE#2,"BAND="+STR$(F5)
300 FOR I=0 TO N1-1
302 IF F3(I)=F5 THEN 310 ELSE330
310 D$=W$(I*10+1,I*10+10)
320 PRINT D$
323 WRITE#2,D$
330 NEXT I
335 RETURN
340 IF F1=1 THEN 410
360 FOR I=0 TO N1-1
370 WRITE #2,W$(I*10+1,I*10+10)
380 NEXT
390 WRITE #2,"-*-"
400 CLOSE#2
410 GOTO 5045
420 GOTO 5550 \ REM GET STNDATA
430 PRINT"STARTING TO SORT"
440 N9=1
450 C=0\FORI=N1-1TO N9 STEP -1
460 IFW$((I-1)*10+1,(I-1)*10+10)<W$(I*10+1,I*10+10)THEN500
470 A$=W$((I-1)*10+1,(I-1)*10+10)
480 W$((I-1)*10+1,(I-1)*10+10)=W$(I*10+1,I*10+10)
490 W$(I*10+1,I*10+10)=A$ \ C=1 \B2=B1(I-1)\B1(I-1)=B1(I)\B1(I)=B2
492 F2=F3(I-1)\F3(I-1)=F3(I)\F3(I)=F2
500 NEXT \IFC=0THEN510 ELSE N9=N9+1\PRINTN9,\IF N9/20=INT(N9/20)THENPRINT\
    GOTO450
510 PRINT"SORT OVER"
520 RETURN
530 FORI=1TO4
540 IF FILE("SYSTEM,"+STR$(I))=2 THEN EXIT 560
550 NEXT \ GOTO 570
560 CHAIN "SYSTEM,"+STR$(I)
570 PRINT"put a system disc in any drive, then hit 'RETURN'"
580 INPUT" ",A$ \ GOTO 530
590ERRSET 590,E1,E2
600 IF E2=15THEN530
610 IFE1=540 THENIFI<0OR^@^@I>4THEN570ELSE550
620 IFE2<>7THEN630ELSEPRINT"FILE ERROR"\GOTO530
630 IFE2<>8THEN640ELSEPRINT"HARD DISC ERROR"\GOTO530
640 PRINT"ERROR ",E2," AT LINE ",E1\GOTO530
5000 REM SUBROUTINE PACKAGE VER 790817
```

Fig. 4-12. Modified version of CKLSTGEN (multiband).

```
10 REM MULTIBAND CKLSTRD VER 820330
20 ERRSET 9060,E1,E2
30 INPUT"Name of CHECK LIST file ? ",L1$\IF L1$=""THEN 30
40 INPUT"Which drive is it on ? ",D1 \ IF D1<1ORD1>4 THEN40
50 D1$=STR$(D1) \ D1$=D1$(2,2)
60 INPUT"Which output device (0-7) ? ",Z\IFZ>7 OR Z<0 THEN60
70 OPEN #1,L1$+","+D1$
80 I=0
90 PRINT#Z,"Check list ......",TAB(20),L1$
100 FOR J=0 TO 5
110 READ #1,C$
120 IF LEN (C$)<3 THEN EXIT230
130 IF C$(1,3)="-*-"THEN EXIT 230
132 IF C$(1,5)="BAND=" THENJ=6
136 IF C$(1,5)="TOTAL" THENJ=6
140 IF C$=C1$THEN 150 ELSE 155
150 PRINT#Z, TAB(J*10),"******", \ GOTO 180
155 IF J<6THEN160ELSE PRINT#Z\PRINT #Z,C$\GOTO200
160 PRINT#Z, TAB(J*10),C$,
170 C1$=C$
180 NEXT
190 PRINT#Z
200 I=I+1 \ IF I<60 THEN 100
210 REM FORM FEED
220 FOR I=1 TO 5 \ PRINT #Z \ NEXT \ GOTO 80
230 PRINT#Z
240 CLOSE#1
9000 FOR I = 1 TO 4
9010 IF FILE("SYSTEM,"+STR$(I))=2 THEN EXIT 9030
9020 NEXT \ GOTO 9040
9030 CHAIN "SYSTEM,"+STR$(I)
9040 PRINT"put a system disc in any drive, then hit 'RETURN'"
9050 INPUT" ",A$ \ GOTO 9000
9060ERRSET 9060,E1,E2
9070 IF E2=15THEN9000
9080 IFE1=9010THENIFI<1ORI>4THEN9040ELSE9020
9090 IFE2<>7THEN9100ELSEPRINT"FILE ERROR"\GOTO9000
9100 IFE2<>8THEN9110ELSEPRINT"HARD DISC ERROR"\GOTO9000
9110 PRINT"ERROR ",E2," AT LINE ",E1\GOTO9000
```

Fig. 4-13. Modified version of CKLSTRD (multiband).

Checklist with 600 calls

1	Y56YF	AC3A	UA1FV	YU7QDT						
2	YU3UPI	UA3AHA								
3	I2SVA	Y37XJ								
4	LZ1BM	W3XU								
5	AB0I	KA1YQ	W1IHN	F6BXQ	HA5KKB	OK1FV				
6	OK3CFA	UA9XWU								
7	AK1A	UK2BBB	UK3AAC	UK5AAA	YU2RVL	DL5NBC				
8	K3EST	EA3WZ	YU7AD	OH3OQ						
9	VE5UF	P42J	Y22OM/A	RA3DKE	OE1WD	UK2AAF	UA3TN	UB5AAL		
10	HA5KKG	YU3EF								
11	N2AA	OK1DLA	G3XBY	HA4XX	UB5EJA	UK2GAB				
12	AI7B	HB9BMY/4X	FC6FPH	R6L	UA3AGL	OK3JW	I2QMU	UK5FAA	JG1ILF	
13	RB5CCO	UG6LQ	UK3IAA	UL7QF						
14	HA5KKC/7	I5MYL	K2BA/4	KC1F	HG5A	UP2NK				
15	UK3DBG	UK3DAH	UA3QAE	UK5FAD	JA9YBA	DJ8UV				
16	KG1D									
17	AH2M	VE3PCA	I0KWX	YU1NZW	VE1BRB	OK1KUR/P	DL6RAI	HB9BLQ	VE7WJ	UK5IBB
18	PJ2VR	UK0QAA	YU7NZR	YU7KWX	UA9MAF	UA3DUA	ZY5EG			
19	IO6FLD	KE3G	VE6OU	OK1AJN	DL6FAW	UK6LAA	I4VOS	YU3EO	I2YKV	HG3KHO
20	SV1MO	YU2JL	I5JUX							
21	YU2CT	UK3QAA	OK3CSC	UK5OAA	LZ1RN	UB5KAN				
22	YU3CAB	OK1ARI	AI9J	HK5BCZ	OK2RZ	UK2GBL	OE3KTA			
23	WA3GSC	UA0ZDD	HB9BUN	YO9HP	UB5DBV	HG1KZC	OZ5EV			
24	HA4KYH	KC3N	AD8R	UK5MAF	OK1ALQ	FR0FLO	VK3AJJ	UK3SAB	UK9AAN	VE1DXA
25	I4ZSQ	YU3DBC	OK3CRH	DL4SAR	UK3QAE	UA3TBK	YU7AU	I5MXX	LU5FGG	YU1DW
26	YU4YA	WD0EWD	YU7AV	UC2OBP	OK2BQL	OK2BTI	UA0WAM	YU1FW	UK0AMM	XK5XK
27	KF2O	4U1ITU	UW3HV	K2NG	VE7UBC					
28	NP4A	UK5QBE	UK4WAA	YO5KAD	UY5YB	UB5MPD	G3VZT	K4KZZ	KI3L	
29	N4KG	KK5I	UK3XAB	YU3EY						
30	HA4KYN	RA3RBU	VK2APK	UA6LGP	JI1QQI	UY5XE	YO6KEA			
31	IZ1ARI	UK6XAA	KI2P	UG6GDS	ZS5IV	WB1GZE				
32	UK4FAV	YU7ECD	KS2G							
33	UK3ADZ	WB4KRH	HA9KPU	UK6YAB						
34	YU3VM	UA3VAS	VE3BMV	UQ2GFN	ZS4SP	UK2PAP				
35	YU2CHI	F8WE	DK5AD							
36	KQ2M	UA3ACY/U9J	UB5MBZ	UC2OBZ	UK5WBG	HG6V	YO6AFP	GW4HSH		
37	UK9FER	K6IR	KA4SUN	UA9CRR	OH2AA					
38	UK2PCR	IS0QDV	PA2TMS	YU1BAU	UK7NAQ	W1NG				
39	4X4VE/5N8	N4XD	LZ13C							
40	N4NO	LZ2KIM	4N6HN	Y23DL	LU8FEU	OK1MSN	KM5R	OZ5EDR		
41	YV3BJL	4X6CJ	YU7MY							
42	KQ8M	UK2GKW	EA2AEA							
43	K3ZJ	Y22JJ	I0JBL	OH6AC	JN1XEV	OE1DH	F5RU			
44	OK1KST	K2SS	YU3DAW	F6BFN	VP2MGQ	UK6LTG	UK5QAV	UW3ZU	HA0DU	4N0SM

Fig. 4-14. Sample checklist (in computer memory) for 600 calls.

```
45      EI7DJ    VE7BTV   KB2MG    VK2ERT   UK5GKW   EA6GP    UB5VAZ   DJ3HJ    VE3JTQ
46      IV3PRK   KJ9W     W8UA     LU7MAY   IP1TTM
47      UK2RDX   IS0WON   YU4EXA   UP2PBW
48      KV5Q     Y27FN    F6GKH    VE1AI
49      N2US     YU3TLA
50      N8AKF    OH1IG    EG3SF    JR1WHW
51      KB8LH    DK0MM    HA4ZB
52      Y35LM    DL0UE    I2JSB    N2WT     6Y5HN
53      CT7AL    Y33ZB    DL8PC    UK5UDX   YT3L     IV3WMP   DJ1BZ
54      DF4NP    K2IJL    JH0BBE   LZ2KRR   I2ARN
55      K3KHL    WB9TIY   LA7JO    HA8ZB    4X4UH
Checklist with 600 calls
56      YU3TDP   Y32ZF    LZ2DB    LZ2CC    DF2AO/A  I6CXD    YU2BST
57      Y34YF    Y24UK    EA1ABT   I1JHS    HB9AGC   N4WW     F6KAW    ED3VM    DL0IV    NU4Y
58      VP2EC    YU7KMN   YZ6G     W9LT     4Z4JS    I0NKN
59      IT9WPO   KL7RA    OH2AW    DK7JQ    DF5IY    UB5JK
60      I0SKK    Y47XF    9H4B
```

Fig. 4-14. Sample checklist (in computer memory) for 600 calls. (continued)

mand was added to display at will the last contact and QSO number information given following a contact. Changes were made in the display routine (lines 1000 and 1002) including adding time (the last time entered), band, and log file data to the information displayed. Corresponding modifications must also be made to the checklist generating and reading programs of Fig. 4-8 and Fig. 4-9. These changes are shown in Fig. 4-12 and Fig. 4-13.

Fig. 4-12 lists the modified CKLSTGEN program. The frequency information for each contact is entered (line 462) in the frequency matrix (F3) which was set up in line 220. Once the calls are loaded they must be sorted (line 492), and printed according to bands worked (line 295). Each set of data is prefixed with a header. Note that the patch was written for the case in which the checklist is stored on the disk.

The modified CKLSTRD program is shown in Fig. 4-13. Here the modifications lie in lines 132, 136 and 155, which control the formatting of the listing and provide the correct paragraphing. Setting the value of J to 6 is a convenient way of exiting from the FOR/NEXT loop and signalling that an extra carriage return line feed sequence is required. An example of the printout from this program is shown in Appendix A.

CONTEST STRATEGIES

Using the programs in a contest is an art. The fudge feature is useful in correcting an entry that has been logged to the disk erroneously. It can also be used to correct an earlier error. For example, a contact with 4X4VI/5N8 is entered. When he is tuned in again later, you realize to your great dismay that he is actually 4X4VE/5N8. What can you do? Well, if a dummy contact for 4X4VE/5N8 is logged and then fudged, both calls will be in the checklist and the dupe routine will recognize them. If a note is made, the original entry can be corrected using the LOGEDIT editor program after the contest, and the LOGFUDGE program will clear any fudged entries as well as the dummy entry.

THE CHECKLIST

The contest programs of this chapter use a hashing technique to distribute the callsigns of stations worked throughout memory so as to speed up the access time to

Checklist with 1000 calls

1	4X6CJ									
2	KQ8M	EA2AEA								
3	K3ZJ	Y22JJ	I0JBL	OH6AC	OE1DH	F5RU				
4	K2SS	F6BFN	HA0DU	4N0SM						
5	EI7DJ	KB2MG	EA6GP	DJ3HJ						
6	KJ9W	W8UA								
7										
8	KV5Q	Y27FN	F6GKH	VE1AI						
9	N2US									
10	N8AKF	OH1IG	EG3SF							
11	KB8LH	DK0MM	HA4ZB							
12	Y35LM	DL0UE	I2JSB	N2WT	6Y5HN					
13	CT7AL	Y33ZB	DL8PC	YT3L	DJ1BZ					
14	DF4NP	K2IJL	JH0BBE	I2ARN						
15	K3KHL	LA7JO	HA8ZB	4X4UH						
16	Y32ZF	LZ2DB	UA3ACY/U9JLZ2CC		DF2AO/A	I6CXD				
17	Y34YF	Y24UK	EA1ABT	I1JHS	HB9AGC	N4WW	F6KAW	ED3VM	DL0IV	NU4Y
18	VP2EC	YZ6G	W9LT	4Z4JS	I0NKN					
19	KL7RA	OH2AW	DK7JQ	DF5IY	UB5JK					
20	I0SKK	Y47XF								
21	Y56YF	UA1FV								
22	UA3AHA									
23	I2SVA	Y37XJ								
24	LZ1BM	W3XU								
25	KA1YQ	W1IHN	F6BXQ	HA5KKB	OK1FV					
26	OK3CFA									
27	UK2BBB	UK3AAC	UK5AAA	DL5NBC						
28	K3EST	EA3WZ	YU7AD	OH3OQ						
29	VE5UF	Y22OM/A	RA3DKE	OE1WO	UK2AAF	UA3TN	UB5AAL			
30	HA5KKG	YU3EF								
31	OK1DLA	G3XBY	HA4XX	UB5EJA	UK2GAB					
32	FC6FPH	UA3AGL	OK3JW	I2QMU	UK5FAA	JG1ILF				
33	RB5CCO	UG6LQ	UK3IAA	UL7QF						
34	HA5KKC/7	I5MYL	UP2NK							
35	UK3DBG	UK3DAH	UA3QAE	UK5FAD	JA9YBA	DJ8UV				
36										
37	VE3PCA	I0KWX	VE1BRB	DL6RAI	HB9BLQ	VE7WJ	UK5IBB			
38	PJ2VR	UK0QAA	UA9MAF	UA3DUA	ZY5EG					
39	IO6FLD	VE6OU	OK1AJN	DL6FAW	UK6LAA	I4VOS	YU3EO	I2YKV	HG3KHO	I5JUX
40	SV1MO	YU2JL								
41	YU2CT	UK3QAA	OK3CSC	UK5OAA	LZ1RN	UB5KAN				
42	YU3CAB	OK1ARI	HK5BCZ	OK2RZ	UK2GBL	OE3KTA				

Fig. 4-15. Sample checklist (in computer memory) for 1000 calls.

Checklist with 1000 calls

```
43  WA3GSC  UA0ZDD  HB9BUN  YO9HP   UB5DBV  HG1KZC  OZ5EV
44  HA4KYH  UK5MAF  OK1ALQ  FR0FLO  VK3AJJ  UK3SAB  UK9AAN  VE1DXA  YU1DW
45  I4ZSQ   YU3DBC  OK3CRH  DL4SAR  UK3QAE  UA3TBK  YU7AU   I5MXX   LU5FGG  XK5XK
46  YU4YA   WD0EWD  YU7AV   UC2OBP  OK2BQL  OK2BTI  UA0WAM  YU1FW   UK0AMM
47  4U1ITU  UW3HV   VE7UBC
48  UK5QBE  UK4WAA  YO5KAD  UY5YB   UB5MPD  G3VZT   K4KZZ
49  UK3XAB  YU3EY
50  HA4KYN  RA3RBU  VK2APK  UA6LGP  JI1QQI  UY5XE   YO6KEA
51  IZ1ARI  UK6XAA  UG6GDS  ZS5IV   WB1GZE
52  UK4FAV  YU7ECD
53  UK3ADZ  WB4KRH  HA9KPU  UK6YAB
54  YU3VM   UA3VAS  VE3BMV  UQ2GFN  ZS4SP   UK2PAP
55  YU2CHI
56  UB5MBZ  UC2OBZ  UK5WBG  YO6AFP  GW4HSH
57  UK9FER  KA4SUN  UA9CRR
58  UK2PCR  IS0QDV  PA2TMS  YU1BAU  UK7NAQ
59  4X4VE/5N8
60  LZ2KIM  LU8FEU  9H4B    OK1MSN  OZ5EDR
61  AC3A    YV3BJL  YU7MY
62  UK2GKW
63  JN1XEV
64  OK1KST  YU3DAW  VP2MGQ  UK6LTG  UK5QAV  UW3ZU   UB5VAZ  VE3JTQ
65  VE7BTV  AB0I    VK2ERT  UK5GKW
66  IV3PRK  LU7MAY  IP1TTM
67  AK1A    UK2RDX  IS0WON  YU4EXA  UP2PBW
68
69  P42J    YU3TLA
70  JR1WHW
71  N2AA
72  AI7B    R6L
73  UK5UDX  IV3WMP
74  K2BA/4  KC1F    LZ2KRR  HG5A
75  WB9TIY
76  YU3TDP  KG1D    YU2BST
77  AH2M
78  YU7KMN
79  KE3G    IT9WPO
80
81  YU7QDT
82  YU3UPI  AI9J
83
84  KC3N    AD8R
85
```

Fig. 4-15. Sample checklist (in computer memory) for 1000 calls. (continued)

```
86  UA9XWU
87  KF2O        K2NG        YU2RVL
88  NP4A        KI3L
89  N4KG        KK5I
90
91  KI2P
92  HB9BMY/4X KS2G
93
94
95  F8WE        DK5AD
96  KQ2M        HG6V
97  K6IR        YU1NZW      OK1KUR/P OH2AA
98  YU7NZR      YU7KWX      W1NG
99  N4XD        LZ13C
100 N4NO        4N6HN       Y23DL       KM5R
```

Fig. 4-15. Sample checklist (in computer memory) for 1000 calls. (continued)

determine if the station has been worked before. The programs dynamically allocate space in memory for storing the calls. Figures 4-14 and 4-15 show how the callsigns of the stations worked during the 1982 WPX contest were stored in memory. Figure 4-14 shows a matrix with space allocated for 600 callsigns, while Fig. 4-14 shows the same calls stored in a matrix of 1000 callsigns. The hashing technique used only computes an entry point into the table. From then on a sequential search is performed. This simple technique significantly reduces the search time, even in the case of Fig. 4-14 in which the matrix is over half full. Even at this stage of the contest, at the worst, 35 entries will have to be checked when testing for a duplicate entry (lines 24 to 27). Otherwise, usually less than ten entries have to be checked. If a thousand spaces are allocated, as shown in Fig. 4-15, wide open spaces appear and the search times are reduced. It is possible to detect the difference in search times during the contest as the response time varies for different callsigns. The difference in the amount of memory used by the arrays is such that in a system which can allocate space for 1000 calls using the program of Fig. 4-1, space can only be allocated for 600 callsigns when using the multiband version of Fig. 4-11. (This was in a Northstar computer with 40K of memory.)

Once a contact has been entered in the log (that is, written to disk), it is gone from sight. If a second terminal is available, the log routine should be modified to write the entry to that other terminal as well. A modification of the code used in the Q command should suffice. This allows the last few entries to be constantly displayed.

These programs were exercised in several contests during 1981 and 1982 and examples of their use are presented in Appendix A. However, the best computer program in the world is no substitute for high power, a good antenna, and good operating practices.

Simulations and Modeling

Simulation and modeling are an exciting area of computer activity often sadly neglected by amateur radio. This chapter presents a simulation of an actual contest in game format; the game incorporates several different models.

Computer games are basically simulations which react with the player via the console. The common "Star Trek" game is a simulation of a universe in which the Starship is directed to search and destroy enemy vessels within the constraints set up by the programmer. These same simulation techniques can be employed in amateur radio as training aids, for sharpening operator skills, or just for having fun.

A SWEEPSTAKES CONTEST GAME

The particular contest discussed here is the 1977 version of the American Radio Relay League (ARRL) Sweepstakes Contest. The rules of the contest are outlined in Fig. 5-1. This particular contest is designed to improve the third-party traffic or message handling skills of the contestants, and the information exchange is thus set up to simulate a real message.

An explanation of the message is as follows:

	Number	*Precedence*	*Call*	*Check*	*Section*
EXCHANGE	Consecutive Serial Number	Power Input Level	Station Callsign	Last two digits of year first licensed	The ARRL section in which the station is located
SAMPLE	003	A (less than 200W)	G3ZCZ/W3	68	MD

Rules

1) **Eligibility:** This contest is open to all radio amateurs in (or officially attached to) the ARRL sections listed in Fig. 5-2. U.S. possessions in the Pacific are part of the Pacific section. KP4, KV4, and KG4 are all part of the West Indies section.

2) **Object:** To exchange QSO information (as explained in Rule 5) with as many amateurs in (or officially attached to) ARRL sections.

3) **Conditions of Entry:** Each entrant agrees to be bound by the provisions as well as the intent of this announcement, the regulations of his licensing authority and the decisions of the ARRL Awards Committee.

4) **Contest Period and Time:** All contacts must be made during the contest period indicated, Saturday November 19 (2100 GMT) to Monday November 21 (0300 GMT). Time spent listening counts as operating time. No more than 24 hours of operation are permitted during the 30-hour period. "Off" periods may not be less than 15 minutes at a time. Times on and off and QSO times must be intered in your log.

5) **Valid QSOs:** Contacts must include certain information sent in the form of a standard message preamble, as shown in the example. Cw stations work only cw stations and phone stations only other phones. Valid points can be earned by contacting stations not working in the contest upon acceptance of your preamble and receipt of a preamble.

6) **Scoring:** Each station from which a preamble is received and to whom a preamble is sent and acknowledged, results in 2 points. Partial QSOs do not count for scoring purposes. No additional points can be earned by recontacting the same station, regardless of the frequency band (i.e., repeat QSOs, even if on different bands, are not allowed). The total number of ARRL sections (plus VE8) worked during the contest is the section multiplier—maximum possible total of 75. The final score equals the total points times the section multiplier.

7) **Entry Classification:** Entries will be classified as single- or multiple-operator stations. Single-operator stations are those in which one person performs all transmitting, receiving, spotting, and logging functions. Multiple-operator stations are those obtaining any form of assistance, such as from spotting or relief operators, or keeping the station log or records. The use of any type of "spotting" or "multiplier" net places the entry in the multiple-operator class.

8) **Reporting:** Every competing entry claiming 200 or more QSOs must have cross-checking sheets attached. All entries become the property of ARRL and none can be returned. Although FCC rules no longer require most log-keeping, each competitive entry submitted must include date, QSO times, times on/off, exchange sent, exchange received, band, and mode.

9) **Misc. Rules:** Yukon-N. W. T. (VE8) counts as a separate multiplier, for a possible total of 75 multipliers.

If your power is 200-watts dc or less, send "A" as your precedence; otherwise send "B."

Fig. 5-1. Rules for the Sweepstakes Contest Game.

This is a little different from the usual 5900 1 or 59-1kW exchange of other contests, but is not too difficult to manage.

The operation of the contest is set up to simulate participation by an amateur radio station located in Silver Spring, Maryland, just outside Washington, D.C. Before going into the mechanics of the contest, consider some of the background information needed to generate the complete simulation. The basic aim of the contest is to contact as many different stations and as many different sections of the United States and Canada as possible. Contacts with sections are governed by two factors: (a) propagation of radio frequencies between Maryland and the rest of the USA, and (b) the number of amateur radio stations active in each section during the contest.

Propagation conditions are reasonably well known. As the general propagation characteristics of each amateur band are different, experienced amateurs by and large know which band to use to contact which part of the world. A model can be created of propagation between one part of the world and any other based on the frequency band, the time of day, and the occupancy of the band.

Consider propagation between Washington, D.C. and the rest of the United States. The United States is divided into 10 call areas, and Canada is divided into 8 by the governments involved. The ARRL has further subdivided the USA and Canada

NMBR	ABBRV.	SECTION
0	CT	Connecticut
1	EMASS	East Massachusetts
2	ME	Maine
3	NH	New Hampshire
4	RI	Rhode Island
5	VT	Vermont
6	WMASS	West Massachusetts
7	ENY	Eastern New York
8	NY - LI	New York City - Long Island
9	NNJ	Northern New Jersey
10	SNJ	Southern New Jersey
11	WNY	Western New York
12	DE	Delaware
13	E.PA	East Pennsylvania
14	M.DC	Maryland - DC.
15	W.PA	West Pensylvania
16	ALA	Alabama
17	GA	Georgia
18	KY	Kentucky
19	NC	North Carolina
20	NFLA	North Florida
21	SC	South Carolina
22	SFLA	South Florida
23	TN	Tennessee
24	VA	Virginia
25	WI	West Indies
26	AK	Arkansas
27	LA	Louisiana
28	MISS	Mississippi
29	NM	New Mexico
30	N.TEX	North Texas
31	OK	Oklahoma
32	S.TEX	South Texas
33	CZ	Canal Zone (Since deleted)
34	E BAY	East Bay
35	LA	Los Angeles
36	ORG	Orange
37	SB	Santa Barbara
38	SCV	Santa Clara Valley
39	SD	San Diego
40	SF	San Francisco
41	SJV	San Joaqim Valley
42	SV	Sacramento Valley
43	PCF	Pacific (Hawaii and Islands)
44	AZ	Arizona
45	ID	Idaho
46	MT	Montana
47	NV	Nevada
48	OR	Oregon
49	UT	Utah
50	WA	Washington
51	AL	Alaska
52	WY	Wyoming
53	MI	Michigan
54	OH	Ohio
55	W VA	West Virginia
56	IL	Illinois
57	IN	Indiana
58	WSC	Wisconsin
59	CO	Colorado
60	IA	Iowa
61	KS	Kansas
62	MN	Minnesota
63	MO	Missouri
64	NB	Nebraska
65	ND	North Dakota
66	SD	South Dakota
67	MAR	Maritime Provinces
68	QB	Quebec
69	ONT	Ontario
70	MAN	Manitoba
71	SK	Saskatchewan
72	AB	Alberta
73	BC	British Columbia
74	NWT	North West Territories

Fig. 5-2. List of sections in game.

into 75 sections for its organizational purposes. These sections are listed in Fig. 5-2. A map showing the sections and call areas is shown in Fig. 5-3. The assignment of numbers to the sections is for modelling purposes only; the numbers have no other significance. Some of the more populated states may contain several sections.

The truth table for a *simple* propagation model is shown in Fig. 5-4. The model predicts the probability of propagation between Maryland and the different call areas of the USA for various times of the day, which is divided into 4-hour periods. For example, the model shows that propagation with W2 is possible on 80m and 40m day or night, is never possible on 15m, and is possible on 20m and 10m only under specific conditions. This model, although simple, can be used to predict propagation to any part of the USA for determining the best time for working into particular states for contests, awards, or setting up schedules.

Thus far we have considered the probability of a successful contact as a function of propagation. In the real world, however, the distribution of amateur radio stations in these sections is not uniform. There are, for example, more radio amateurs in Colorado than in North Dakota, so that even if propagation is possible to the W0 call area, the probability of a contact with Colorado is greater than the probability of a contact with North Dakota.

Statistical data on the distribution of active amateurs is available as part of the results of the contest. Figure 5-5 lists the various sections in the different call areas, and the number of amateurs in each section sending entries in for the CW and SSB parts

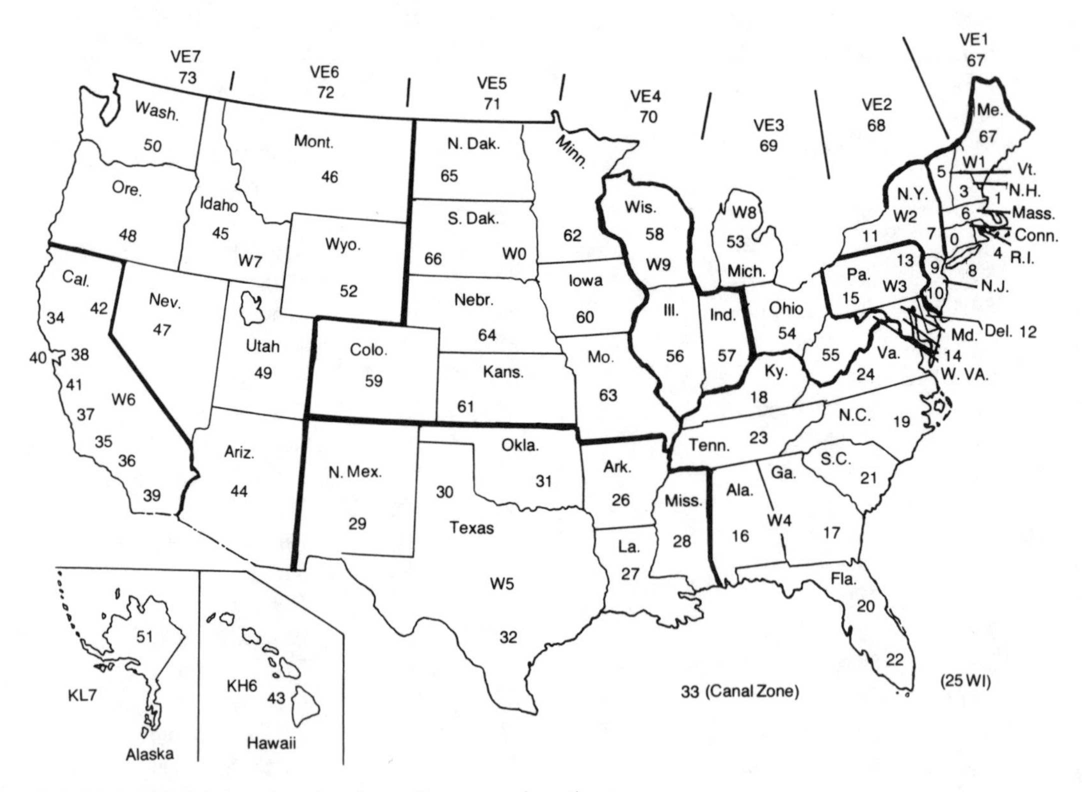

Fig. 5-3. Map of North America showing call areas and sections.

Time Z	L	W1 1015204080					W2 1015204080					W3 1015204080					W4 1015204080				
4	0	0	0	0	1	1	0	0	0	1	1	0	0	0	5	1	0	0	0	1	1
8	4	0	.1	.5	1	1	0	.5	1	1	0		0	0	1	1	0	0	1	1	1
12	8	2	0	1	1	0	1.5	0	0	1	1	0	0	0	0	1	3.5	.5	1	1	5
16	12	2	0	1	1	0	0	0	0	1	1	0	0	0	1	1	3.5	.5	1	1	0
20	16	2.5	.1	.5	1	.5	0	0	.5	1	1	0	0	0	1	1	3.5	0	1	1	.5
0	20	0	0	0	1	1	0	0	0	1	1	0	0	0	5	1	0	0	0	1	1

Z	L	W5					W6					W7					W8				
4	0	0	0	0	1	1	0	0	0	1	1	0	0	0	1	1	0	0	0	.5	1
8	4	.5	.5	.5	.5	.5	.1	.1	.1	.1	.1	0	0	0	.5	0	0	0	0	1	1
12	8	1	1	1	0	0	1	1	1	0	0	1	1	1	0	0	0	0	0	1	1
16	12	1	1	1	0	0	1	1	1	0	0	1	1	1	0	0	0	0	0	1	1
20	16	.5	.5	.5	.5	.5	.5	.5	1	.1	0	0	.5	.5	.5	0	0	0	0	1	1
0	20	0	0	0	1	.5	0	0	0	1	.5	0	0	0	1	.5	0	0	0	.5	1

Z	L	W9					W0				
4	0	0	0	0	1	1	0	0	0	1	1
8	4	0	0	0	1	.5	0	0	0	.5	.5
12	8	0	0	1	1	0	5.5	.5	1	.0	0
16	12	0	0	1	1	0	5.5	.5	1	0	0
20	16	0	0	0	1	.5	5.5	0	.5	.5	.1
0	20	0	0	0	1	1	0	0	0	1	1

P > 1 = (Short Skip + N)

Z	L	KV4					KZ5					KL7/KH6				
4	0	0	0	0	1	1	0	0	0	1	1	0	0	0	1	.5
8	4	1	1	1	0	0	.5	.5	0	.5	0	0	0	.1	0	0
12	8	1	1	1	0	0	1	1	0	0	0	1	.5	.5	0	0
16	12	1	1	1	0	0	1	1	1	0	0	1	1	1	0	0
20	16	.5	1	1	0	0	0	.5	1	.5	0	.5	.1	.5	0	0
0	20	0	0	0	.1	1	0	0	0	1	.5	0	0	0	.5	0

Fig. 5-4. Truth table for simple propagation model.

of the 1977 contest. The total CW and SSB entries have been added together, and the percentage of amateurs in each call area in the section is listed. Thus, if the band is open to the W1 call area, there is a 30% probability of a contact with Connecticut, but only a 4% probability of a contact with Vermont based on the number of amateurs active in the contest. Figure 5-6 summarizes the percentage of stations in the various call areas active in the contest.

The actions of the operator during the contest must then be modelled. The operator can tune the band, call cq, request repeat transmissions of information missed, check his status, change bands, take a break, etc. The options considered in this simulation also happen to be the commands available in the game, and are as follows:

BAND—Switch to a new band or look at setting of band switch. Allowable bands are 80, 40, 20, 15, and 10 meters.

#	AREA	CW	SSB	TOT	PROBABILITY	ΣP
	W1					
0	CT	43	30	73	0.30165	0.30165
1	EMASS	37	31	68	0.28099	0.58264
2	ME	12	9	21	0.08677	0.66941
3	NH	12	10	22	0.09090	0.76031
4	RI	8	12	20	0.08264	0.84295
5	VT	7	4	11	0.04545	0.88840
6	WMASS	15	12	27	0.11157	0.99997
	W2					
7	ENY	35	24	59	0.17507	0.17507
8	NY-LI	28	28	56	0.16617	0.34124
9	NNJ	34	32	66	0.19584	0.53708
10	SNJ	37	47	84	0.24925	0.78633
11	WNY	38	34	72	0.21364	0.99997
	W3					
12	DE	5	6	11	0.04135	0.04135
13	E.PA	51	48	99	0.37218	0.41353
14	MDC	65	53	118	0.44360	0.85713
15	WPA	22	16	38	0.14285	0.99998
	W4					
16	ALA	7	7	14	0.04590	0.04590
17	GA	13	16	29	0.09508	0.14098
18	KY	15	12	27	0.08852	0.22950
19	NC	17	12	29	0.09508	0.32458
20	NFL	13	15	28	0.09180	0.41638
21	SC	7	3	10	0.03278	0.44916
22	SFL	12	7	19	0.06229	0.51145
23	TN	11	15	26	0.08524	0.59669
24	VA	58	58	116	0.38032	0.97701
25	WI	4	3	7	0.02215	0.99996
	W5					
26	AK	11	6	17	0.11038	0.11038
27	LA	6	3	9	0.05844	0.16882
28	MS	9	4	13	0.08441	0.25323
29	NM	4	6	10	0.06493	0.31816
30	NTX	25	23	48	0.31168	0.62984
31	OK	4	7	11	0.07142	0.70126
32	STX	26	19	45	0.29220	0.99346
33	CZ	1	0	1	0.00649	0.99995

Fig. 5-5. Probability of working particular sections during the contest.

#	AREA	CW	SSB	TOT	PROBABILITY	ΣP
	W6					
34	EB	17	18	35	0.11945	0.11945
35	LA	23	21	44	0.15017	0.26962
36	ORG	11	8	19	0.06484	0.33446
37	SB	8	8	16	0.05460	0.38906
38	SCV	45	44	89	0.30375	0.69281
39	SD	13	7	20	0.06825	0.76106
40	SF	11	10	21	0.07167	0.83273
41	SJV	14	16	30	0.10238	0.93511
42	SV	7	6	13	0.04436	0.97947
43	PCF	3	3	6	0.02047	0.99994
	W7					
44	AZ	20	11	31	0.17613	0.17613
45	ID	4	7	11	0.06250	0.23863
46	MT	2	7	9	0.05113	0.28976
47	NV	7	3	10	0.05681	0.34657
48	OR	15	10	25	0.14204	0.48861
49	UT	2	4	6	0.03409	0.52270
50	WA	34	42	76	0.43181	0.95451
51	AL	2	3	5	0.02840	0.98291
52	WY	0	3	3	0.01704	0.99995
	W8					
53	MI	126	113	239	0.61280	0.61280
54	OH	73	61	134	0.34358	0.95638
55	WVA	11	6	17	0.04358	0.99996
	W9					
56	IL	68	58	126	0.48837	0.48837
57	IN	31	48	79	0.30620	0.79457
58	WSC	30	23	53	0.20542	0.99999
	W0					
59	CO	16	21	37	0.16299	0.16299
60	IA	20	11	31	0.13656	0.29955
61	KS	14	11	25	0.11013	0.40968
62	MN	30	21	51	0.22466	0.63434
63	MO	24	20	44	0.19383	0.82817
64	NB	6	10	16	0.07048	0.89865
65	ND	2	3	5	0.02202	0.92067
66	SD	8	10	18	0.07929	0.99996

Fig. 5-5. Probability of working particular sections during the contest. (continued)

#	AREA	CW	SSB	TOT	PROBABILITY	ΣP
	VE					
67	MAR	4	3	7	0.11864	0.11864
68	QB	6	2	8	0.13559	0.25423
69	ONT	10	6	16	0.27118	0.52541
70	MAN	6	3	9	0.15254	0.67795
71	SK	2	1	3	0.05084	0.72879
72	AB	2	5	7	0.11864	0.84743
73	BC	4	4	8	0.13559	0.98302
74	NWT	0	1	1	0.01694	0.99996

Fig. 5-5. Probability of working particular sections during the contest. (continued)

BREAK—Take a break. A minimum of 15 minutes is set by the contest rules. This can be used to simulate sleep or rest periods, since only 24 hours of operation are allowed during the total contest time of 30 hours.

CALL—Call another station that has been heard in an effort to establish contact.

CHECK—Check to see if a station or section heard has been worked before or if it's a new one.

CQ—Call CQ and see who responds.

DATA—Determine what data is missing from the information sent by the other station. This command may be used in the presence of interference (QRM).

DUPLICATE—Tell another station that a contact has already been made during the contest.

HELP—Prints a brief summary of the commands at the system console (used in the game only, not in the real contest).

LOG—Enter the contact in the log.

QRT—Terminate the game and print a final summary.

QSL—Confirm a message sent by the other station.

REPEAT—Request a repeat of an incoming transmission which was garbled due to interference.

SECTION—List the sections still not contacted at the console. This can be used as

#	AREA	TOT	P	Σ
0	W1	242	0.08939	0.08939
1	W2	337	0.12449	0.21388
2	W3	266	0.09826	0.31214
3	W4	305	0.11267	0.42481
4	W5	154	0.05688	0.48169
5	W6	293	0.10820	0.58989
6	W7	176	0.06501	0.65490
7	W8	390	0.14407	0.79897
8	W9	258	0.09530	0.89427
9	W0	227	0.08385	0.97812
10	VE	59	0.02179	0.99991
		2707		

Fig. 5-6. Statistical distribution of contestants by call areas.

an aid in deciding when to switch bands to take advantage of differences or changes in propagation.

SEND—Send your station information to the other station.

STATUS—Obtain a display of your current status, including time of day, elapsed time (24 hours maximum), number of contacts, sections worked, score, and scoring rate at that time.

Only the first two letters of the command need to be entered (except SECTION—you must enter three letters.) Either upper or lowercase may be used.

THE GAME MATRIX

The game matrix stores the calls of the contestants in the game C$(K), the sections that they are located in S$(K), and the flags associated with working them C(K) and S(K). The game matrix can be generated and used in a number of different ways which fall into two basic categories, disk-based and memory-based.

In a disk-based system, the matrix is stored on disk and all accesses to the matrix have to read and write to the disk. This technique, which uses the disk as a virtual memory, slows down the access time and increases the delay time noticed by the player when waiting for the computer to respond to an input.

A memory-based technique, however, stores all the data in RAM. Response time is faster, but large amounts of memory are required to play the game. For example, in this version about 480 callsigns fit into a 48K CP/M system, and about 1000 fit into a 62K CP/M system. Since the probability of a contact is also a function of the numbers involved, contacts would be much faster with fewer duplicates if a disk-based file containing thousands of calls were used. In the real world, however, the number of contestants are finite. Since only about 3000 stations actually took part in (rather sent in logs for) the contest, and memory is getting cheaper all the time, the memory-based technique was chosen. (It also saves wear and tear on disks and disk drives.)

The callsigns can be placed into the matrix in two ways: read in from a floppy disk file, or computed on the spot. If they are read in from a disk data file, the same calls and sections will show up each time the game is played. If they are computed each game, the calls will be different and rare sections may show up. The technique used makes no difference. This version of the game computes them at the beginning of the game. This results in the slight disadvantage that it takes a few minutes to set up the game each time it is played.

THE PROPAGATION MODEL

The propagation model of Fig. 5-4 can easily be converted to software. Standard Boolean algebra logic techniques are used to reduce each table to a set of equations. These equations are then written in a form that the computer can understand. See Figs. 5-7 and 5-8.

Consider the table for the W2 as an example. It can be seen that contacts are always possible on 40 Meters and 80 Meters any time of day. There is a 50% probability of a QSO on 20 Meters from 0800-01200 UTC and 2000-2400 UTC. Contacts are not possible on 15 Meters because either the band is dead at night or skip is too long by day (not strictly true in the real world, due to backscattering, but close enough for a first-generation model). Contacts are possible on 10 Meters either if short skip sporadic E is present between the hours of 1200-1600 UTC. There is also a 50% probability of a QSO at that time if short skip is not present. To summarize, contacts are possible if:

- □ the band is 40 Meters or 80 Meters
- □ the band is 20 Meters and (the time is 0800-1200 or 2000-2400) and the probability is 50%

```
2070 Y = 0:IF C(P) = 14 OR C(P) = T6 THEN 2810
2080 IF C(P)<Z8 THEN 2230
2090 IF C(P)<T3 THEN 2280
2100 IF C(P)<T4 THEN 2310
2110 IF C(P) = 25 THEN 2710
2120 IF C(P)<25 THEN 2340
2130 IF C(P) = 33 THEN 2750
2140 IF C(P)<33 THEN 2380
2150 IF C(P) = 43 THEN 2640
2160 IF C(P)<43 THEN 2420
2170 IF C(P) = 51 THEN 2640
2180 IF C(P)<53 THEN 2480
2190 IF C(P)<56 THEN 2530
2200 IF C(P)<59 THEN 2550
2210 IF C(P)<67 THEN 2580
2220 ON C(P)-67 GOTO 2280,2530,2550,2580,2480,2480,2480
2230 IF B = B4 OR B = B5 AND H<T3 OR B = B3 AND H>= T5 AND RND(Z1)>P5 THEN 2810
2240 IF B = B3 AND (H<T5 AND H>= T3 OR RND(Z1)>P5 AND H>= Z8) THEN 2810
2250 IF B = B2 AND (H>= T5 OR H>= Z8 AND H<T3) AND RND(Z1)<P1 THEN 2810
2260 IF B = B1 AND (H>= T3 AND Q = Z2 AND H<T5 OR H>= T5 AND RND(Z1)>P5) THEN 2810
2270 RETURN
2280 IF B = B3 AND (H>= Z8 AND H<T3 OR H>= T5) AND RND(Z1)>P5 THEN 2810
2290 IF B>= B4 OR B = B1 AND Q = Z1 AND RND(Z1)<P5 AND H<T4 AND H>= T3 THEN 2810
2300 RETURN
2310 IF B = B5 OR B = B4 AND H<Z8 AND RND(Z1)>P5 THEN 2810
2320 IF B = B4 AND (H>= T5 OR H>= Z8 AND H<T3) THEN 2810
2330 RETURN
2340 IF B = B4 OR B = B3 AND H>= Z8 OR B = B1 AND Q = Z3 AND H>= T3 AND RND(Z1)>P5 THEN 2810
2350 IF B = B2 AND H<T5 AND H>= T3 AND RND(Z1)>P5 THEN 2810
2360 IF B = B5 AND H<T4 OR (H>= T5 OR H<T3) AND RND(Z1)>P5 THEN 2810
2370 RETURN
2380 IF B<B4 AND H>= T3 AND H<T5 OR B = B4 AND H<Z8 THEN 2810
2390 IF B = B5 AND (H<Z8 AND H>= Z4 OR H<Z4 AND RND(Z1)>P5) THEN 2810
2400 IF RND(Z1)>P5 AND (H>= T5 OR H>= Z8 AND H<T3) THEN 2810
2410 RETURN
2420 IF H<Z8 AND H>= T3 AND RND(Z1)<P1 THEN 2810
2430 IF B<B4 AND H>= T3 AND H<T5 OR B>= B3 AND H<Z8 AND H>= 4 THEN 2810
2440 IF H>= T5 AND B<B3 AND RND(Z1)>P5 THEN 2810
2450 IF H>= T5 AND (B = B3 OR B = B4 AND RND(Z1)<P1) THEN 2810
2460 IF H<Z4 AND (B = B4 OR B = B5 AND RND(Z1)>P5) THEN 2810
2470 RETURN
2480 IF B<B4 AND H<T5 AND H>= T3 THEN 2810
2490 IF H>= T5 AND B>B1 AND B<B5 AND RND(Z1)>P5 THEN 2810
2500 IF B = B4 AND (H<Z8 OR H>= T5 AND RND(Z1)>P5) THEN 2810
2510 IF B = B5 AND (H<Z8 AND H>= Z4 OR H<Z4 AND RND(Z1)>P5) THEN 2810
```
Fig. 5-7. The propagation model program.

```
2520 RETURN
2530 IF B = B5 OR B = B4 AND (H>= Z8 OR RND(Z1)>P5) THEN 2810
2540 RETURN
2550 IF B = B4 OR B = B5 AND (H<Z8 OR RND(Z1)>P5 AND H<T3 OR H>= T5) THEN 2810
2560 IF B = B3 AND H>= T3 AND H<T5 THEN 2810
2570 RETURN
2580 IF B>B3 AND H<Z8 OR H<T3 AND RND(Z1)>P5 THEN 2810
2590 IF B = B1 AND Q = Z5 AND H>= T3 AND RND(Z1)>P5 THEN 2810
2600 IF B = B2 AND H<T5 AND H>= T3 AND RND(Z1)>P5 THEN 2810
2610 IF B = B3 AND (H<T5 AND H>= T3 OR H>= T5 AND RND(Z1)>P5) THEN 2810
2620 IF H>= T5 AND (B = B4 AND RND(Z1)>P5 OR B = B5 AND RND(Z1)<P1) THEN 2810
2630 RETURN
2640 IF B<B4 AND H<T5 AND H>= T4 THEN 2810
2650 IF B = B3 AND (H>= T3 AND RND(Z1)>P5 OR H>= Z8 AND H<T3 AND RND(Z1)<P1) THEN 2810
2660 IF B = B1 AND H>= T3 AND (H<T5 OR RND(Z1)>P5) THEN 2810
2670 IF B = B2 AND (H<T4 AND H>= T3 AND RND(Z1)>P5 OR H>= T5 AND RND(Z1)<P1) THEN 2810
2680 IF B = B4 AND H<Z8 AND (H>= Z4 OR RND(Z1)>P5) THEN 2810
2690 IF B = B5 AND H>= Z4 AND H<Z8 AND RND(Z1)>P5 THEN 2810
2700 RETURN
2710 IF B = B5 AND H<Z8 OR B<B4 AND H>= Z8 AND H<T5 THEN 2810
2720 IF H>= T5 AND (B = B1 AND RND(Z1)>P5 OR B = B2 OR B = B3) THEN 2810
2730 IF B = B4 AND (H>= Z4 AND H<Z8 OR H<Z4 AND RND(Z1)<P1) THEN 2810
2740 RETURN
2750 IF B<B3 AND H<T5 AND (H>= T3 AND H<T4 OR H>= Z8 AND RND(Z1)>P5) THEN 2810
2760 IF H>= T5 AND (B = B2 OR B = B4) AND RND(Z1)>P5 THEN 2810
2770 IF H>= T4 AND B = B3 OR B = B4 AND H<Z8 THEN 2810
2780 IF B = B4 AND RND(Z1)>P5 AND H<T3 THEN 2810
2790 IF B = B5 AND H<Z8 AND (H>= Z4 OR RND(Z1)>P5) THEN 2810
2800 RETURN
2810 Y = 1
2820 RETURN
```

Fig. 5-7. The propagation model program. (continued)

- the band is 10 Meters and (short skip is present or the probability is 50%)

The probability value is generated by a random-number generator in this model.

THE SECTION MODEL

The section model generates the callsigns and allocates the sections to the call areas. Figure 5-9 shows a summary of the distribution of contestants by call areas. In this contest, there were 2707 reporting participants. Based on this total, the P column shows the percentage of participants in each call area. Thus, 8.9% were in the W0 call area, 12.4% were in the W1 area, etc. The total is summed in the Σ column as a check. The difference between 0.99991 and 1 in the final line (VE entry) is due to round-off error.

For the purposes of the model, the percentage of contestants in the call area is the same as the probability that any contact will be made with that call area. This means that when setting up the game matrix to determine who is in the game, a random number can be generated. This number is

```
25 K9 = 13:B5 = 500
30 Q4 = 18:K7 = 74:K6 = 66:P1 = .1:P5 = .5:Z0 = 0:Z1 = 1:Z2 = 2:Z3 = 3:
40 N1 = Z1:Z2 = 2:Z3 = 3:Z4 = 4:Z5 = 5:Z6 = 6:Z7 = 7:Z8 = 8:Z9 = 9:B1 = 100:B2 = 200:B3 = 300:T3 = 12
50 DIM S(B5+B1),B$(Z4),C(K7),C1(K9),C$(K9)
60 B$(Z0) = '10':B$(Z1) = '15':B$(Z2) = '20':B$(Z3) = '40'
80 B$(Z4) = '80':
100 B4 = 400:B5 = 500:T4 = 16:T5 = 20:T6 = 24:T7 = 60
140  LPRINT'COPYRIGHT (C) JOE KASSER [G3ZCZ] 1979 '
145 C$(0) = 'W1' :REM CALL AREAS
155 C$(1) = 'W2':C$(2) = 'W3':C$(3) = 'W4':C$(4) = 'W5':C$(5) = 'W6'
165 C$(6) = 'W7':C$(7) = 'W8':C$(8) = 'W9':C$(9) = 'W0':C$(10) = 'KL7'
175 C$(11) = 'CARRIBEAN':C$(12) = 'KZ5':C$(13) = 'PACIFIC'
195 FOR  I = 0 TO K9: READ C1(I):NEXT
205 DATA 0,7,12,16,26,34,44,53,56,59,51,25,33,43
615 REM MAIN LOOP
635 FOR  I = 0  TO  K9
645 C(0) = C1(I):REM SET UP C(P)
655 LPRINT :LPRINT :LPRINT C$(I)+ '   CALL AREA':LPRINT :LPRINT ' HR    ';
665 FOR  B = B1 TO B5 STEP B1:LPRINT  B$(B/100-Z1),:NEXT:LPRINT
675 FOR  H = 0 TO  23 STEP 4
685 LPRINT H;:LPRINT '    ';
695 FOR  B = B1 TO  B5 STEP B1
696 IF I = 0 THEN Q = 2:REM SET SHORT SKIP VALUES ON
697 IF I = 1 THEN Q = 1
698 IF I = 3 THEN Q = 3
699 IF I = 9 THEN Q = 5
705 FOR  J = 1 TO  100:GOSUB 2070
706 PRINT  ' * ',
715 X3 = X3+Y : NEXT J
716 PRINT  'RUNNING A SIMULATION PLEASE STANDBY'
717 PRINT : PRINT  'JOE KASSER ' : PRINT  : PRINT :
725 LPRINT  X3,:X3 = Z0 :REM RESET FOR  NEXT TIME AROUND
735 NEXT B
736 LPRINT
745 NEXT H,I
755 GOTO 615
```

Fig. 5-8. Program for the propagation model.

then compared against the P values in Fig. 6-6 and the call area allocated to that station can be found. This is done by the use of a FOR/NEXT loop in line 480 of the routine shown in Fig. 5-10, where the P values for each call area are stored in the P matrix where P(75) = value for W0, P(76) = value for W1, etc.

Once the call area is identified, a section within that call area is allocated to that station. There use is made of the information in Fig. 5-4. The total number of reporting CW and SSB entrants in section in each call are summarized. The percentage of entrants in each section are then calculated. Thus, for example, 30% of the W1 partici-

W1 CALL AREA

HR	10	15	20	40	80
0	0	0	0	100	100
4	0	0	0	100	100
8	0	12	50	100	100
12	100	0	100	100	0
16	100	0	100	100	0
20	55	16	81	100	0

W2 CALL AREA

HR	10	15	20	40	80
0	0	0	0	100	100
4	0	0	0	100	100
8	0	12	53	100	100
12	0	0	100	100	0
16	0	0	100	100	0
20	55	8	76	100	0

W3 CALL AREA

HR	10	15	20	40	80
0	0	0	0	43	100
4	0	0	0	49	100
8	0	0	0	100	100
12	0	0	0	0	100
16	0	0	0	0	100
20	0	0	0	100	100

W4 CALL AREA

HR	10	15	20	40	80
0	47	62	48	100	100
4	42	47	56	100	100
8	48	54	100	100	100
12	54	56	100	100	100
16	51	45	100	100	0
20	76	54	100	100	50

W5 CALL AREA

HR	10	15	20	40	80
0	0	0	0	100	49
4	0	0	0	100	100
8	46	37	45	50	52
12	100	100	100	0	0
16	100	100	100	0	0
20	51	56	51	53	54

W6 CALL AREA

HR	10	15	20	40	80
0	0	0	0	100	47
4	0	0	100	100	100
8	0	0	0	0	0
12	100	100	100	0	0
16	100	100	100	0	0
20	47	46	100	11	0

W7 CALL AREA

HR	10	15	20	40	80
0	0	0	0	100	44
4	0	0	0	100	100
8	0	0	0	0	0
12	100	100	100	0	0
16	100	100	100	0	0
20	0	55	47	75	0

W8 CALL AREA

HR	10	15	20	40	80
0	0	0	0	52	100
4	0	0	0	52	100
8	0	0	0	100	100
12	0	0	0	100	100
16	0	0	0	100	100
20	0	0	0	100	100

Fig. 5-9. Results obtained by exercising propagation model.

W9 CALL AREA

HR	10	15	20	40	80
0	0	0	0	100	100
4	0	0	0	100	100
8	0	0	0	100	64
12	0	0	100	100	0
16	0	0	100	100	0
20	0	0	0	100	100

CARRIBEAN CALL AREA

HR	10	15	20	40	80
0	0	0	0	11	100
4	0	0	0	100	100
8	100	100	100	0	0
12	100	100	100	0	0
16	100	100	100	0	0
20	51	100	100	0	0

W0 CALL AREA

HR	10	15	20	40	80
0	44	42	48	100	100
4	53	46	54	100	100
8	52	57	39	52	49
12	53	46	100	0	0
16	54	49	100	0	0
20	52	0	57	45	11

KZ5 CALL AREA

HR	10	15	20	40	80
0	0	0	0	100	48
4	0	0	0	100	100
8	45	55	0	57	0
12	100	100	0	0	0
16	42	43	100	0	0
20	0	56	100	48	0

KL7 CALL AREA

HR	10	15	20	40	80
0	0	0	0	54	0
4	0	0	0	100	53
8	0	0	10	0	0
12	100	43	59	0	0
16	100	100	100	0	0
20	46	19	46	0	0

PACIFIC CALL AREA

HR	10	15	20	40	80
0	0	0	0	52	0
4	0	0	0	100	55
8	0	0	6	0	0
12	100	58	51	0	0
16	100	100	100	0	0
20	45	4	46	0	0

Fig. 5-9. Results obtained by exercising propagation model. (continued)

pants were in Connecticut (CT), 28% in East Massachusetts (EMASS), 8.6% in Maine (ME), 9% in New Hampshire (NH), 8.2% in Rhode Island (RI), 4.5% in Vermont (VT), and 11% in West Massachusetts (WMASS).

The table can also be used to compare the levels of activity based on the actual participants, but not on the probability, or percentage figures. This is because the CW or SSB numbers reflect actual contestants, while the P values reflect the percentage in the call area. Thus, the 61.2% participation of Michigan in the W8 call area reflects the fact that 61.2% of the W8 calls were in Michigan, not that 61% of the contestants were there. Only 239 out 2707 or 8.83% were actually there. Summing the CW and SSB entrants compensates for the fact that some stations took part but never sent in an entry and, hence, do not appear in the results because other stations took part in both modes and are counted twice. The assumption here is that the distribution is about the same.

Interesting facts can be read into the summary, such as the "rarity of sections."

For example, out of 2707 recorded entrants, NWT and CZ only had 1 participant each, WY and SK had 3 each, AL and ND had 5 each, LA only had 9, DE had 11, while MI, OH, IL, MD and VA had the largest amounts of entrants.

The section model can be investigated on a stand-alone basis. The listing of a program that does this is shown in Fig. 5-11. When this program was exercised some tabular data was obtained as shown in Fig. 5-12A. A summary of six runs in which occurrences of each section were computed 2707 times (the number of occurrences in original data) is summarized in Fig. 5-12B.

An examination of these results shows some interesting phenomena, including the fact that the occurrences of W1's, W6's, and W0's are consistently below the desired probability, while the occurrences of W2's, W9's and VE's are consistently above the desired probability. The first run gave a call

Time		W1					W2					W3					W4				
Z	L	10	15	20	40	80	10	15	20	40	80	10	15	20	40	80	10	15	20	40	80
4	0	0	0	0	1	1	0	0	0	1	1	0	0	0	.5	1	.5	.5	.5	1	1
8	4	0	.1	.5	1	1	0	.1	.5	1	1	0	0	0	1	1	.5	.5	1	1	1
12	8	1	0	1	1	0	0	0	1	1	0	0	0	0	0	1	.5	.5	1	1	1
16	12	1	0	1	1	0	0	0	1	1	0	0	0	0	0	1	.5	.5	1	1	0
20	15	.5	.1	.5	1	0	.5	.1	.5	0	0	0	0	0	1	1	.5	.5	1	1	.5
0	20	0	0	0	1	1	0	0	0	1	1	0	0	0	.5	1	.5	.5	.5	1	1

		W5					W6					W7					W8				
Z	L	10	15	20	40	80	10	15	20	40	80	10	15	20	40	80	10	15	20	40	80
4	0	0	0	0	1	1	0	0	1	1	1	0	0	0	1	1	0	0	0	.5	1
8	4	.5	.5	.5	.5	.5	0	0	0	0	0	0	0	0	0	0	0	0	0	1	1
12	8	1	1	1	0	0	1	1	1	0	0	1	1	1	0	0	0	0	0	1	1
16	12	1	1	1	0	0	1	1	1	0	0	1	1	1	0	0	0	0	0	1	1
20	16	.5	.5	.5	.5	.5	.5	.5	1	.1	0	0	.5	.5	.5	0	0	0	0	1	1
0	20	0	0	0	1	.5	0	0	0	1	.5	0	0	0	1	.5	0	0	0	.5	1

		W9					W0				
Z	L	10	15	20	40	80	10	15	20	40	80
4	0	0	0	0	1	1	.5	.5	.5	1	1
8	4	0	0	0	1	.5	.5	.5	.5	.5	.5
12	8	0	0	1	1	0	.5	.5	1	0	0
16	12	0	0	1	1	0	.5	.5	1	0	0
20	16	0	0	0	1	1	.5	0	.5	.5	.1
0	20	0	0	0	1	1	.5	.5	.5	1	1

		KV4					KZ5					KL7/KH6				
Z	L	10	15	20	40	80	10	15	20	40	80	10	15	20	40	80
4	0	0	0	0	1	1	0	0	0	1	1	0	0	0	1	.5
8	4	1	1	1	0	0	.5	.5	0	.5	0	0	0	.1	0	0
12	8	1	1	1	0	0	1	1	0	0	0	1	.5	.5	0	0
16	12	1	1	1	0	0	.5	.5	1	0	0	1	1	1	0	0
20	16	.5	1	1	0	0	0	.5	1	.5	0	.5	.1	.5	0	0
0	20	0	0	0	.1	1	0	0	0	1	.5	0	0	0	.5	0

Differences Between Predicted and Actual Values are Circled

Fig. 5-10. Propagation model verification results.

```
10 REM SECTOR DISTRIBUTION MODEL
20 K = Z1 : REM DUMMY
40 Q4 = 18 : K7 = 74 : K6 = 66 : P1 = .1 : P5 = .5 : Z0 = 0 : Z1 = 1 : Z2 = 2 : Z3 = 3 : Q = (RND(Z1) * Z2)
50 N1 = Z1: Z2 = 2: Z3 = 3: Z4 = 4: Z5 = 5: Z6 = 6: Z7 = 7: Z8 = 8: Z9 = 9: B1 = 100: B2 = 200: B3 = 300: T3 = 12
60 DIM L(Z7),L$(Z5),C$(K),C(K),S(K7),S$(K7),P(B5),P$(Z4),Z$(Q4),B$(Z4)
65 DIM C1(K7),C2(K7)
90 B$(Z4) = "B0" : Z$(Z0) = "CALL" : Z$(Z1) = "LOG" : Z$(Z2) = "B AND " : Z$(Z3) = "CHECK" : X$ = ""
110 B4 = 400 : B5 = 500 : T4 = 16 : T5 = 20 : T6 = 24 : T7 = 60 : Z$(Z8) = "REPEAT" : Z$(Z9) = "QRT"
120 DIM M$(Z5) : M$(Z0) = "NUMBER" : M$(Z1) = "POWER" : M$(Z4) = Z$(Z3) : M$(Z5) = "SECTION"
140 K$ = "CK" : Z$(7) = "SEND" : Z$(10) = "TUNE" : Z$(11) = "DATA " : Z$(T3) = "TIME" : M$(3) = Z$(3)
160 FOR I = Z0 TO 85 : READ P(I) : NEXT
170 DATA .3,.58,.66,.76,.84,.88,1,.17,.34,.53,.78,1,.04,.41,.85,1,.04
180 DATA .14,.22,.32,.41,.44,.51,.59,.97,1,.11,.16,.25,.31,.62,.7,.99,1
190 DATA .11,.26,.33,.38,.69,.76,.83,.93,.97,1,.17,.23,.28,.34,.48,.52
200 DATA .95,.98,1,.61,.95,1,.48,.79,1,.16,.29,.4,.63,.82,.89,.92,1
210 DATA .11,.25,.52,.67,.72,.84,.98,1,.08,.21,.31,.42,.48,.58,.65,.79
220 DATA .89,.97,1
240 S$(Z0) = "CT" : S$(Z1) = "EMASS" : S$(Z2) = "ME" : S$(Z3) = "NH" : S$(Z4) = "RI"
250 S$(Z5) = "VT" : S$(Z6) = "WMASS" : S$(Z7) = "ENY" : S$(Z8) = "NY - LI" : S$(Z9) = "NNJ"
260 S$(10) = "SNJ" : S$(11) = "WNY" : S$(12) = "DE" : S$(13) = "E.PA" : S$(14) = "M.DC"
270 S$(15) = "W.PA" : S$(16) = "ALA" : S$(17) = "GA" : S$(18) = "KY" : S$(19) = "NC"
280 S$(20) = "NFLA" : S$(21) = "SC" : S$(22) = "SFLA" : S$(23) = "TN" : S$(24) = "VA"
290 S$(25) = "WI" : S$(26) = "AK" : S$(27) = "LA" : S$(28) = "MISS" : S$(29) = "NM"
300 S$(30) = "N.TEX" : S$(31) = "OK" : S$(32) = "S.TEX" : S$(33) = "CZ" : S$(34) = "E BAY"
310 S$(35) = "LA" : S$(36) = "ORG" : S$(37) = "SB" : S$(38) = "SCV" : S$(39) = "SD" : S$(40) = "SF"
320 S$(41) = "SJV" : S$(42) = "SV" : S$(43) = "PCF" : S$(44) = "AZ" : S$(45) = "ID" : S$(46) = "MT"
330 S$(47) = "NV" : S$(48) = "OR" : S$(49) = "UT" : S$(50) = "WA" : S$(51) = "AL" : S$(52) = "WY"
340 S$(53) = "MI" : S$(54) = "OH" : S$(55) = "W VA" : S$(56) = "IL" : S$(57) = "IN" : S$(58) = "WSC"
350 S$(59) = "CO" : S$(60) = "IA" : S$(61) = "KS" : S$(62) = "MN" : S$(63) = "MO" : S$(64) = "NB"
360 S$(65) = "ND" : S$(66) = "SD" : S$(67) = "MAR" : S$(68) = "QB" : S$(69) = "ONT" : S$(70) = "MAN"
370 S$(71) = "SK" : S$(72) = "AB" : S$(73) = "BC" : S$(74) = "NWT"
375 I = Z0 : REM RESET IT
395 FOR Q5= Z1 TO 30
405 FOR J = Z0 TO K7 : C1(J) = Z0 : NEXT
415 I = I + 1
425 REM MAIN LOOP
480   X = RND(Z1) : FOR J = 75 TO 86 : IF X > P(J) THEN NEXT J
490 X = RND(Z1) : N = J - K7 : ON J - 75 GOTO 510,520,530,540,550,560,570,580,590,600
500 Y = Z0 : Z = Z6 : GOTO 610
510 Y = Z7 : Z = 11 : GOTO 610
520 Y = T3 : Z = 15 : GOTO 610
530 Y = 16 : Z = 25 : GOTO 610
540 Y = 26 : Z = 33 : GOTO 610
550 Y = 34 : Z = 43 : GOTO 610
560 Y = 44 : Z = 52 : GOTO 610
570 Y = 53 : Z = 55 : GOTO 610
580 Y = 56 : Z = 58 : GOTO 610
590 Y = 59 : Z = K6 : GOTO 610
600 Y = 67 : Z = K7
610 FOR T = Y TO Z : IF X > P(T) THEN NEXT T
```

Fig. 5-11. Program to test the section model.

```
620 REM
675 C1(T) = C1(T) + Z1
680 PRINT I , S$(T) , C1(T)
685 IF I < 2707 THEN 415 : REM LOOP
695 LPRINT "SUMMARY" : LPRINT
705 FOR I = Z0 TO K7 : C2(I) = C2(I) + C1(I) : LPRINT I + Z1, S$(I),C1(I) : NEXT
715 NEXT Q5
718 LPRINT "RUNS = ";Q5 : LPRINT
725 FOR I = Z0 TO K7 : LPRINT I + Z1,S$(I),C2(I) : NEXT
795 STOP
```

Fig. 5-11. Program to test the section model. (continued)

area distribution identical to the model but varied the distribution of sections within the area.

The probabilities in the model could be changed to vary the distribution (i.e., lowering the VE threshold would decrease the occurrence of VE's and at the same time increase the occurrence of W0's—a desired condition). The sample, however, is small (only 6 runs) and the original data contains assumptions that may or may not be valid, so the values in the model are probably as good as any others. Variations of the probabilities within each call area will also change the distribution. Try it in the VE area to redistribute the section occurrence. Computing the Section distribution each time the main program is invoked means that there will be a slightly different distribution each time, which will add to the interest.

THE FLOW OF THE GAME PROGRAM

The propagation model and section distribution model can only determine if a contact with a particular section is possible at any time of the day. The actual number of contacts is governed by the action that the operator or player takes. Consider what can happen as a result of each of the Commands.

SUMMARY

1	CT	77
2	EMASS	59
3	ME	17
4	NH	19
5	RI	17
6	VT	12
7	WMASS	34
8	ENY	63
9	NY - LI	59
10	NNJ	79
11	SNJ	89
12	WNY	64
13	DE	12
14	E.PA	103
15	M.DC	114
16	W.PA	51
17	ALA	9
18	GA	25
19	KY	21
20	NC	38
21	NFLA	25
22	SC	6
23	SFLA	21
24	TN	19
25	VA	103
26	WI	15
27	AK	18
28	LA	6
29	MISS	16
30	NM	8
31	N.TEX	39
32	OK	18
33	S.TEX	45
34	CZ	1
35	E BAY	32
36	LA	39
37	ORG	14
38	SB	16
39	SCV	76
40	SD	11
41	SF	13
42	SJV	29
43	SV	11
44	PCF	7
45	AZ	41
46	ID	11
47	MT	11
48	NV	13
49	OR	33
50	UT	7
51	WA	71
52	AL	4
53	WY	4
54	MI	236
55	OH	126
56	W VA	17
57	IL	126
58	IN	89
59	WSC	65
60	CO	41
61	IA	28
62	KS	31
63	MN	49
64	MO	39
65	NB	12
66	ND	3
67	SD	19
68	MAR	8
69	QB	11
70	ONT	23
71	MAN	13
72	SK	3
73	AB	13
74	BC	10
75	NWT	0

Fig. 5-12. Section distribution model test results.

W1		Individual Run Results						Average	Real
1	CT	79	63	64	47	63	76	65.3	73
2	EMASS	51	54	73	71	44	73	61.0	68
3	ME	17	15	17	20	19	18	17.6	21
4	NH	23	16	21	20	17	17	19.0	22
5	RI	16	21	14	23	21	11	17.6	20
6	VT	7	7	10	14	11	6	9.1	11
7	WMASS	28	24	25	31	30	18	26.0	27
		242	200	224	226	205	219	219.3	242

W2									
8	ENY	58	63	49	58	68	62	59.6	59
9	NY-LI	60	70	57	53	67	77	64.0	56
10	NNJ	75	52	61	62	61	63	62.3	66
11	SNJ	93	77	101	91	98	82	90.3	84
12	WNY	76	82	94	68	76	66	77.0	72
		337	344	362	332	370	350	349.1	337

W3									
13	DE	12	18	16	13	10	15	14.0	11
14	EPA	112	102	99	90	92	85	96.6	99
15	MD	123	122	124	122	128	119	123.0	118
16	WPA	35	29	31	42	32	38	34.5	38
		266	271	270	267	262	257	265.5	266

W4									
17	ALA	13	18	10	12	10	12	12.5	14
18	GA	29	25	22	35	29	29	28.1	29
19	KY	26	22	20	16	31	29	24.0	27
20	NC	45	26	31	39	35	33	34.8	29
21	NFLA	21	30	22	30	23	19	24.1	28
22	SC	8	7	12	12	10	8	9.5	10
23	SFLA	16	21	18	29	26	23	22.1	19
24	TN	19	29	21	21	14	26	21.6	26
25	VA	102	111	101	112	100	138	110.6	116
26	WI	14	6	4	13	7	7	8.5	7
		305	295	261	319	285	324	296.8	305

Fig. 5-12. Section distribution model test results. (continued)

W5		Individual Run Results						Average	Real
27	AK	13	13	14	18	20	20	16.3	17
28	LA	6	13	5	6	6	6	7.0	9
29	MS	15	12	8	19	19	20	15.5	13
30	NM	5	15	8	9	8	10	9.16	10
31	NTEX	47	46	48	61	54	52	51.3	48
32	OK	14	17	11	20	17	11	15.0	11
33	STEX	42	50	47	42	58	44	47.1	45
34	CZ	1	0	1	0	2	3	1.1	1
		154	66	142	175	184	166	164.5	154
W6									
35	EBAY	31	33	33	22	29	22	28.3	35
36	LA	43	40	43	41	47	36	41.6	44
37	ORG	12	14	26	18	17	18	17.5	19
38	SB	13	13	9	14	10	9	11.3	16
39	SCV	91	70	97	78	83	89	84.6	89
40	SD	7	21	22	16	21	15	17.0	20
41	SF	13	21	20	20	30	19	20.5	21
42	SJV	28	29	36	20	27	25	27.5	30
43	SV	14	9	11	10	8	7	9.8	13
44	PCF	6	4	4	8	8	10	6.6	6
		293	254	301	247	280	250	270.8	293
W7									
45	AZ	39	41	35	30	37	36	36.3	31
46	ID	11	14	12	7	21	11	12.6	11
47	MT	14	15	17	6	14	5	11.8	9
48	NV	16	16	10	9	8	14	12.1	10
49	OR	26	32	27	24	29	25	27.1	25
50	UT	5	9	8	8	10	4	7.3	6
51	WA	72	85	74	65	89	63	74.6	76
52	AL	4	5	2	6	7	5	4.8	5
53	WY	4	5	8	4	3	5	4.8	3
		176	222	193	159	218	168	189.3	176
W8									
54	MI	219	207	241	265	213	239	230.6	239
55	OH	135	142	138	129	119	127	131.6	134
56	WVA	19	23	23	13	20	17	19.1	17
		390	372	402	407	352	383	384.3	390

Fig. 5-12. Section distribution model test results. (continued)

W9		Individual Run Results						Average	Real
57	IL	126	138	135	144	142	128	135.5	126
58	IN	89	83	78	93	85	109	89.5	79
59	WSC	65	59	79	52	43	61	56.8	53
		258	271	283	289	270	298	278.1	258
W0									
60	CO	42	28	35	37	34	35	35.1	37
61	IA	26	33	23	31	30	22	27.5	31
62	KS	33	27	23	27	19	19	24.6	25
63	MN	48	47	37	40	58	52	47.0	51
64	MO	35	38	37	33	32	36	35.1	44
65	NB	13	13	14	18	10	10	13.0	16
66	ND	2	7	11	2	2	4	4.6	5
67	SD	18	21	14	25	21	17	19.3	18
		227	214	194	213	206	195	208.1	227
VE									
68	MAR	5	11	6	10	9	8	8.1	7
69	QB	11	13	17	9	6	19	12.5	8
70	ONT	24	28	25	17	21	24	23.1	16
71	MAN	18	20	5	12	13	11	13.1	9
72	SK	5	3	3	4	3	4	3.6	3
73	AB	14	10	8	8	14	13	11.1	7
74	BC	10	12	9	10	6	15	10.3	8
75	NWT	0	1	2	3	3	3	2.0	1
		59	98	75	73	75	97	83.6	59

Fig. 5-12. Section distribution model test results. (continued)

Call

Call can be invoked in a number of situations. A station can be heard on a frequency as a result of the TUNE or CQ commands. He can be calling CQ himself, or calling QRZ, or be in QSO with you or even with someone else. The result of CALL depends on what status the station heard is in (Q1 flag). You may even call someone with no one being on frequency (error conditon). A flowchart analyzing the call sequence is shown in Fig. 5-13.

Log

The log sequence enters the QSO data into the log. In this model, the exchange data is verified to ensure that all necessary data has been received. The callsign and section are then flagged as having been worked. Since this is a model designed to be played as a game, the computer checks to see that rules are not infringed or that it is not logging a duplicate QSO. After logging the contact, if the player is in the CQ mode, there is also a low probability that someone else will call anyhow. A flowchart analyzing the logging sequence is shown in Fig. 5-14.

Band

Switching to a new band is simple: the operator just enters a new band number. It has to be valid, of course, so the input is checked to make sure it's allowed. If a new band is selected, the QSO buffers are cleared. A flowchart describing the operation is shown in Fig. 5-15.

Check

The check operation determines if a station or a section has been worked before. In the computer model, the first thing that has to be done is to verify that a section or a call are loaded in the QSO buffer.

If the call is in the log (i.e., worked), a message is printed. Then the section is tested. If it has not been worked, no call in that section has been worked, and a message is printed out accordingly. If it has been worked before, a message is printed out saying that the section has been worked but that that particular callsign has not. The flowchart for this sequence is shown in Fig. 5-16.

Status

The status sequence shown in Fig. 5-17 displays the state of the game at any time. A date, time, QSO count, and section worked summary is given. The score and scoring rate are also computed and displayed.

Break

Taking a break should be simple enough. The computer prompts for the amount of time to break, verifies that the input is valid, and updates the clocks. Breaks for less than 15 minutes are against the rules. A flowchart for the BREAK sequence is shown in Fig. 5-18. In this model, the status is displayed after the break and the QSO buffers are cleared.

Breaks are used in the game to advance to a time with better desired propagation characteristics. Note that the rules state that the contest lasts 30 hours, but only 24 hours of actual operating are allowed.

QSL

The QSL sequence is used to tell the other station that his data has been received. If he has sent data he can send a "73" or "ROGER." If QRM is present he can request a repeat. He can also move off frequency without repeating any data you lost in the QRM. A typical sequence is shown in Fig. 5-19.

Send

The send operation is shown in Fig. 5-20. After checking that there really is someone to send to, the message is sent or displayed at the console. Once it has been sent, there is a probability that the other station will request a repeat, or QSL the data. There is also a good probability that you will get a reply in kind directly without having to request it.

Repeat

When a repeat or a report is requested, it can either be received or QRM can interfere with the transmission. This is shown in Fig. 5-21. Note that the computer first checks that a QSO is in progress.

QRT

QRT means that the game is over. A final report is printed out and control is returned to the operating system. A flowchart is shown in Fig. 5-22.

Tune

What happens when an operator is tuning around is a function of the propagation at any particular time. For example, the

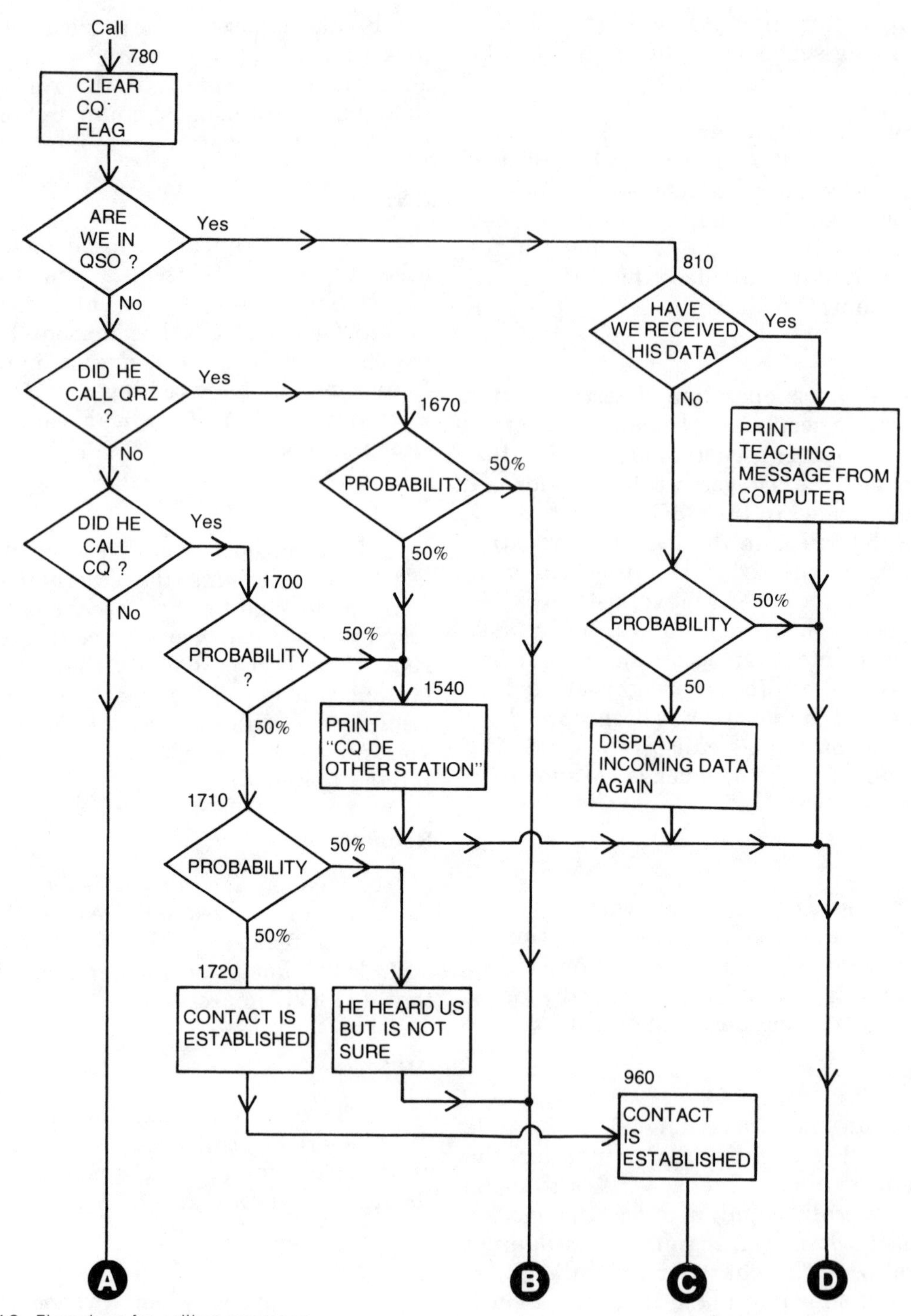

Fig. 5-13. Flowchart for calling sequence.

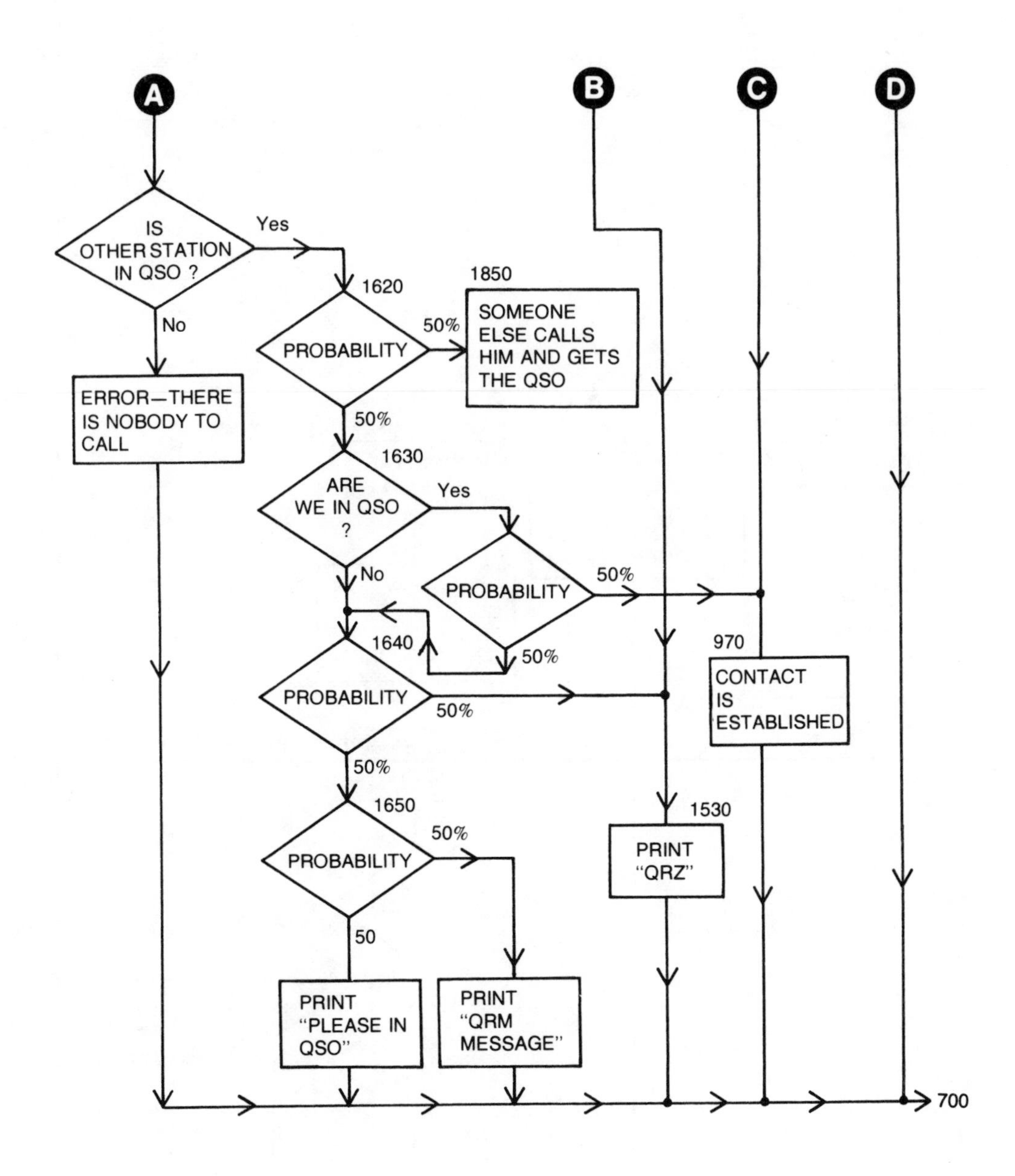

Fig. 5-13. Flowchart for calling sequence. (continued)

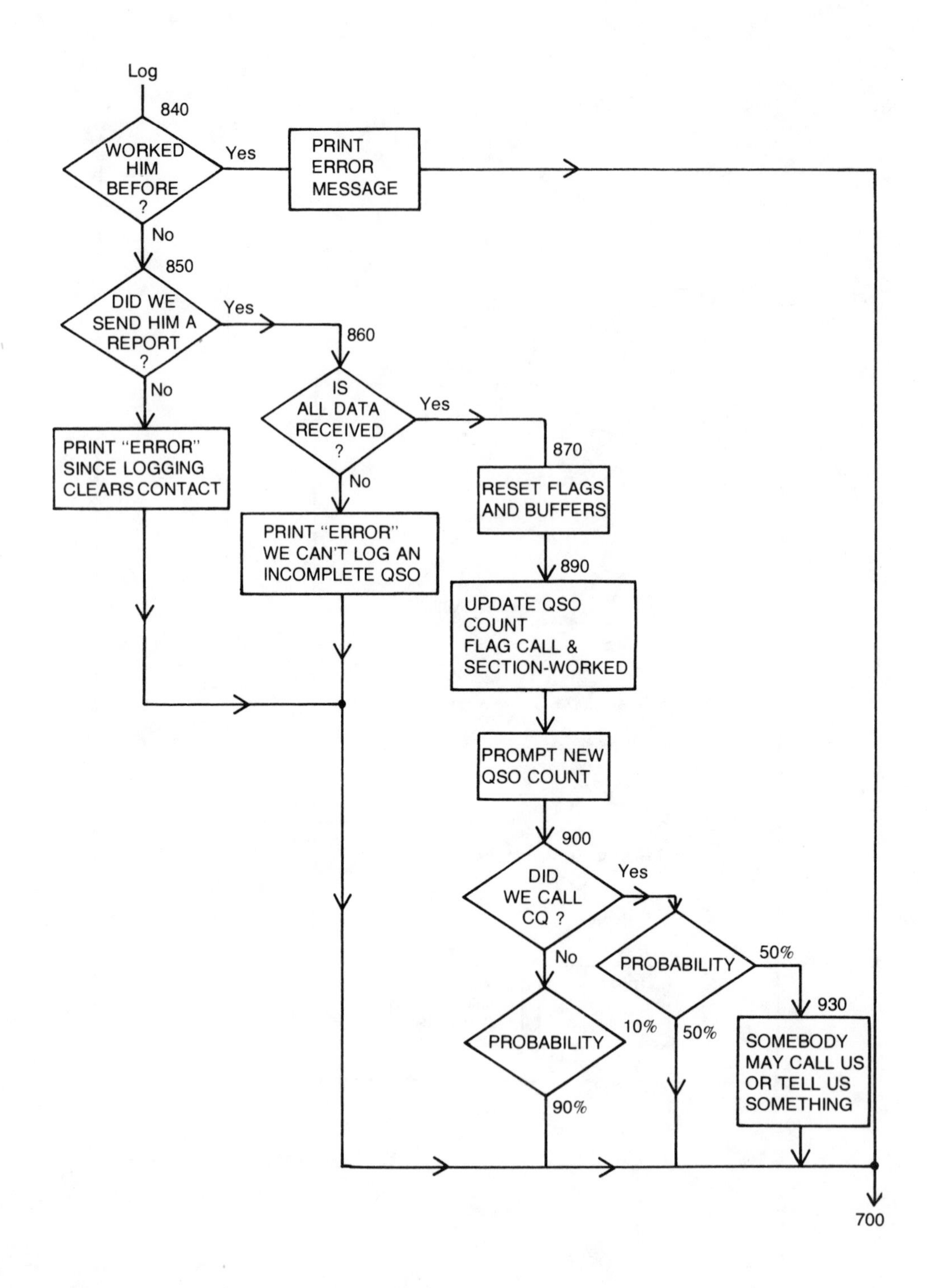

Fig. 5-14. Flowchart for logging sequence.

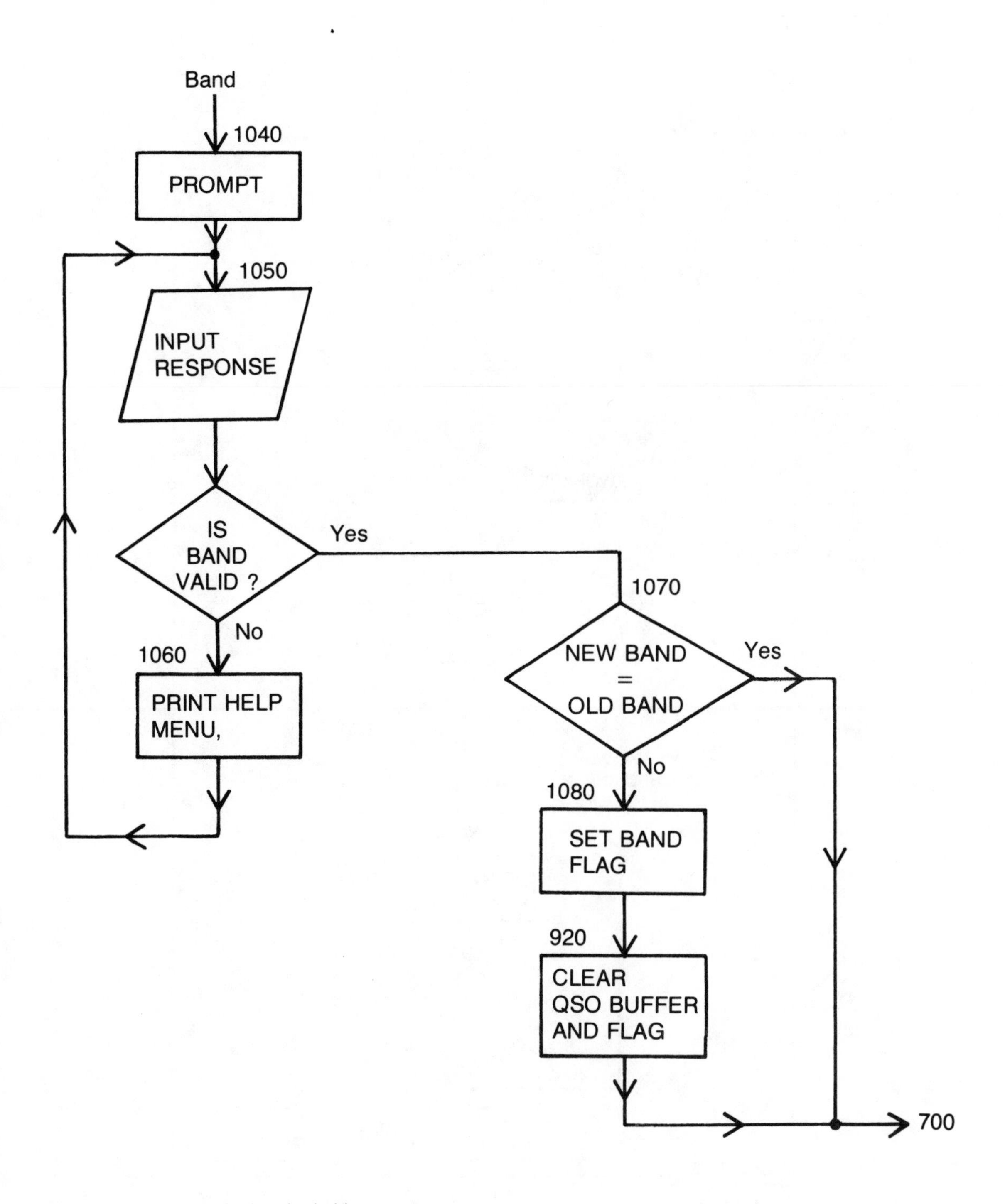

Fig. 5-15. Flowchart for bandswitching.

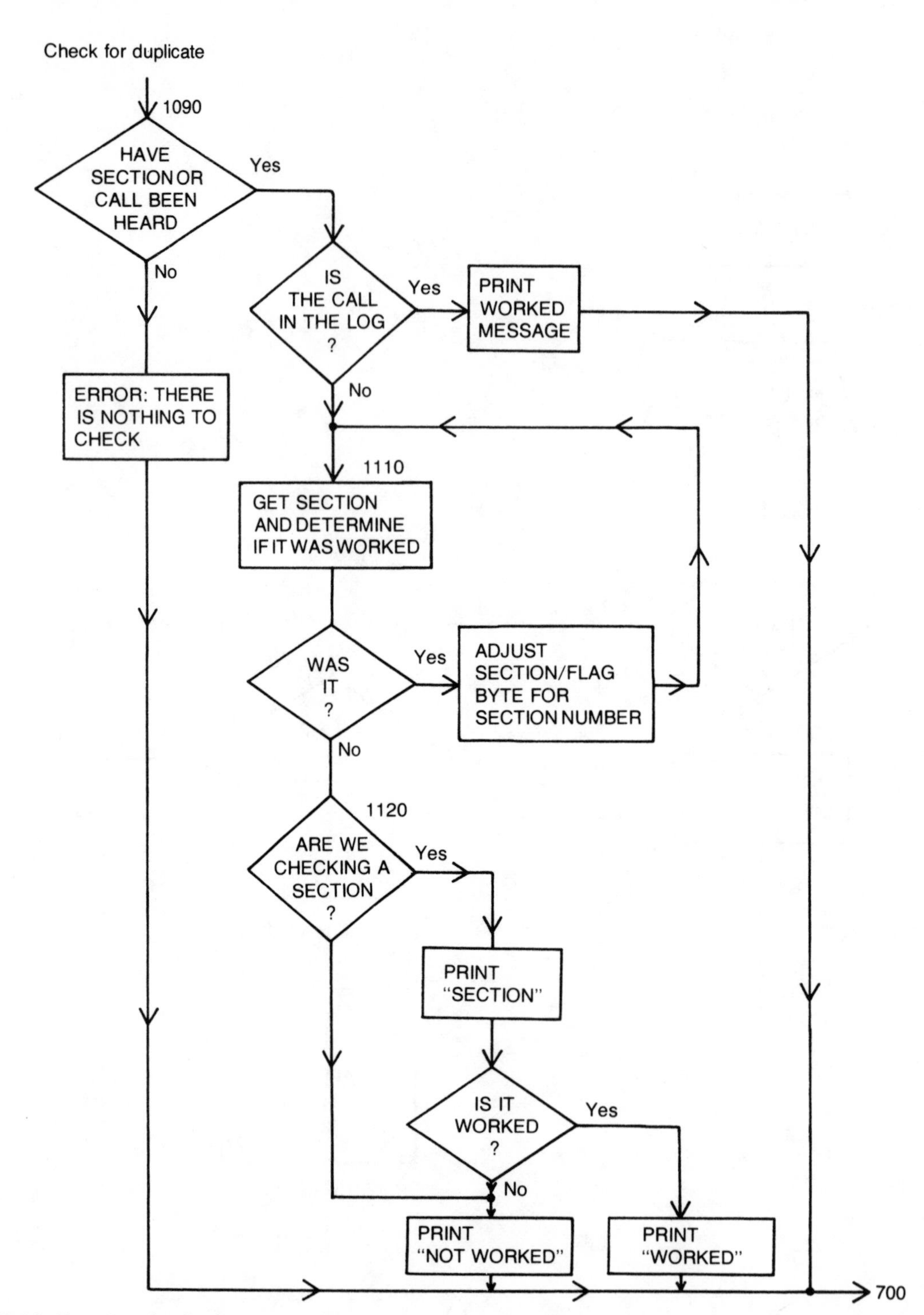

Fig. 5-16. Flowchart for checking for duplicate contact.

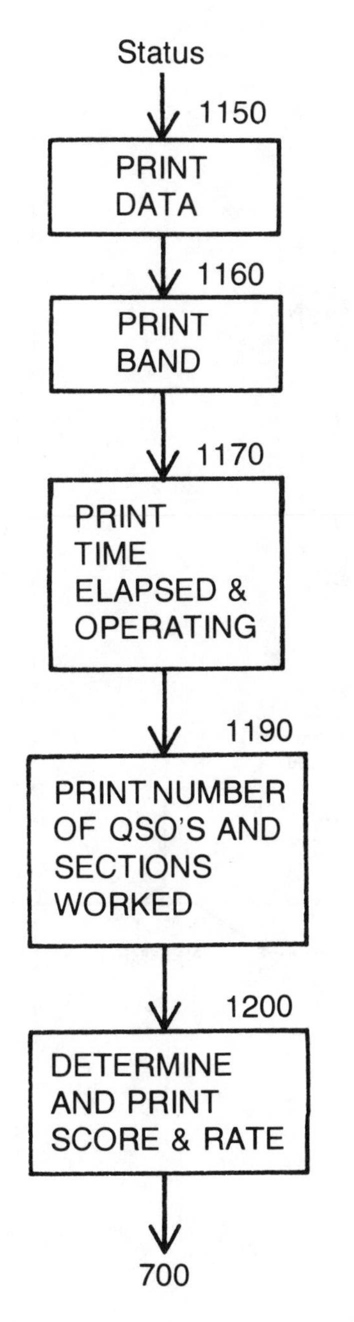

Fig. 5-17. Flowchart for the state of the game information display.

40 Meter band in the evening (local time) will contain a number of Region 1 broadcast stations. The 10, 15, and 20 Meter bands in the late afternoon (local time) may contain South American stations looking for phone patches into the U.S.A. These are incorporated into the game model as shown in Fig. 5-23.

Once these "ringers" have been taken care of, a plunge is made into the game matrix at random and a callsign is identified. The elapsed tuning time is then incremented and a check made to see if propagation is available to that station at the present time on the present band. If it is, the parameters for the contact are set up and a QSO is then possible. Now the station heard may be calling CQ, QRZ, or may be already in QSO with someone else. The flowchart of Fig. 5-23 also outlines the sequence of determining if a QSO will be possible.

Data

DATA takes a quick glance at the log sheet to see if any data is missing. It is flow charted in Fig. 5-24.

Time

TIME looks at the station clock. Its simple flowchart is shown in Fig. 5-25.

Duplicate

This command tells the other station that a contact has been made previously. There is a low probability that your message will be ignored or interfered with and a good probability that he will send a "sorry" message before tuning off. There is then a probability that someone else may call you. The flowchart for this sequence is shown in Fig. 5-26.

Section

This is an assist command that lists the sections still needed. It demonstrates the usefulness of the computer to the user.

Help

This is a simulation command only. It lists the options available to the player.

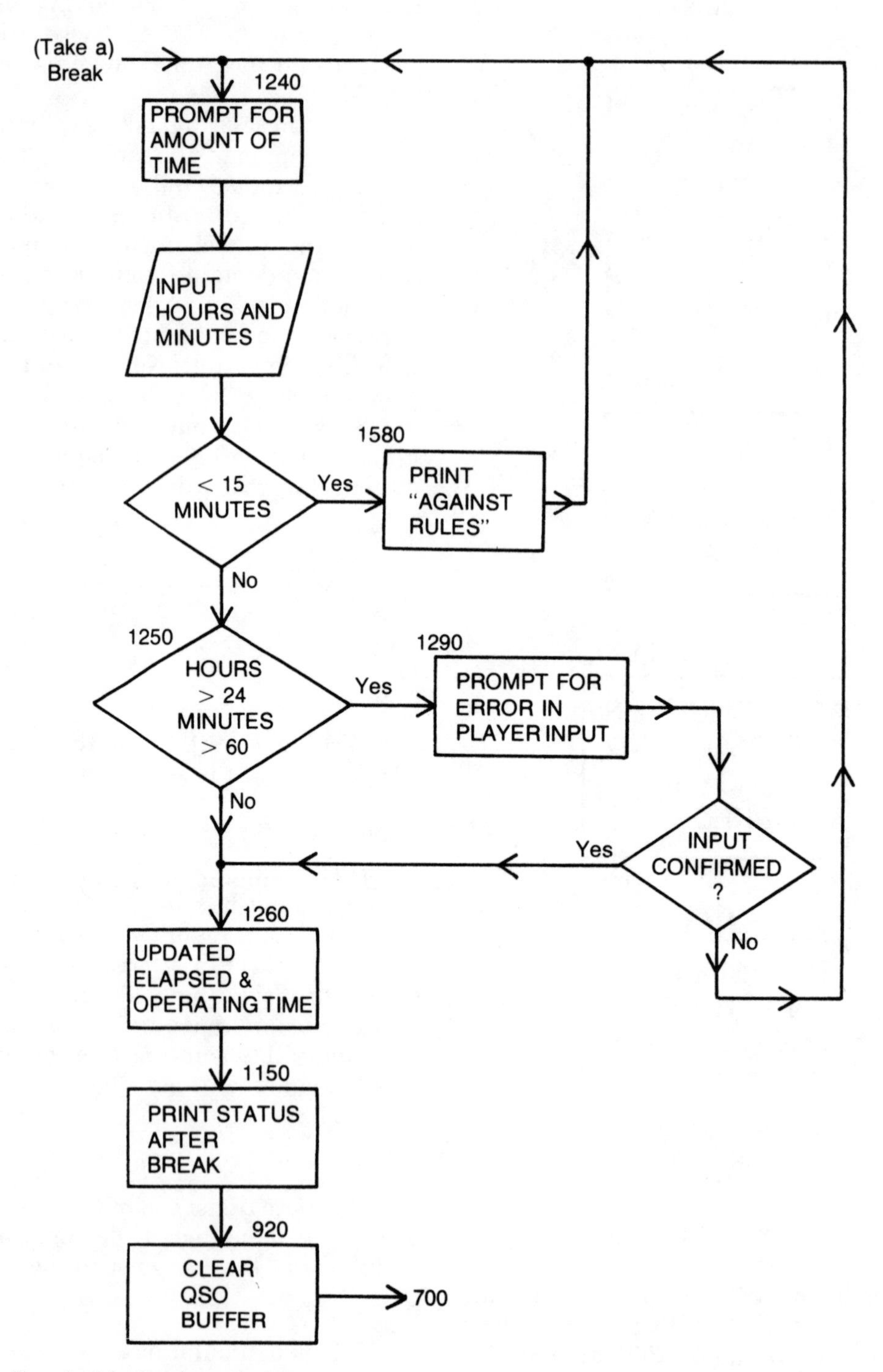

Fig. 5-18. Flowchart for breaks or timeouts.

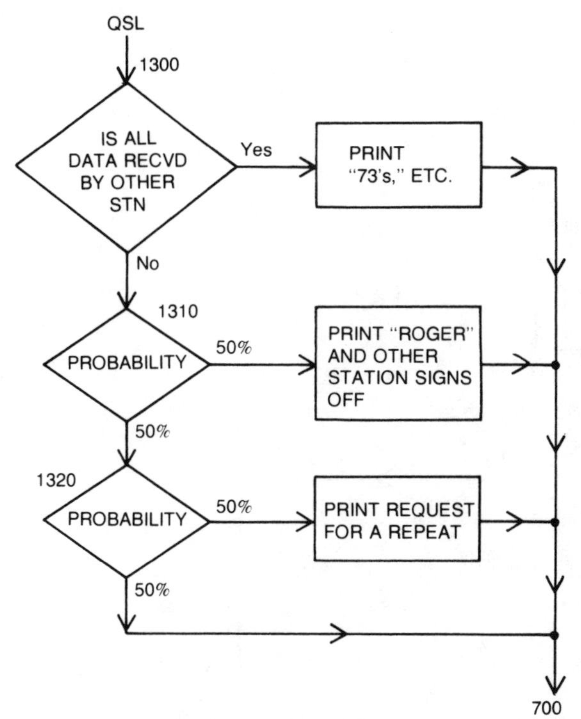

Fig. 5-19. Flowchart for QSL sequence.

CQ

This section models what happens as a result of a CQ. Two things usually happen as a result of calling CQ: either a reply or no reply. If no reply is heard, then a "Frequency in use" message could be received, or there could be QRM on frequency masking any incoming traffic.

Once called by another station, the possibility arises that he has not checked carefully and this could be a duplicate QSO. He could also call and pass his data at the same time. A flowchart for the model used in the game is shown in Fig. 5-27.

Worked

This command scans the callsigns in the game matrix and lists any that have been worked.

Debug

This command lists all callsigns in the game matrix. It was used to check that the callsign generation routine did not generate duplicates. It can be used to get an idea of the distribution of the players.

AN EXAMPLE OF THE PLAYER DIALOG

An example of the dialog in the game follows. The output of the computer is presented in upper case (capitals), the player input is in lowercase, and the author's comments are in parentheses. The standard prompt that the computer offers when awaiting an input is

QRU ? .

All the callsigns used in this example, except G3ZCZ/W3, have been set up by the random number generator in the program. Operating procedures given here bear no resemblance to the operating habits of the real owner of the callsign.

A BASIC listing for the propagation model with comment or REM statements is shown in Fig. 5-7. Each table has been reduced to a set of equations in the manner already described. This propagation model is incorporated in the game. The subroutine signals that a QSO is possible by setting the value of Y to 1. If a return from the subroutine occurs where Y = 0, propagation is not possible at that time.

This and subsequent listings are not explained at this time, since they are part of the total program listed in Fig. 5-28. The whole program is described and explained later in this chapter.

The propagation model can also be tested on a stand-alone basis. The listing shown in Fig. 5-8 contains the necessary statements to exercise the propagation model. The program goes through each call area testing propagation on each band at each time of day. One hundred tests are

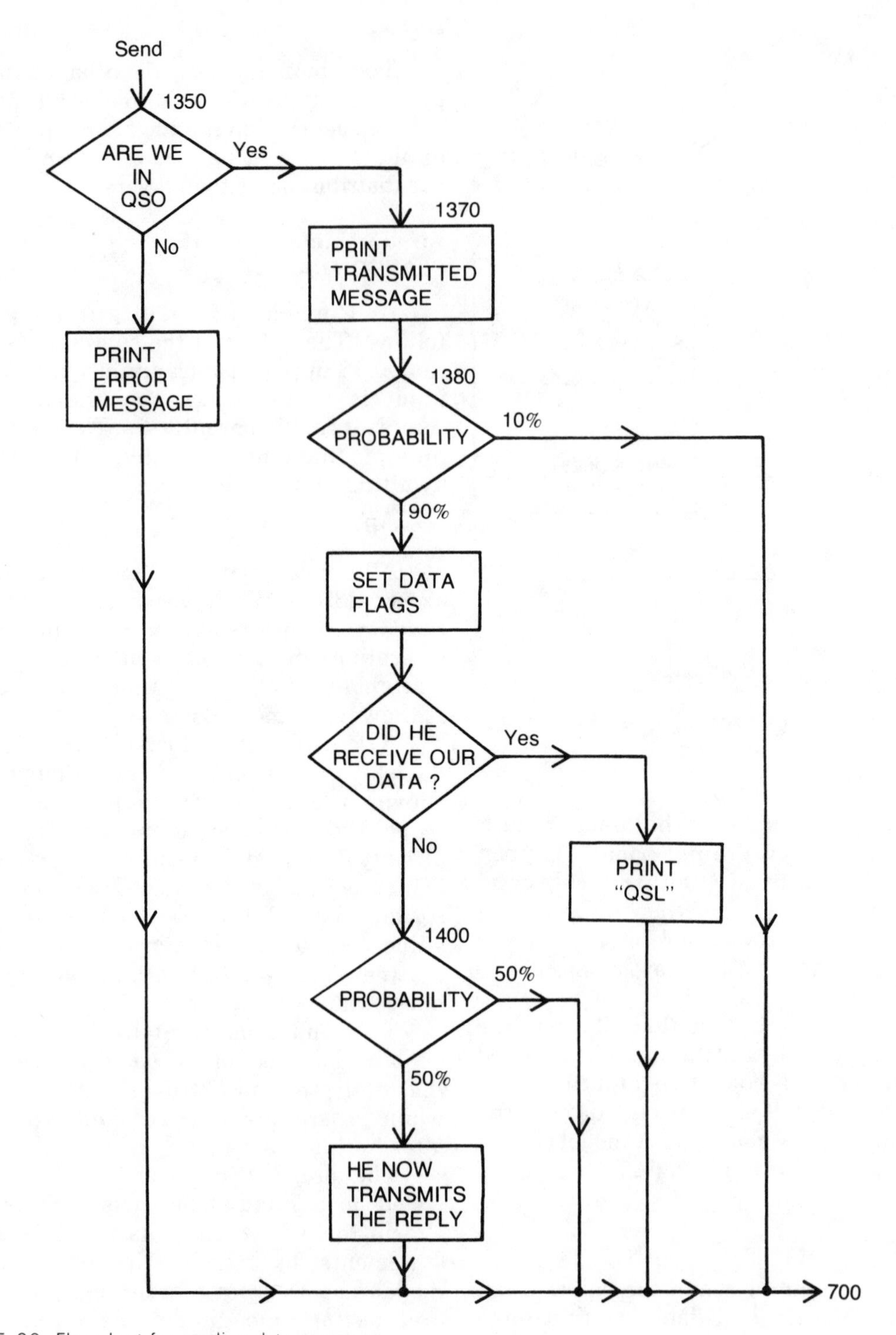

Fig. 5-20. Flowchart for sending data sequence.

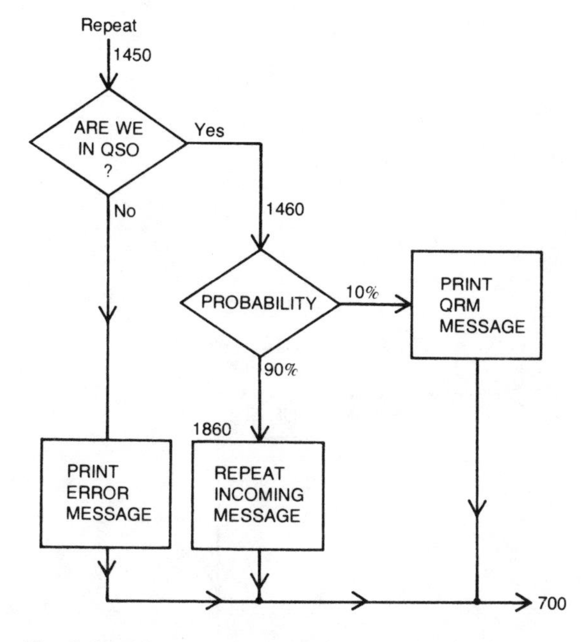

Fig. 5-21. Flowchart for tuning sequence.

performed on each band at each time and the results are shown in Fig. 5-9. They are also tabulated and summarized in Fig. 5-10. If the results in Fig. 5-10 are compared to the desired data shown in Fig. 5-4 it can be seen that coding errors crept into the model for the W1, W2, W4, W7, W9, W0 and KZ5 call areas, because the results differ from the desired values. The differences are probably due to errors in converting the truth table to logic statements in BASIC. Since you may decide to change the propagation model in your version of the game, the changes are left up to you. You may verify your changes by running this program.

The program is doing a lot of number crunching and the time intervals between printouts are relatively long. Statements were inserted in lines 706, 716, and 717 of the program (Fig. 5-8) to reassure the user that the system is active and performing correctly.

The lines of code used in the truth table or model verification that are not used in the game have unique line numbers; otherwise, the line numbers correspond in each of the listings.

ARRL SWEEPSTAKES CONTEST SIMULATION VERSION 2.1

COPYRIGHT (C) JOE KASSER (G3ZCZ) 1979

ARE YOU A HAM RADIO OPERATOR ? no

OK THE STATION CALL IS G3ZCZ/W3
THE QTH IS SILVER SPRING, MARYLAND

SETTING UP THE CALLS OF THE CONTESTANTS

THIS IS GOING TO TAKE ABOUT 5 MINUTES

WHY DON'T YOU GET A CUP OF COFFEE OR SOMETHING?

(The player is notified that a time delay is to be expected.)

(Several minutes later . . . . .)

IT IS 2100 HRS ON 19 NOVEMBER
THE CONTEST HAS STARTED

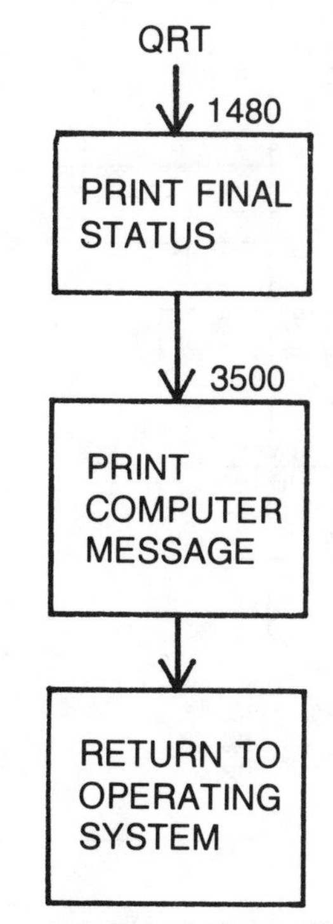

Fig. 5-22. Flowchart for end of game (QRT) sequence.

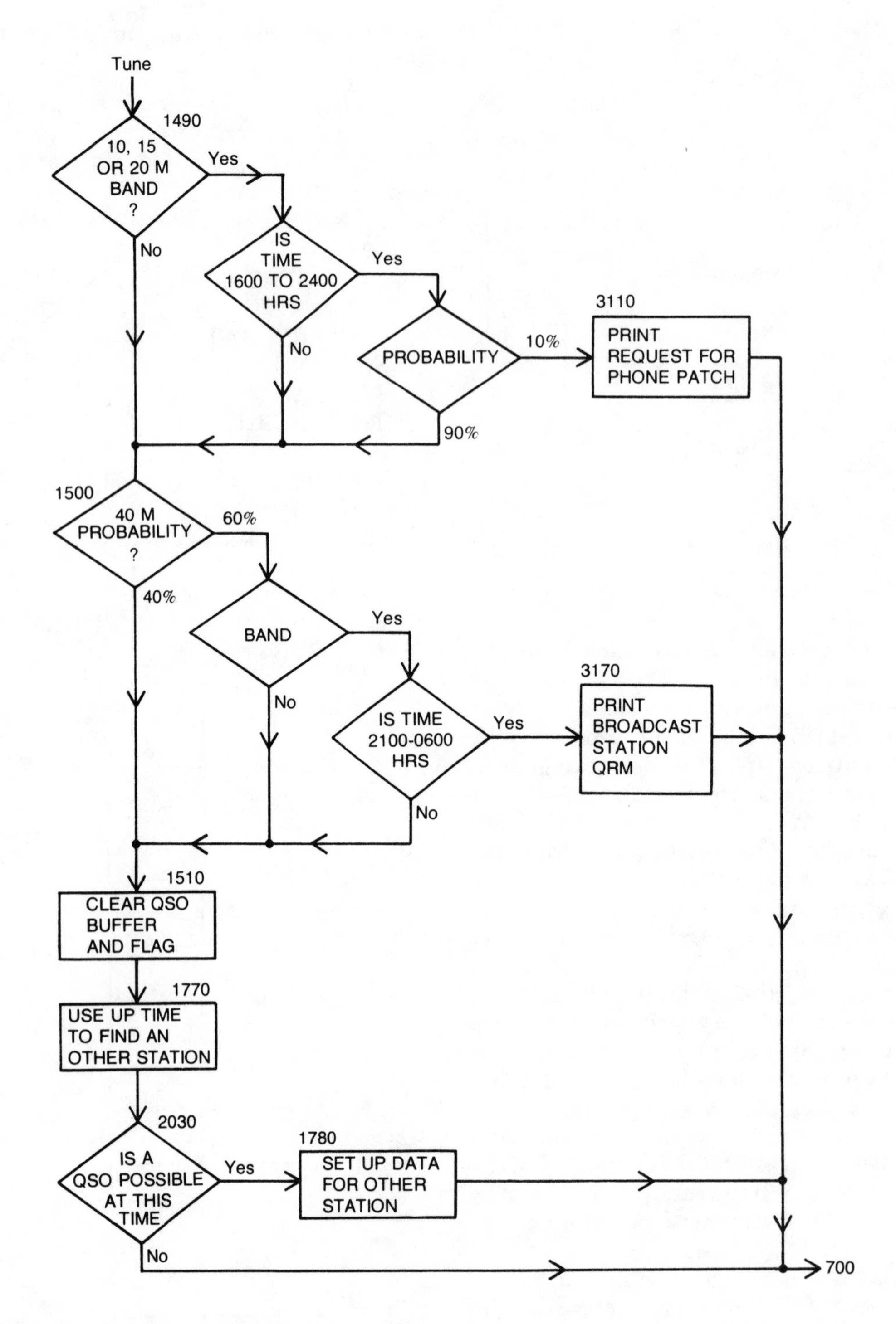

Fig. 5-23. Flowchart for tune sequence.

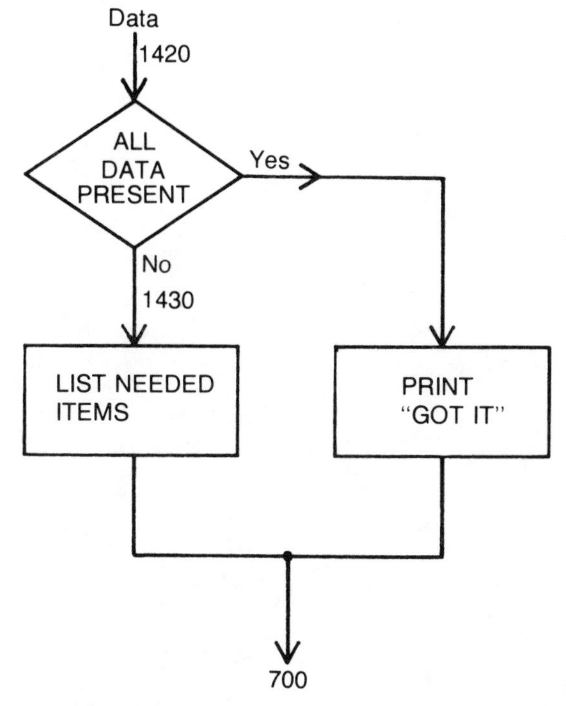

Fig. 5-24. Flowchart for data sequence.

WHICH BAND ? 40

QRU ? (Standard prompt) tu (Answer: tune around.)

CQ CQ CQ DE WB8CHF CQ CQ CQ DE WB8CHF

(Here we hear a CQ call.)

QRU ? ca (Now let's call him.)

QRZ DE WB8CHF (No response to our call, but he heard something.)

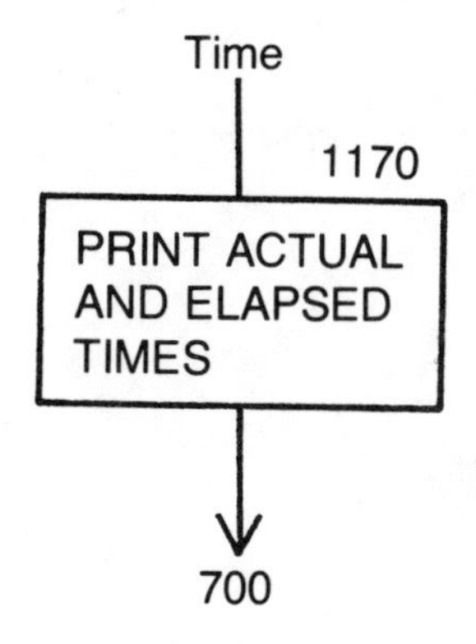

Fig. 5-25. Flowchart for time of day (clock) sequence.

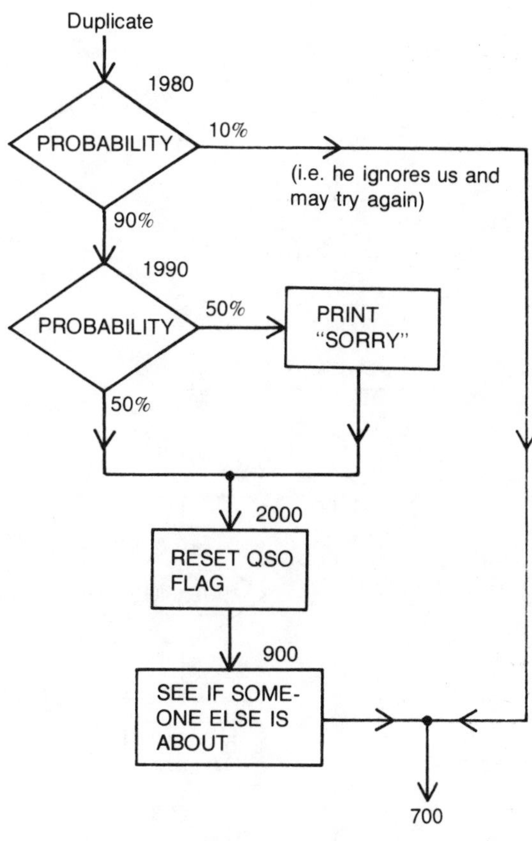

Fig. 5-26. Flowchart to tell other station that he has been worked.

QRU ? ca (If at first you don't succeed, try again!)

G3ZCZ/W3 DE WB8CHF (Got him!)

QRU ? se (Now send him our information.)

WB8CHF UR 1 A DE G3ZCZ/W3 CK 68 MD K

G3ZCZ/W3 UR 1 B WB8CF CK 47 OH DO YOU QSL?

(He got our information and sent his right back.)

QRU ? qsl

73, SEE YOU LATER

QRU ? lo (Now that we worked him, let's put him in the log.)

NEXT QSO = 2

QRU ? tu (Let's look around for someone else.)

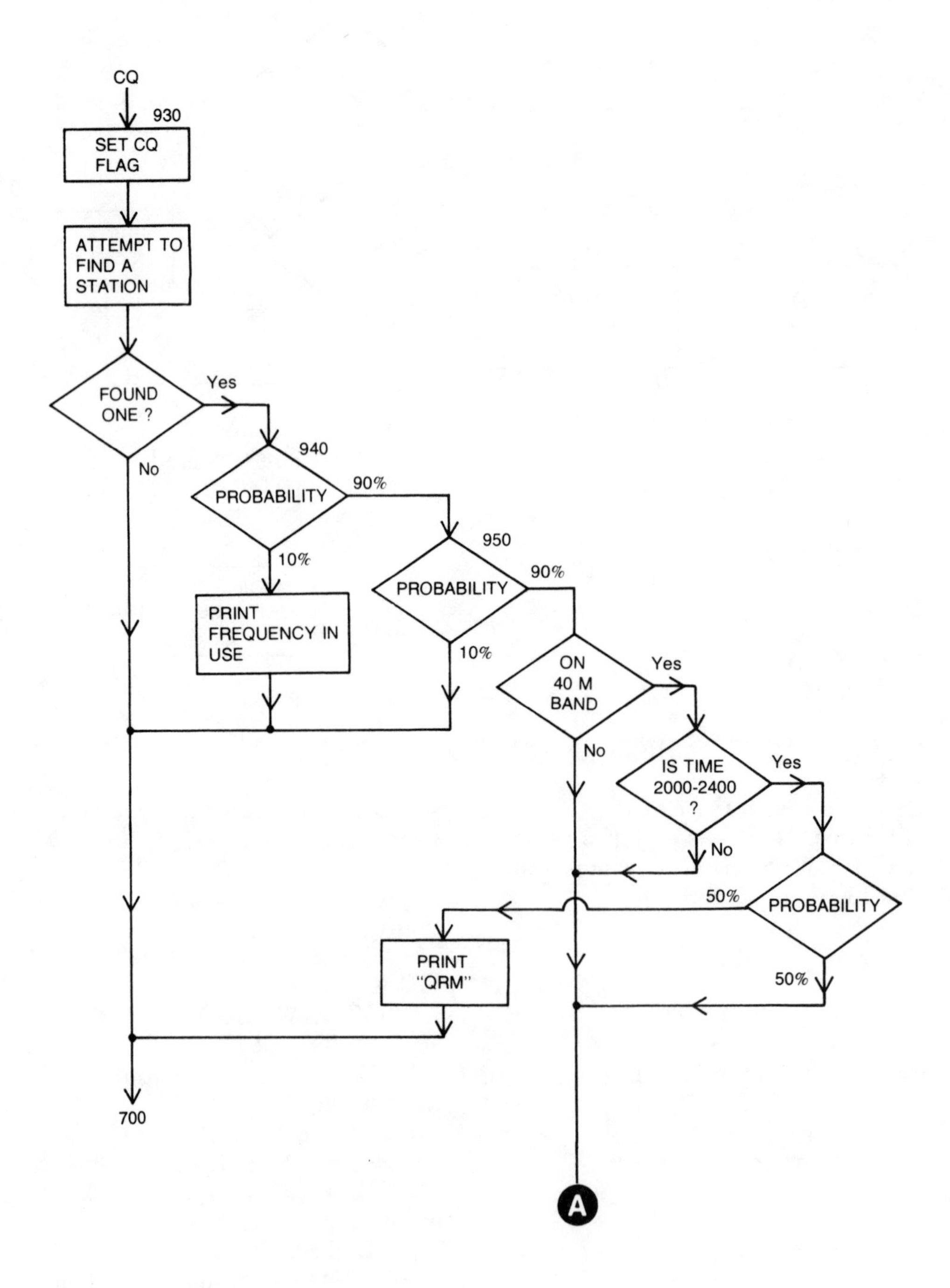

Fig. 5-27. Flowchart for calling CQ.

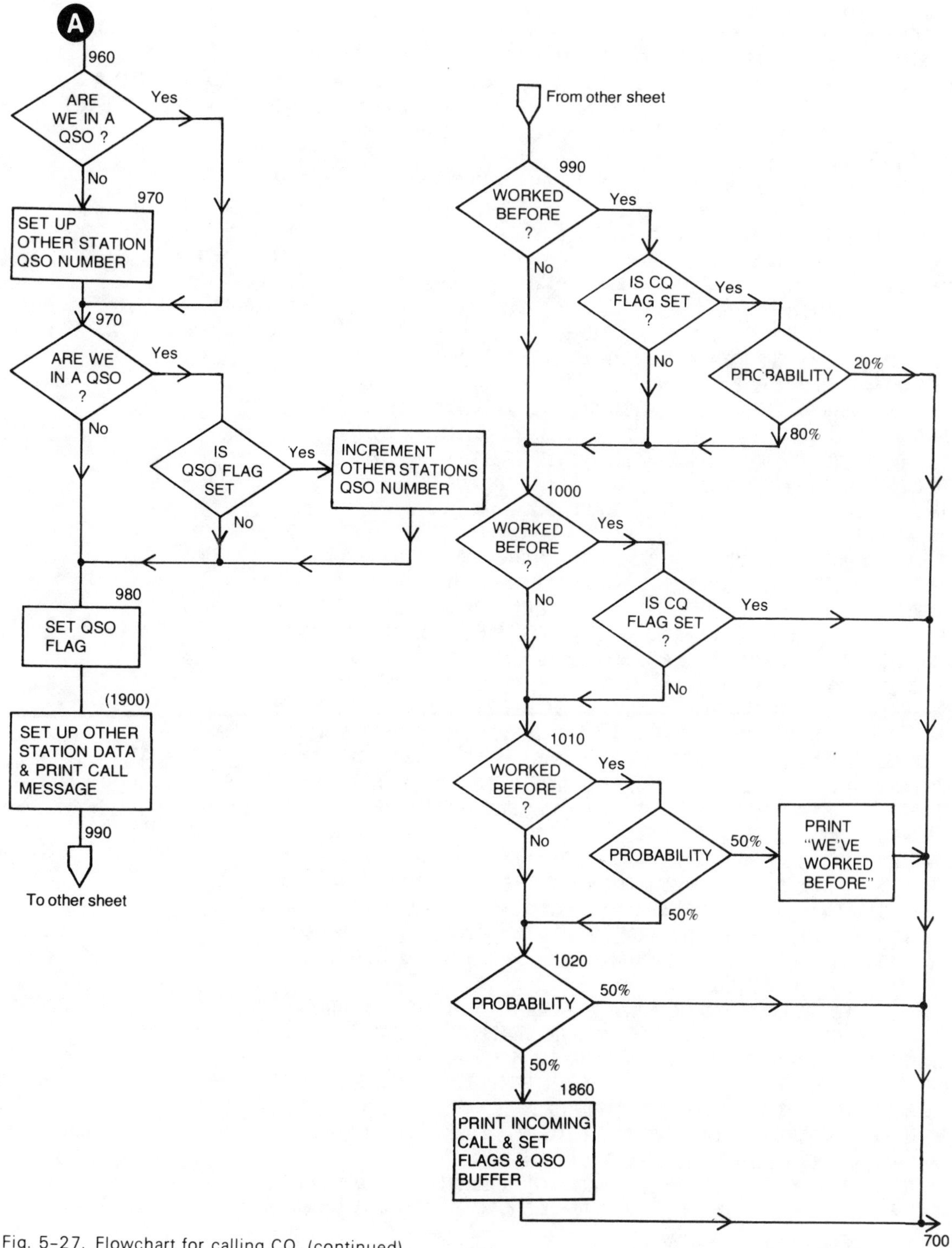

Fig. 5-27. Flowchart for calling CQ. (continued)

```
10 REM SET LINE 20 FOR YOUR SYSTEM
20 CLEAR 5000 : K = 480 : REM 48K CP/M
30 PRINT "ARRL SWEEPSTAKES CONTEST SIMULATION VERSION 2.1"
40 Q4 = 18 : K7 = 74 : K6 = 66 : P1 = .1 : P5 = .5 : Z0 = 0 : Z1 = 1 : Z2 = 2 : Z3 = 3 : Q = (RND(Z1) * Z2)
50 N1 = Z1: Z2 = 2: Z3 = 3: Z4 = 4: Z5 = 5: Z6 = 6: Z7 = 7: Z8 = 8: Z9 = 9: B1 = 100: B2 = 200: B3 = 300: T3 = 12
60 DIM L(Z7),L$(Z5),C$(K),C(K),S(K7),S$(K7),P(85),P$(Z4),Z$(Q4),B$(Z4)
70 Z$(13) = "DUPLICATE" : B$(Z0) = "10" : B$(Z1) = "15" : B$(Z2) = "20" : B$(Z3) = "40"
80 P$(Z0) = "W" : P$(Z1) = "K" : P$(Z2) = "WA" : Q$ = "QWERTYUIOPASDFGHJKLZXCVBNM"
90 B$(Z4) = "80" : Z$(Z0) = "CALL" : Z$(Z1) = "LOG" : Z$(Z2) = "BAND " : Z$(Z3) = "CHECK" : X$ = ""
100 P$(Z3) = "WB" : Z$(Z4) = "STATUS" : Z$(Z5) = "BREAK" : Z$(Z6) = "QSL" : E$ = " DE "
110 B4 = 400 : B5 = 500 : T4 = 16 : T5 = 20 : T6 = 24 : T7 = 60 : Z$(Z8) = "REPEAT" : Z$(Z9) = "QRT"
120 DIM M$(Z5) : M$(Z0) = "NUMBER" : M$(Z1) = "POWER" : M$(Z4) = Z$(Z3) : M$(Z5) = "SECTION"
130 M$(Z2) = Z$(Z0) : A$ = "G3ZCZ/W3" : H = 21 : D = 19 : R$ = " !# * " : D$ = "NOVEMBER" : N = 1968
140 K$ = "CK" : Z$(7) = "SEND" : Z$(10) = "TUNE" : Z$(11) = "DATA " : Z$(T3) = "TIME" : M$(3) = Z$(3)
150 P$(Z4) = "N" : PRINT "COPYRIGHT (C) JOE KASSER [G3ZCZ] 1979 "
160 FOR I = Z0 TO 85 : READ P(I) : NEXT
170 DATA .3,.58,.66,.76,.84,.88,1,.17,.34,.53,.78,1,.04,.41,.85,1,.04
180 DATA .14,.22,.32,.41,.44,.51,.59,.97,1,.11,.16,.25,.31,.62,.7,.99,1
190 DATA .11,.26,.33,.38,.69,.76,.83,.93,.97,1,.17,.23,.28,.34,.48,.52
200 DATA .95,.98,1,.61,.95,1,.48,.79,1,.16,.29,.4,.63,.82,.89,.92,1
210 DATA .11,.25,.52,.67,.72,.84,.98,1,.08,.21,.31,.42,.48,.58,.65,.79
220 DATA .89,.97,1
230 Z$(14) = M$(Z5) : Z$(15) = "HELP" : Z$(T4) = "CQ" : Z$(17) = "WORKED" : Z$(18) = "DEBUG"
240 S$(Z0) = "CT" : S$(Z1) = "EMASS" : S$(Z2) = "ME" : S$(Z3) = "NH" : S$(Z4) = "RI"
250 S$(Z5) = "VT" : S$(Z6) = "WMASS" : S$(Z7) = "ENY" : S$(Z8) = "NY - LI" : S$(Z9) = "NNJ"
260 S$(10) = "SNJ" : S$(11) = "WNY" : S$(12) = "DE" : S$(13) = "E.PA" : S$(14) = "M.DC"
270 S$(15) = "W.PA" : S$(16) = "ALA" : S$(17) = "GA" : S$(18) = "KY" : S$(19) = "NC"
280 S$(20) = "NFLA" : S$(21) = "SC" : S$(22) = "SFLA" : S$(23) = "TN" : S$(24) = "VA"
290 S$(25) = "WI" : S$(26) = "AK" : S$(27) = "LA" : S$(28) = "MISS" : S$(29) = "NM"
300 S$(30) = "N.TEX" : S$(31) = "OK" : S$(32) = "S.TEX" : S$(33) = "CZ" : S$(34) = "E BAY"
310 S$(35) = "LA" : S$(36) = "ORG" : S$(37) = "SB" : S$(38) = "SCV" : S$(39) = "SD" : S$(40) = "SF"
320 S$(41) = "SJV" : S$(42) = "SV" : S$(43) = "PCF" : S$(44) = "AZ" : S$(45) = "ID" : S$(46) = "MT"
330 S$(47) = "NV" : S$(48) = "OR" : S$(49) = "UT" : S$(50) = "WA" : S$(51) = "AL" : S$(52) = "WY"
340 S$(53) = "MI" : S$(54) = "OH" : S$(55) = "W VA" : S$(56) = "IL" : S$(57) = "IN" : S$(58) = "WSC"
350 S$(59) = "CO" : S$(60) = "IA" : S$(61) = "KS" : S$(62) = "MN" : S$(63) = "MO" : S$(64) = "NB"
360 S$(65) = "ND" : S$(66) = "SD" : S$(67) = "MAR" : S$(68) = "QB" : S$(69) = "ONT" : S$(70) = "MAN"
370 S$(71) = "SK" : S$(72) = "AB" : S$(73) = "BC" : S$(74) = "NWT"
380 INPUT "ARE YOU A HAM RADIO OPERATOR" ; I$ : IF LEFT$(I$,Z1) <> "Y" THEN 420
390 INPUT "YOUR CALL" ; A$ : IF RIGHT$(A$,Z2) <> "/3" THEN A$ = A$ + "/3"
400 INPUT "YEAR FIRST LICENSED " ; N
410 IF N < 1900 THEN PRINT "YOU MUST HAVE BEEN LICENSED BEFORE 1900" : GOTO 400
420 F$ = STR$(N - 1900) : PRINT : PRINT "OK","THE STATION CALL IS",A$
430 PRINT ,"THE QTH IS SILVER SPRING, MARYLAND " : PRINT
440 INPUT "DO YOU NEED INSTRUCTIONS " ; I$ : IF LEFT$(I$,Z1) = "Y" THEN GOSUB 3300
450 PRINT : PRINT "SETTING UP CALLS OF CONTESTANTS"
460 PRINT "THIS IS GOING  TO TAKE ABOUT" ; INT(K / (T7 * Z2)) + Z1 ; " MINUTES"
470 PRINT "WHY DON'T YOU GET A CUP OF COFFEE OR SOMETHING"
480 FOR I = Z0 TO K - Z1 : X = RND(Z1) : FOR J = 75 TO 86 : IF X > P(J) THEN NEXT J
490 X = RND(Z1) : N = J - K7 : ON J - 75 GOTO 510,520,530,540,550,560,570,580,590,600
```

Fig. 5-28. Program for sweepstakes contest game.

```
500 Y = Z0 : Z = Z6 : GOTO 610
510 Y = Z7 : Z = 11 : GOTO 610
520 Y = T3 : Z = 15 : GOTO 610
530 Y = 16 : Z = 25 : GOTO 610
540 Y = 26 : Z = 33 : GOTO 610
550 Y = 34 : Z = 43 : GOTO 610
560 Y = 44 : Z = 52 : GOTO 610
570 Y = 53 : Z = 55 : GOTO 610
580 Y = 56 : Z = 58 : GOTO 610
590 Y = 59 : Z = K6 : GOTO 610
600 Y = 67 : Z = K7
610 FOR T = Y TO Z : IF X > P(T) THEN NEXT T
620 Y = INT(RND(Z1) * Z5) : J$ = P$(Y) : C(I) = T : IF T > K6 THEN J$ = "VE" : N = T - K6
630 IF T = 51 THEN J$ = "KL" : GOTO 670
640 IF T = 33 THEN J$ = "KZ" : GOTO 670
650 IF T = 43 THEN J$ = "KH" : IF RND(Z1) < P1 THEN J$ = "KG" : IF RND(Z1) > P5 THEN J$ = "KM"
660 IF T = 25 THEN J$ = "KP" : IF RND(Z1) > P5 THEN J$ = "KV" : IF RND(Z1) > P5 THEN J$ = "KC"
670 C$(I) = J$ + RIGHT$(STR$(N),Z1) : GOSUB 2920 : GOSUB 2920 : GOSUB 2920
680 PRINT K - I : NEXT I : N = Z1 : PRINT "IT IS 2100 HRS ON" ; D ; " " + D$
690 PRINT "THE CONTEST HAS STARTED" + X$ : GOTO 1050
700 PRINT : IF N >= K THEN 2010
710 IF Q1 = Z0 AND RND(Z1) < P1 THEN 2930
720 IF H1 > T6 OR D >= 21 AND H > T5 THEN PRINT "CONTEST IS OVER" : GOTO 1480
730 INPUT "QRU" ; I$ : IF LEN(I$) < Z2 THEN PRINT "WHAT" + X$ : GOTO 730
740 FOR I = Z0 TO Q4 : IF I$ = LEFT$(Z$(I),LEN(I$)) THEN 760
750 NEXT : FOR I = Z0 TO Q4 - Z1 : PRINT Z$(I), : NEXT : PRINT : GOTO 730
760 IF I > Z8 THEN ON I - Z8 GOTO 1480,1490,1420,1170,1980,3270,1340,930,1750,1730
770 ON I GOTO 840,1040,1090,1140,1240,1300,1350,1450
780 Q6 = Z0 : ON Q1 GOTO 800,1670,1700,1620
790 PRINT "CALL WHOM ?" + X$ : PRINT "TRY TUNING OR PUT OUT A CQ" : GOTO 700
800 IF L(Z7) = Z0 AND RND(Z1) > P5 THEN 1030
810 PRINT "TRY SEND OR REPEAT, DON'T WASTE TIME" ; X$ : GOTO 700
820 IF L(Z6) = Z0 AND RND(Z1) < P1 THEN PRINT "REPEAT PLEASE" : GOTO 700
830 Q1 = Z0 : GOTO 700
840 IF C(P) > K7 THEN PRINT "IN THE LOG ALREADY" : GOTO 700
850 IF Q5 = Z0 THEN PRINT X$ + "WHY DON'T YOU SEND HIM HIS REPORT FIRST" : GOTO 730
860 FOR I = Z0 TO Z7 : IF L(I) = Z0 THEN I = Z1 : GOSUB 1580 : GOTO 700
870 Q7 = Z0 : NEXT : Y = C(P) : N = N + Z1
880 IF Y > K7 THEN Y = Y - B : GOTO 880
890 S(Y) = Z1 : C(P) = C(P) + B : GOSUB 920 : Q1 = Z0 : PRINT "NEXT QSO = " ; N
900 IF Q6 = Z1 AND RND(Z1) > P5 OR RND(Z1) < P1 THEN 930
910 GOTO 700
920 Q5 = Z0 : FOR I = Z0 TO Z7 : L(I) = Z0 : NEXT : RETURN
930 Q6 = Z1 : Y = T6 : GOSUB 1770 : IF Y = Z0 THEN 700
940 IF RND(Z1) < P1 THEN PRINT "FREQUENCY IS IN USE OM" : GOTO 700
950 IF RND(Z1) < P1 OR B = B4 AND H >= T5 AND RND(Z1) > P5 THEN PRINT "QRM" : GOTO 700
960 IF Q7 = Z0 THEN GOSUB 1970
970 IF Q7 = Z1 AND Q1 = Z1 THEN N1 = N1 + Z1
980 GOSUB 1900 : Q1 = Z1 : PRINT A$ ; E$ ; C$(P) : L(Z2) = Z1
```

Fig. 5-28. Program for sweepstakes contest game. (continued)

```
990 IF C(P) > K7 AND Q6 = Z1 AND RND(Z1) < Z2 * P1 THEN 700
1000 IF C(P) > K7 AND Q6 = Z1 THEN 1020
1010 IF C(P) > K7 AND RND(Z1) > P5 THEN PRINT "WE'VE WORKED OM !" : GOTO 700
1020 IF RND(Z1) > P5 THEN PRINT : GOTO 700
1030 J$ = A$ : GOSUB 1860 : PRINT : GOTO 700
1040 PRINT Z$(Z2) ; " = " ; B$(INT(B - Z1) / B1)
1050 INPUT "WHICH BAND " ; I$ : FOR I = Z0 TO Z4 : IF I$ = B$(I) THEN 1070
1060 NEXT : FOR I = Z0 TO Z4 : PRINT B$(I), : NEXT : PRINT : GOTO 1050
1070 IF B = (Z1 + I) * B1 THEN 700
1080 B = (Z1 + I) * B1 : GOSUB 920 : GOTO 700
1090 IF L(Z2) = Z0 AND L(Z5) = Z0 THEN PRINT "CHECK WHAT" : GOTO 700
1100 IF C(P) >= B1 THEN PRINT C$(P) ; " WORKED ON " ; B$(INT(C(P) / B1) - Z1) : GOTO 700
1110 Y = C(P) : IF Y >= B1 THEN Y = Y - B1 : GOTO 1110
1120 IF L(Z5) = Z1 THEN PRINT "SECTION " ; : IF S(Y) = Z1 THEN PRINT "WORKED" : GOTO 700
1130 PRINT "NOT WORKED YET" : GOTO 700
1140 GOSUB 1150 : GOTO 700
1150 PRINT Z$(Z4) ; : PRINT TAB(T6) ; D ; D$ : PRINT Z$(Z2) ; : PRINT TAB(T6) ;
1160 PRINT B$(INT(B - Z1) / B1)
1170 PRINT "GMT. TIME" ; : PRINT TAB(T6) ; : PRINT H ; " HRS",M ; " MIN"
1180 PRINT "ELAPSED TIME" ; : PRINT TAB(T6) ; H1 ; " HRS",M1 ; " MIN" : IF I = T3 THEN 700
1190 PRINT "QSO'S" ; : PRINT TAB(T6) ; N - Z1 : PRINT M$(Z5) + "S WORKED" ;
1200 PRINT TAB(T6) ; : Y = Z0 : FOR J = Z0 TO K7 : IF S(J) = Z1 THEN Y = Y + Z1
1210 NEXT : PRINT Y : PRINT "SCORE" ; : PRINT TAB(T6) ; Z2 * (N - Z1) * Y : PRINT "RATE" ;
1220 IF H1 = Z0 AND M1 = Z0 THEN PRINT TAB(T6) ; " - - - " : RETURN
1230 PRINT TAB(T6) ; (N - Z1) * T7 / (H1 * T7 + M1) ; "QSO'S PER HOUR" : RETURN
1240 INPUT "HRS" ; X : INPUT "MIN" ; Y : IF X = Z0 AND Y < 15 THEN GOSUB 1580 : GOTO 1240
1250 IF X >= T6 OR Y >= T7 THEN GOSUB 1290 : IF LEFT$(I$,Z1) <> "Y" THEN 1240
1260 M = M + Y : IF M >= T7 THEN M = M - T7 : H = H + Z1
1270 H = H + X : IF H >= T6 THEN H = H - T6 : D = D + 1
1280 GOSUB 1150 : GOSUB 920 : Q1 = Z0 : GOTO 700
1290 INPUT "ARE YOU SURE YOU MEANT THAT" ; I$ : RETURN
1300 IF L(Z7) = Z1 THEN PRINT "73, SEE YOU LATER" : GOTO 700
1310 IF RND(Z1) > P1 THEN L(Z7) = Z1 : PRINT "ROGER" : GOTO 700
1320 IF RND(Z1) >= P1 THEN PRINT "DO YOU QSL ?"
1330 GOTO 700
1340 GOSUB 3300 : GOTO 700
1350 IF L(Z2) = Z0 OR Q1 = Z0 THEN PRINT X$ + "TO WHOM" : GOTO 700
1360 IF Q1 <> Z1 THEN PRINT "TUT TUT - CALL HIM FIRST" ; X$ : GOTO 700
1370 PRINT L$(Z2) ; " UR" ; N ; : PRINT "A" ; E$ ; A$ ; : PRINT " " ; K$ ; " " ; F$ ; " MD K"
1380 IF RND(Z1) < P1 THEN PRINT A$ ; " PLEASE REPEAT" ; E$ ; C$(P) : GOTO 700
1390 Q5 = Z1 : L(Z6) = Z1 : IF L(Z7) = Z1 THEN PRINT "QSL" : GOTO 700
1400 IF RND(Z1) > P5 THEN J$ = A$ : GOSUB 1860 : PRINT "DO YOU QSL ?"
1410 GOTO 700
1420 FOR I = Z0 TO Z5 : IF L(I) = Z1 THEN NEXT : PRINT " GOT IT ALL" : GOTO 700
1430 PRINT "STILL NEED", : FOR J = I TO Z5 : IF L(J) = Z0 THEN PRINT " " ; M$(J) ;
1440 NEXT : PRINT : GOTO 700
1450 IF Q1 <> Z1 THEN PRINT "CALL HIM FIRST" + X$ : GOTO 700
1460 IF RND(Z1) > P1 THEN J$ = A$ : GOSUB 1860 : GOTO 700
1470 PRINT "SORRY OM, QRM ... TRY AGAIN" : GOTO 700
```

Fig. 5-28. Program for sweepstakes contest game. (continued)

```
1480 PRINT "FINAL " ; : GOSUB 1150 : GOTO 3500
1490 IF B < 400 AND H > T4 AND RND(Z1) < P1 THEN GOSUB 3110 : GOTO 700
1500 IF RND(Z1) >= P5 + P1 AND B = B4 AND (H >= T5 OR H < Z6) THEN GOSUB 3170 : GOTO 700
1510 GOSUB 920 : Q1 = Z0 : Y = T7 : GOSUB 1770 : IF Y = Z0 THEN 700
1520 L(Z2) = Z1 : Q1 = INT(RND(Z1) * Z3) + Z2 : ON Q1 GOTO 1530,1540,1550
1530 PRINT "QRZ" ; E$ ; C$(P) : GOTO 700
1540 FOR T = 0 TO INT(RND(1) * 3) : PRINT "CQ CQ CQ" ; E$ ; C$(P) ; " " ; : NEXT : PRINT "K" : GOTO 700
1550 GOSUB 1810 : PRINT C$(K) ; E$ ; C$(P) : IF RND(Z1) > P5 THEN 700
1560 Q7 = Z1 : GOSUB 1970
1570 GOSUB 1900 : J$ = C$(K) : GOSUB 1860 : GOTO 700
1580 PRINT "THAT'S AGAINST THE RULES" : PRINT X$
1590 IF I = Z1 THEN PRINT "QSO NOT COMPLETED"
1600 IF I = Z5 THEN PRINT "15 MINUTE MINIMUM"
1610 RETURN
1620 IF RND(Z1) > P5 THEN 1850
1630 IF Q7 = Z1 AND RND(Z1) > P5 THEN 970
1640 IF RND(Z1) > P5 THEN Q1 = Z2 : GOTO 1530
1650 IF RND(Z1) < P5 THEN PRINT "PLEASE, I'M IN QSO" : GOTO 700
1660 PRINT C$(K) ; E$ ; C$(P) ; " SORRY QRM, REPEAT" : GOTO 700
1670 IF RND(Z1) > P5 THEN 1530
1680 IF RND(Z1) < P1 THEN Q1 = Z3 : GOTO 1540
1690 Q1 = Z1 : GOTO 960
1700 IF RND(Z1) > P5 THEN 1540
1710 IF RND(Z1) > P5 THEN Q1 = Z2 : GOTO 1530
1720 Q1 = Z1 : GOTO 960
1730 FOR J = Z0 TO K - Z1 : PRINT C$(J) ; : IF C(J) > K7 THEN PRINT " * " ;
1740 PRINT , : NEXT : GOTO 700
1750 FOR J = Z0 TO K - Z1 : IF C(J) > K7 THEN PRINT C$(J),
1760 NEXT : GOTO 700
1770 Y = RND(Z1) * Y + Z4 : FOR T = Z0 TO Z2 : GOSUB 2860 : GOSUB 2820 : GOSUB 2030
1780 IF Y = Z0 OR RND(Z1) < P1 THEN NEXT : Y = Z0
1790 IF Y = Z1 THEN GOSUB 1900
1800 RETURN
1810 I = K : T = INT(RND(Z1) * Z3 + Z2) : J$ = P$(INT(RND(Z1) * Z3))
1820 C$(I) = J$ + RIGHT$(STR$(T),Z1) : GOSUB 2920 : GOSUB 2920 : GOSUB 2920
1830 IF C$(I) = A$ THEN 1810
1840 RETURN
1850 GOSUB 1810 : J$ = C$(K) : GOSUB 1860 : N1 = N1 + Z1 : Q7 = Z1 : GOTO 700
1860 L$(Z0) = STR$(N1) : L(Z7) = Z1 : PRINT J$ ; " UR" ; :
1870 FOR J = Z0 TO Z5 : IF L(J) = Z1 OR RND(Z1) >= P1 THEN L(J) = Z1 : PRINT " " ; L$(J) ; : GOTO 1890
1880 PRINT R$ ;
1890 NEXT : PRINT : RETURN
1900 FOR J = Z1 TO 26 : IF RIGHT$(C$(P),Z1) <> MID$(Q$,J,Z1) THEN NEXT
1910 L$(Z1) = "A" : IF J >= T3 THEN L$(Z1) = "B"
1920 Y = Z0 : FOR J = Z1 TO Z3 : Y = Y + ASC(MID$(RIGHT$(C$(P),Z3),J,Z1)) : NEXT
1930 L$(Z4) = STR$(INT(Y / Z2 - 57)) : L$(Z0) = STR$(N1)
1940 L$(Z2) = C$(P) : L$(Z3) = K$ : Y = C(P)
1950 IF Y > K7 THEN Y = Y - B1 : GOTO 1950
1960 L$(Z5) = S$(Y) : RETURN
```

Fig. 5-28. Program for sweepstakes contest game. (continued)

```
1970 N1 = Z1 + INT(N * T3 * P1 * RND(Z1)) : RETURN
1980 IF RND(Z1) < P1 THEN 700
1990 IF RND(Z1) > P5 THEN PRINT "SORRY"
2000 Q1 = Z0 : GOTO 900
2010 PRINT "ANTENNA BLEW DOWN IN WIND"
2020 PRINT " NO WAY  TO FIX IT - QRT" : GOTO 1480
2030 IF C(P) < B1 THEN 2070
2040 X = C(P)
2050 C(P) = C(P) - B1 : IF C(P) > B1 THEN 2050
2060 GOSUB 2070 : C(P) = X : RETURN
2070 Y = 0 : IF C(P) = 14 OR C(P) = T6 THEN 2810
2080 IF C(P) < Z8 THEN 2230
2090 IF C(P) < T3 THEN 2280
2100 IF C(P) < T4 THEN 2310
2110 IF C(P) = 25 THEN 2710
2120 IF C(P) < 25 THEN 2340
2130 IF C(P) = 33 THEN 2750
2140 IF C(P) < 33 THEN 2380
2150 IF C(P) = 43 THEN 2640
2160 IF C(P) < 43 THEN 2420
2170 IF C(P) = 51 THEN 2640
2180 IF C(P) < 53 THEN 2480
2190 IF C(P) < 56 THEN 2530
2200 IF C(P) < 59 THEN 2550
2210 IF C(P) < 67 THEN 2580
2220 ON C(P) - 67 GOTO 2280,2530,2550,2580,2480,2480,2480
2230 IF B = B4 OR B = B5 AND H < T3 OR B = B3 AND H >= T5 AND RND(Z1) > P5 THEN 2810
2240 IF B = B3 AND (H < T5 AND H >= T3 OR RND(Z1) > P5 AND H >= Z8) THEN 2810
2250 IF B = B2 AND (H >= T5 OR H >= Z8 AND H < T3) AND RND(Z1) < P1 THEN 2810
2260 IF B = B1 AND (H >= T3 AND Q = Z2 AND H < T5 OR H >= T5 AND RND(Z1) > P5) THEN 2810
2270 RETURN
2280 IF B = B3 AND (H >= Z8 AND H < T3 OR H >= T5) AND RND(Z1) > P5 THEN 2810
2290 IF B >= B4 OR B = B1 AND Q = Z1 AND RND(Z1) < P5 AND H < T4 AND H >= T3 THEN 2810
2300 RETURN
2310 IF B = B5 OR B = B4 AND H < Z8 AND RND(Z1) > P5 THEN 2810
2320 IF B = B4 AND (H >= T5 OR H >= Z8 AND H < T3) THEN 2810
2330 RETURN
2340 IF B = B4 OR B = B3 AND H >= Z8 OR B = B1 AND Q = Z3 AND H >= T3 AND RND(Z1) > P5 THEN 2810
2350 IF B = B2 AND H < T5 AND H >= T3 AND RND(Z1) > P5 THEN 2810
2360 IF B = B5 AND H < T4 OR (H >= T5 OR H < T3) AND RND(Z1) > P5 THEN 2810
2370 RETURN
2380 IF B < B4 AND H >= T3 AND H < T5 OR B = B4 AND H < Z8 THEN 2810
2390 IF B = B5 AND (H < Z8 AND H >= Z4 OR H < Z4 AND RND(Z1) > P5) THEN 2810
2400 IF RND(Z1) > P5 AND (H >= T5 OR H >= Z8 AND H < T3) THEN 2810
2410 RETURN
2420 IF H < Z8 AND H >= T3 AND RND(Z1) < P1 THEN 2810
2430 IF B < B4 AND H >= T3 AND H < T5 OR B >= B3 AND H < Z8 AND H >= 4 THEN 2810
2440 IF H >= T5 AND B < B3 AND RND(Z1) > P5 THEN 2810
2450 IF H >= T5 AND (B = B3 OR B = B4 AND RND(Z1) < P1) THEN 2810
```

Fig. 5-28. Program for sweepstakes contest game. (continued)

```
2460 IF H < Z4 AND (B = B4 OR B = B5 AND RND(Z1) > P5) THEN 2810
2470 RETURN
2480 IF B < B4 AND H < T5 AND H >= T3 THEN 2810
2490 IF H >= T5 AND B > B1 AND B < B5 AND RND(Z1) > P5 THEN 2810
2500 IF B = B4 AND (H < Z8 OR H >= T5 AND RND(Z1) > P5) THEN 2810
2510 IF B = B5 AND (H < Z8 AND H >= Z4 OR H < Z4 AND RND(Z1) > P5) THEN 2810
2520 RETURN
2530 IF B = B5 OR B = B4 AND (H >= Z8 OR RND(Z1) > P5) THEN 2810
2540 RETURN
2550 IF B = B4 OR B = B5 AND (H < Z8 OR RND(Z1) > P5 AND H < T3 OR H >= T5) THEN 2810
2560 IF B = B3 AND H >= T3 AND H < T5 THEN 2810
2570 RETURN
2580 IF B > B3 AND H < Z8 OR H < T3 AND RND(Z1) > P5 THEN 2810
2590 IF B = B1 AND Q = Z5 AND H >= T3 AND RND(Z1) > P5 THEN 2810
2600 IF B = B2 AND H < T5 AND H >= T3 AND RND(Z1) > P5 THEN 2810
2610 IF B = B3 AND (H < T5 AND H >= T3 OR H >= T5 AND RND(Z1) > P5) THEN 2810
2620 IF H >= T5 AND (B = B4 AND RND(Z1) > P5 OR B = B5 AND RND(Z1) < P1) THEN 2810
2630 RETURN
2640 IF B < B4 AND H < T5 AND H >= T4 THEN 2810
2650 IF B = B3 AND (H >= T3 AND RND(Z1) > P5 OR H >= Z8 AND H < T3 AND RND(Z1) < P1) THEN 2810
2660 IF B = B1 AND H >= T3 AND (H < T5 OR RND(Z1) > P5) THEN 2810
2670 IF B = B2 AND (H < T4 AND H >= T3 AND RND(Z1) > P5 OR H >= T5 AND RND(Z1) < P1) THEN 2810
2680 IF B = B4 AND H < Z8 AND (H >= Z4 OR RND(Z1) > P5) THEN 2810
2690 IF B = B5 AND H >= Z4 AND H < Z8 AND RND(Z1) > P5 THEN 2810
2700 RETURN
2710 IF B = B5 AND H < Z8 OR B < B4 AND H >= Z8 AND H < T5 THEN 2810
2720 IF H >= T5 AND (B = B1 AND RND(Z1) > P5 OR B = B2 OR B = B3) THEN 2810
2730 IF B = B4 AND (H >= Z4 AND H < Z8 OR H < Z4 AND RND(Z1) < P1) THEN 2810
2740 RETURN
2750 IF B < B3 AND H < T5 AND (H >= T3 AND H < T4 OR H >= Z8 AND RND(Z1) > P5) THEN 2810
2760 IF H >= T5 AND (B = B2 OR B = B4) AND RND(Z1) > P5 THEN 2810
2770 IF H >= T4 AND B = B3 OR B = B4 AND H < Z8 THEN 2810
2780 IF B = B4 AND RND(Z1) > P5 AND H < T3 THEN 2810
2790 IF B = B5 AND H < Z8 AND (H >= Z4 OR RND(Z1) > P5) THEN 2810
2800 RETURN
2810 Y = 1 : RETURN
2820 IF I = Z0 THEN P = INT(RND(Z1) * K)
2830 P = P + INT(RND(Z1) * Z9) : IF P >= K THEN P = P - K
2840 IF B = B1 AND RND(Z1) > P1 * P1 THEN Q = Q + INT(RND(Z1) + Z4) : IF Q > Z5 THEN Q = Q - Z5
2850 RETURN
2860 S1 = S1 + Y : IF S1 < T7 THEN RETURN
2870 S1 = S1 - T7 : M1 = M1 + Z1 : M = M + Z1 : IF S1 >= T7 THEN 2870
2880 IF M >= T7 THEN M = M - T7 : H = H + Z1 : GOTO 2880
2890 IF M1 >= T7 THEN M1 = M1 - T7 : H1 = H1 + Z1 : GOTO 2890
2900 IF H < T6 THEN RETURN
2910 H = H - T6 : D = D + Z1 : RETURN
2920 I$ = MID$(Q$,INT(RND(Z1) * 26) + Z1,Z1) : C$(I) = C$(I) + I$ : RETURN
2930 IF RND(Z1) > P1 * Z2 THEN 700
2940 X = INT(RND(Z1) * 10) + Z1
```

Fig. 5-28. Program for sweepstakes contest game. (continued)

```
2950 ON X GOTO 2960,2970,2980,2960,2960,2960,2960,2960,3070,3020,2960
2960 PRINT "TIMEOUT IN THE BATHROOM" : Y = T5 : GOTO 2990
2970 PRINT "TIMEOUT  TO CLEAR UP SPILT COFFEE" : Y = Z5 : GOTO 2990
2980 PRINT "KIDS ARE FIGHTING..TIMEOUT" : Y = T4
2990 Y = INT(RND(Z1) * Y) + Z3 : PRINT "OFF THE AIR FOR " ; Y ; " MIN"
3000 IF Y > 14 THEN X = Z0 : GOTO 1260
3010 Y = Y * T7 : GOSUB 2860 : GOTO 700
3020 IF RND(Z1) * 10 < P5 THEN 2010
3030 PRINT "POWER LINE BLACKOUT" : X = INT(RND(Z1) * Z5) : Y = INT(RND(Z1) * T6) + Z8
3040 PRINT "POWER CAME BACK ON AFTER " ; X ; " HRS",Y," MIN"
3050 IF X = Z0 AND Y < 15 THEN Y =  Y * T7 : GOSUB 2860 : GOTO 700
3060 GOTO 1260
3070 PRINT "FINAL TUBES BLEW OUT"
3080 IF H < T3 THEN PRINT "WHERE ARE YOU GOING  TO GET SPARES AT THIS TIME"
3090 PRINT "LUCKY YOU, SPARES ARE  TO HAND "
3100 PRINT "REPLACING THEM NOW" : Y = T7 : GOTO 2990
3110 I$ = "YV4XYZ" : IF RND(Z1) > P5 THEN I$ = "HK3ZZZ"
3120 J$ = "MIAMI" : IF RND(Z1) > P5 THEN J$ = "LOS ANGELES"
3130 PRINT "CQ CQ CQ DE " ; I$
3140 PRINT "THIS IS " ; I$ ; " LOOKING FOR A PHONE PATCH IN TO " ; J$
3150 IF RND(Z1) < P5 - P1 THEN 3140
3160 RETURN
3170 X = INT(RND(Z1) * Z4) : IF RND(Z1) > P5 THEN 3190
3180 PRINT "YOU ARE LISTENING  TO " ; : ON X + Z1 GOTO 3200,3210,3220,3230
3190 PRINT "THIS IS " ; : ON X GOTO 3210,3220,3230
3200 PRINT "RADIO MOSCOW" : GOTO 3240
3210 PRINT "RADIO PEKING" : GOTO 3240
3220 PRINT "RADIO ALBANIA" : GOTO 3240
3230 PRINT "THE BBC" : IF RND(Z1) < P1 * Z2 THEN PRINT "THIS IS RADIO NEWSREEL" : RETURN
3240 IF RND(Z1) > P5 THEN PRINT "A NEWS BROADCAST FOLLOWS" : RETURN
3250 IF RND(Z1) > P5 THEN PRINT "YOU HAVE BEEN LISTENING  TO A COMMENTARY ON THE NEWS"
3260 RETURN
3270 PRINT Z$(14) + "S NEEDED" : J = K7 : FOR I = Z0 TO K7
3280 IF S(I) = Z0 THEN PRINT S$(I), : J = J - Z1
3290 NEXT : PRINT : PRINT J + Z1,Z$(14) + "S WORKED" : PRINT : GOTO 700
3300 PRINT Z$(Z0),"CALL OTHER STATION"
3310 PRINT Z$(Z1),"ENTER CONTACT IN LOG"
3320 PRINT Z$(Z2),"CHANGE BANDS"
3330 PRINT Z$(Z3),"CHECK  TO SEE IF WORKED"
3340 PRINT Z$(Z4),"SEE HOW WELL YOU ARE DOING"
3350 PRINT Z$(Z5),"TAKE A BREAK (TIMEOUT)"
3360 PRINT Z$(Z6),"CONFIRM RECEPTION OF CONTEST DATA "
3370 PRINT Z$(Z7),"SEND YOUR DATA  TO OTHER STATION"
3380 PRINT Z$(Z8),"REQUEST REPEAT OF OTHER STATION'S DATA "
3390 PRINT Z$(Z9),"GIVE UP AND SWITCH OFF EQUIPMENT"
3400 PRINT Z$(10),"TUNE ACROSS BAND "
3410 PRINT Z$(11),"DETERMINE WHICH DATA ARE MISSING FROM REPORT"
3420 PRINT Z$(T3),"LOOK AT TIME OF DAY"
3430 PRINT Z$(13),"TELL OTHER STATION THAT YOU HAVE WORKED BEFORE"
```

Fig. 5-28. Program for sweepstakes contest game. (continued)

```
3440 PRINT Z$(14),'LIST SECTIONS STILL NEEDED'
3450 PRINT Z$(15),'RECEIVE SUMMARY OF INSTRUCTIONS'
3460 PRINT Z$(T4),'CALL ' ; Z$(T4)
3470 PRINT Z$(17),'LIST OF STATIONS WORKED (UNSORTED)'
3480 PRINT 'YOU MAY TYPE FIRST TWO LETTERS FOR ALL ENTRIES, EXCEPT',
3490 PRINT Z$(14) ; ' (TYPE 3)' : RETURN
3500 PRINT 'HOPE YOU ENJOYED THE CONTEST'
```

Fig. 5-28. Program for sweepstakes contest game. (continued)

QRU ? tu (Nothing, so try again.)

CQ CQ CQ DE W2CJR CQ CQ CQ DE W2CJR

QRU ? ca (We heard another CQ call, so let's give him a call.)

QRZ DE W2CJU (Did he hear something?)

QRU ? ca

QRZ DE W2CJU (Why can't he hear us?)

QRU ? ca

G3ZCZ/W3 DE W2CJU (Got him!)

QRU ? se

W2CJR UR 2 A DE G3ZCZ/W3 CK 68 MD K

QRU ? RE (We sent our data, and now are requesting him to send his.)

G3ZCZ/W3 UR !#* A W2CJR CK 54 NNJ (Notice the QRM while blocked his number.)

WRU ? re (Request a repeat to get the number.)

G3ZCZ/W3 UR 3 A W2CJR CK 54 NNJ

QRU ? lo (All ok, log him in.)

NEXT QSO = 3

QRU ? re (Request a repeat to get the more.)

YOU ARE LISTENING TO THE BBC

A NEWS BROADCAST FOLLOWS

(It is evening and propagation from Europe is good, so there are lots of broadcast stations on 40 Meters at this time.)

QRU ? tu

YOU ARE LISTENING TO RADIO ALBANIA

YOU HAVE BEEN LISTENING TO A COMMENTARY ON THE NEWS

QRU ? tu (Continue tuning.)

W4RDK DE K3VZG

W4RDK UR 1 B K3VZG !#* 66 E.PA

QRU ? ch (We heard a QSO in progress; let's see if we've worked him. I know that we haven't because his QSO number is 1 showing that that is his first QSO, but let's see if the computer knows it also.)

SECTION NOT WORKED YET (The computer checks the section first to tell us about multipliers. If we haven't worked the section, we can safely assume that we haven't worked him.)

QRU ? ca

CQ CQ CQ DE K3VZG CQ CQ CQ DE K3VZG

(He didn't hear us, so call again.)

QRU ? ca

QRZ DE K3VZG (No luck—keep trying.)

QRU ? ca

QRZ DE K3VZG

QRU ? ca

QRZ DE K3VZG

QRU ? ca (We'll get him in the end.)

QRZ DE K3VZG

QRU ? ca

G3ZCZ/W3 DE K3VZG (He heard us.)

G3ZCZ/W3 UR 2 B K3VZG CK 66 E.PA

(He sent his data first.)

QRU ? se (Reply)

K3VZG UR 3 A DE G3ZCZ/W3 CK 68 MD K

QSL (He copied it.)

QRU ? lo

NEXT QSO = 4

QRU ? tu

CQ CQ CQ DE WB7ZKJ

QRU ? ch

NOT WORKED YET
QRU ? ca
G3ZCZ/W3 DE WB7ZKJ
G3ZVZ/W3 UR 2 B WB7ZKJ CK 62 WA (Propagation is good.)
QRU ? se
WB7ZKJ UR 4 A DE G3ZCZ/W3 CK 68 MD K
G3ZCZ/W3 PLEASE REPEAT DE WB7ZKJ
QRU ? lo (We worked him, let's log him.)
WHY DON'T YOU SEND HIM HIS REPORT FIRST
(The computer is teaching us to play the game.)
QRU ? se
WB7ZKJ UR 4 A DE G3ZVZ/W3 CK 68 MD K
QSL
QRU ? LOG
NEXT QSO = 5 (Now the computer took the log data.)
QRU ? st (Let's see how we are doing.)
STATUS 19 NOVEMBER
BAND 40
GMT. TIME 21 HRS 3 MIN
ELAPSED TIME 0 HRS 3 MIN
QSO'S 4
SECTIONS WORKED 4
SCORE 32
RATE 80 QSO'S PER HOUR
QRU ? tu
W3VRK DE N3QXC
W3VRK UR 3 b N3QXC CK 61 E.PA
QRU ? ca
G3ZCZ/W3 DE N3QXC
G3ZCZ/W3 UR 4 B N3QXC CK 61 E.PA
QRU ? lo
WHY DON'T YOU SEND HIM HIS REPORT FIRST?
(Again we tried to log him without sending him his report first and the computer caught it, so . . .)
QRU ? se
N3QXC UR 5 A DE G3ZCZ/W3 CK 68 MD K
QSL
QRU ? lo
NEXT QSO = 6
QRU ? tu
THIS IS RADIO MOSCOW
QRU ? ba (40 Meters is terrible. Let's switch to 10 Meters.)
BAND = 40 ? 10
QRU ? cq
G3ZCZ/W3 DE K6KGG
G3ZCZ/W3 UR !#* B K6KGG CK 51 LA
QRU ? se
K6KGG UR 6 A DE G3ZCZ/W3 CK 68 MD K
qru ? lo
THAT'S AGAINST THE RULES
QSO NOT COMPLETED
(We are not supposed to log the QSO if we don't have all the data. The number was lost in the QRM, and the computer is reminding us.)
QRU ? re (We request a repeat.)
G3ZCZ/W3 UR 3 B K6KGG CK 51 LA (Got it now!)
QRU ? lo (So its ok to log it.)
NEXT QSO = 7
(After awhile, we QSY to 20 Meters and tune around.)
QRU ? tu
THIS IS YV4XYZ LOOKING FOR A PHONE PATCH INTO LOS ANGELES
(He'll be lucky if he gets one during the contest.)
QRU ? he (Let's see what commands are available.)

CALL	CALL OTHER STATION
LOG	ENTER CONTACT IN LOG
BAND	CHANGE BANDS
CHECK	CHECK TO SEE IF WORKED
STATUS	SEE HOW WELL YOU ARE DOING
BREAK	TAKE A BREAK
QSL	CONFIRM RECEPTION OF CONTEST DATA
SEND	SEND

	YOUR DATA TO OTHER STATION
REPEAT	REQUEST REPEAT OF OTHER STATION'S DATA
QRT	GIVE UP AND SWITCH OFF EQUIPMENT
TUNE	TUNE ACROSS BAND
DATA	DETERMINE WHICH DATA ARE MISSING FROM REPORT
TIME	LOOK AT TIME OF DAY
DUPLICATE	TELL OTHER STATION THAT YOU HAVE WORKED BEFORE
SECTION	LIST SECTIONS STILL NEEDED
HELP	RECEIVE SUMMARY OF INSTRUCTIONS
CQ	CALL CQ
WORKED	LIST OF STATIONS WORKED (UNSORTED)

YOU MAY TYPE FIRST TWO LETTERS FOR ALL ENTRIES, EXCEPT SECTION (TYPE 3)

QRU ? ti (what time is it?)
GMT.TIME 21 HRS 9 MIN
ELAPSED TIME 0 HRS 9 MIN

(We've been in now the contest for nine minutes.)

QRU ? qrt (We've had enough for now. Let's see how we did.)

FINAL STATUS 20 NOVEMBER
BAND 80
GMT.TIME 10 HRS 17 MIN
ELAPSED TIME 0 HRS 21 MIN
QSO'S 16
SECTION'S WORKED 11
SCORE 353
RATE 45.7143 QSO'S PER HOUR
HOPE YOU ENJOY THE CONTEST.

By the way, some interesting things can happen during the game. Here are some samples;

FINAL TUBES BLEW OUT
LUCKY YOU. SPARES ARE TO HAND
REPLACING THEM NOW
OFF THE AIR FOR 41 MIN

POWER LINE BLACKOUT
POWER CAME BACK ON AFTER 0 HRS 26 MIN

ANTENNA BLEW DOWN IN WIND
NO WAY TO FIX IT—QRT

KIDS ARE FIGHTING . . TIMEOUT
OFF THE AIR FOR 7 MIN

TIMEOUT TO CLEAR UP SPLIT COFFEE
OFF THE AIR FOR 3 MIN

TIMEOUT IN THE BATHROOM
OFF THE AIR FOR 19 MIN

THE GAME PROGRAM

A listing of a program written in Microsoft BASIC is presented in Fig. 5-28. The variables and constants used in the program are shown in Fig. 5-29. Constants are defined as parameters to save memory space. A summary of the commands and their relationship to the program is shown in Fig. 5-30. DEBUG is shown in parenthesis because it is not normally used during a game. Program flow is as follows:

Lines 10-150. Program initialization, constants, and strings.

Lines 160-220. Setting up QSO probabilities for each section.

Lines 230-370. Strings continued.

Lines 380-470. Customization for individual player.

Lines 480-680. Setting up the callsigns of the contestants in the game (simulation).

Lines 680-690. Startup message. X$ alerts operator that the game has started.

A. Variables and Flags

B	Band
C(K)	Section of station and worked flag
D	Date
H	Hours (actual)
H1	Hours (elapsed contest time)
I	For/next loop variable (reused)
J	For/next loop variable (reused)
K	Number of calls in the game
L(0)	Number of station being contacted received flag
L(1)	Power
L(2)	Call
L(3)	Last 2 digits of year first licensed
L(4)	Check
L(5)	Section of station being contacted received flag
L(6)	Outgoing data received flag
L(7)	Incoming data sent
M	Minutes (acutal)
M1	Minutes (elasped contest time)
N	QSO number
N1	Other station's QSO number
P	Index into matrix
Q	Short strip
Q1	QSO Flag
Q4	Number of commands
Q5	Send flag
Q6	CQ flag
Q7	In contact flag
S1	Seconds (elapsed contest time)
S(74)	Section worked flag
X	Local variable (reused)
Y	Local variable (reused)
Z	Local variable (reused)

B. Constants

B1 = 100
B2 = 200
B3 = 300
B4 = 400
B5 = 500
K7 = Sections
N = Year first licensed (then reused)
P1 = 0.1
P5 = 0.5
T3 = 12
T4 = 16
T5 = 20
T6 = 24
T7 = 60
Z0 = 0
Z1 = 1
Z2 = 2
Z3 = 3
Z4 = 4
Z5 = 5
Z6 = 6
Z7 = 7
Z8 = 8
Z9 = 9

C. Strings

A$	= Call sign of player
B$(4)	= Band
C$(K)	= Call sign of others in the game
D$	= Section November
E$	= " = "
F$	= String for year first licensed (N)
I$	= General purpose (reused)
J$	= Temporary general purpose (reused)
K$	= "CK"
L$(0)	=
L$(1)	=
L$(2)	=
L$(3)	=
L$(4)	=
L$(5)	=
M$	= Exchange title
M$(0)	= "Number"
M$(1)	= "Power"
M$(2)	= Z$(Z0) = "CALL"
M$(3)	= K$ = "CK"
M$(4)	= Z$(Z3) = "CHECK"
M$(5)	= "SECTION"
P$(4)	= Prefixes
Q$	= Letters in alphabet
R$	= QRM
S$(74)	= Section names
X$	= Bell character
Z$(18)	= Commands

Fig. 5-29. Variables and constants used in the game.

Lines 700-770. Main command loop. The command associated with each value if I is described by the value of Z$(I). The DEBUG command does not show up in the help list. The command subroutines are as shown in the flowcharts of Figures 5-13 to 5-20.

Lines 780-830. Sequence to call other station. An error message is generated if the QSO flag (Q1) is 0.

Lines 840-910. Sequence to enter contact data in log.

Line 920. Subroutine to clear L( ) flags.

Lines 930-1030. Sequence to call CQ.

Lines 1040-1080. Sequence to change bands.

Lines 1090-1130. Sequence to check to see if other station has been worked.

Line 1040. Sequence to print out status of how well the player is doing.

Lines 1150-1230. Subroutine to print out status or score at any one time.

Lines 1240-1280. Sequence to take a break.

Line 1290. Subroutine: User input error watchdog.

Lines 1300-1330. Sequence to QSL contact.

Line 1340. Sequence to print command summary (HELP).

Lines 1350-1410. Sequence to send your data to the other station.

Lines 1420-1440. Sequence to determine what data was lost in the QRM.

Lines 1450-1470. Sequence to request a repeat or a report from the other station.

I	Command	Description	Line
0	CALL	Call other station	780
1	LOG	Enter QSO in log	840
2	BAND	Change bands	1040
3	CHECK	Check to see if worked	1090
4	STATUS	See how well you are doing	1140
5	BREAK	Take a break or time out	1240
6	QSL	Confirm reception of exchange	1300
7	SEND	Send data to other station	1350
8	REPEAT	Request a repeat or report from other station	1450
9	QRT	Give up and close down	1480
10	TUNE	Tune across band	1490
11	DATA	Determine data missed in QRM	1420
12	TIME	Check time of day	1170
13	DUPLICATE	Tell other station that you have worked him	1980
14	SECTION	List sections still needed	3270
15	HELP	Help - lists all commands usable	1340
16	CQ	Call CQ	930
17	WORKED	Prints unsorted list of stations worked	1750
(18)	(DEBUG)	(Debug listing of all stations in contest. Those worked are flagged with an "*.")	1730

Fig. 5-30. Player options in game.

Line 1480. Sequence to terminate game, printing out status/score at the end.

Lines 1490-1570. Sequence to tune across band.

Lines 1580-1610. Subroutine: Checking for infringement of rules of ungentlemanly conduct.

Lines 1620-1720. Sequence that follows QSO to its conclusion. The QSO flag (Q1) determines the starting points; i.e., Q1 = 1 (in QSO) is at line 800, Q1 = 2 (QRZ) is at line 1670, Q1 = 3 (CQ) is at line 1700, Q1 = 4 (station is in QSO with someone else) is at line 1620.

Lines 1730-1740. Sequence to implement DEBUG command. Prints out everybody in the game and flags those already contacted.

Lines 1750-1760. Sequence to implement worked command. Prints out the calls of stations in the log in the order worked.

Lines 1770-1800. Subroutine to determine if other station is found. Time is spent in waiting.

Lines 1810-1840. Subroutine to set up a callsign on a random basis. Line 1830 checks to ensure that you are not working yourself.

Line 1850. Sequence to set up the "in QSO" flag.

Lines 1860-1890. Subroutine to print incoming report and set flags to verify reception of parts of message.

Lines 1900-1960. Subroutine to set up other stations parameters. Note that the "check" is a function of the letters in the suffix (see lines 1920 and 1930).

Line 1970. Subroutine to set up number of the other stations QSO (N1).

Lines 1980-2000. Sequence to tell other station that this is a duplicate QSO.

Lines 2010-2020. Sequence to terminate game when antenna blows away.

Lines 2030-2810. Subroutine to determine if a QSO with the other station previously located in the contest matrix C(P) is possible on the band at that time of day. Y is used as a QSO flag. Upon exit from the subroutine, if Y = 0 the QSO is not possible; if Y = , then a QSO is possible. The logic in lines 2070–2810 reflects the truth tables shown in Fig. 5-10 and Fig. 5-12.

Lines 2820-2850. Subroutine to find a call in the game contest matrix by setting up a random value for the index pointer into the matrix (P).

Lines 2860-2910. Subroutine to increment elapsed time.

Line 2920. Subroutine to set up one letter in suffix of callsign.

Lines 2930-3100. Subroutine to implement random events (ringers).

Lines 3110-3160. Subroutine to printout signals heard while running across the band.

Lines 3170-3260. Subroutine to print out broadcast stations on 40M on evening.

Lines 3270-3290. Sequence to list sections still needed.

Lines 3300-3490. Subroutine to list help sequence.

Lines 3500-3510. Sequence to close out the game.

The Probability Matrix (P) is set up using a FOR/NEXT loop (line 160) that reads the values stored in the data statements (lines 170 to 220). The Sections are stored in a section matrix (S$) which is set up when writing the program.

Since the contest simulates a station located in Maryland which is in the 3 call area, any ham who is playing the game has his call transferred to portable 3 (line 390). If the player is not a ham, he can use my call sign (G3ZCZ) in this simulation only.

The player is told that some time is going to elapse before the game will start (line 460). The "PRINT K-I" statement (line 680) shows the player that the computer is still active and is setting up the game, and that it has not blown its stack and gone off into never-never land.

Note the error-checking (line 730). Here the input is checked to see that it is at least two characters long. If this was not checked, the Command identification test (the FOR/NEXT loop in lines 740 to 750) could detect an error and abort the game. The trap (line 1220) catches a divided by zero error, if a score rating be requested with zero elapsed time.

To set up the callsigns based on the probability model, the callarea is determined first. Then once it is known, a second random number is generated. This is compared with the probabilities of being in each section again using a FOR/NEXT loop (line 610 of Fig. 5-28) in which the probabilities are stored in the P matrix and the search boundaries associated with each call area are set up (lines 500 to 600).

Once a section is identified, the prefix is set up (J$). It can be either W, K, WA, WB, or N. If, however, it is a Canadian call, the prefix is then changed to VE, and the section number is converted to call area. Checks are made so that the prefix for Alaska is allocated to prefixes in the AL section, calls in the Pacific sections are allocated KH, KG or KM, while stations in the West Indies are allocated KP, KV or KC. The relative rarity of the prefixes are considered when doing the allocations (lines 620 to 660). The complete callsign (prefix, call area, and suffix) is then generated (line 670) using a subroutine (line 2920) that generates a random letter for each of the three letters in the suffix.

This operation is performed for each of the stations in the game, an approach which places the players over geographic North America with the same distribution as the original model, regardless of how many players are in the game. This means that the probability of any callsign being in a particular section is the same in a 48K system having 450 to 500 calls in the game matrix as on a 64K system having 1000 calls stored in the game matrix.

Note that in this version of BASIC, the IF/THEN/ELSE statement works in the following manner: If the test is true, continue on the same line. If the test fails, go to the next line and ignore anything on the line following the test stement. Line 1120 states

```
1120 IF L(5) = Z1 THEN PRINT
  "SECTION" ;:IF S(Y) = Z1 THEN
  PRINT " WORKED ": GOTO 700
```

Its logical flow is

```
1120 IF L (5) = Z1
THEN PRINT " SECTION ";:
  IF S(Y) = Z1
  THEN PRINT " WORKED ":
      GOTO 700
      ELSE
    ELSE
```

If L(5) is not equal to Z1 or if S(Y) is not equal to Z1 the program flow skips everything else and skips to line 1130. Different dialects of BASIC may behave differently. For example, NORTHSTAR BASIC will continue at the next : if the test fails. It will not automatically ignore the rest of the line.

(Northstar BASIC also uses the backslash (\) instead of the colon (:) as the statement delimiter).

MODIFICATIONS TO THE PROGRAM

If you are at all interested in Amateur Radio Contests, you'll enjoy this game. It's a good way to get an idea of what real contesting is like. It contains complex models for propagation activity and operator action. Why not try to modify it? Change the propagation model to reflect the truth table or to make it more "real." Change the probabilities of various events occurring, and see what happens.

You could make a start in lines 1870–1890. Change the QRM probability to something else; you might increase it on 20 Meters, increase it on 40 Meters in the evening, or decrease it at night.

Perhaps in the future a standard contest program might be used. Everyone could take part with their computer running the simulation. There will then be no need for the QRM that the contesters generate to interfere with the non-contest radio amateurs, and the contesters will still get their training and enjoyment.

Antenna Positioning and Pointing

Conventional maps that give a flat or two dimentional representation of the surface of the Earth cannot be used to calculate directions and distances, because straight lines on the surface of the Earth are really curves. The computer has to use spherical geometry to compute directions and distances rather than regular geometry. The Earth is close enough in shape to a sphere to allow the differences to be ignored.

DIRECTION AND DISTANCE COMPUTATIONS

In the globe depicted in Fig. 6-1, the distance D is given by the equation

$$\text{Cos } D = \text{Sin } A \cdot \text{Sin } B + \text{Cos } A \cdot \text{Cos } B \cdot \text{Cos } (L1-L2)$$

where

A is the latitude of the station
L1 is its longitude
B is the latitude of the other station
L2 is its latitude

The distance D is in minutes of arc which is directly related to nautical miles.

The range bearing R is given by the equation:

$$\text{Sin } R = \frac{\text{Cos } B \cdot \text{Sin } (L1-L2)}{\text{Sin } D}$$

As most computers available in the Amateur Radio Station do not contain inverse trig functions, the values of the inverse Sine and Cosines are converted to the inverse tangent function using the following identities, as required.

SQR((1-Cos X)/Cos X)
SQR((Sin X)/(1-Sin X))

The equations that solve D and R can now be written in BASIC using the inbuilt ATN

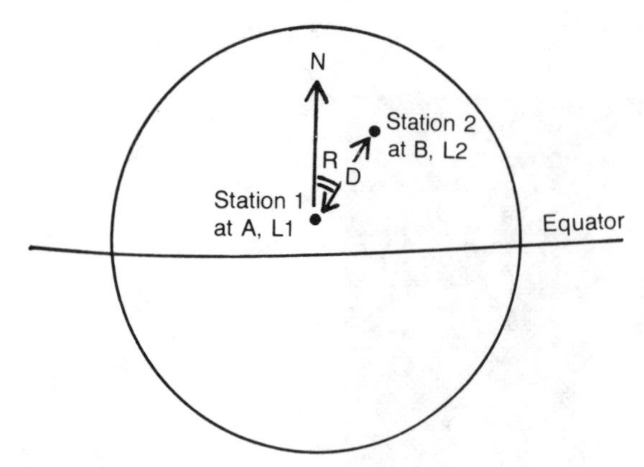

Fig. 6-1. Distances on the surface of the earth.

function to restore the angles to their conventional form.

An equation that converts latitude and longitude to direction and distance is not much use in itself because DX stations very rarely give their lattitude and longitude during a transmission. What is useful is a map, chart, or table that shows roughly where to point the antenna to optimize signals for a given contact. Most antennas have beam widths in excess of 30 degrees, so exact accuracy is not required. As long as the desired station is located within the main lobe of the antenna and not off the side or rear, all will be well.

A second use for such a table is VHF/UHF. Here a list can be kept of stations usually contacted and their antenna bearings. Any time a contact is established, the computer can advise the operator where to point the beam antenna, and, given the required hardware, can actually do the pointing.

This chapter discusses a set of programs that can be used to set up tables of directions and distances for pointing antennas and discusses the requirements for antenna-pointing hardware.

QTHLIB

A subroutine to perform direction and distance calculations is shown in Fig. 6-2. The subroutine begins at line 5000. It requires the values of latitude and longitude of both stations as parameters. Latitudes south of the equator are expressed as negative numbers, while longitudes are expressed in degrees EAST of greenwhich.

The radians to degrees conversion factor (P8) is defined (line 5010). The difference in longitude (L) between the stations is computed (line 5020). The value of L is then adjusted to fall within the range of ± 180 degrees (lines 5030 and 5040). The value of

```
5000 REM SUBROUTINE TO PERFORM CALCULATIONS
5005 REM Based on a routine used by WA3VQD and N3NN, 73 Magazine
     DEC 1978
5010 P8 = 3.14159 / 180 \ REM DEG -> RAD CONV FACTOR
5020 L = L2 - L1
5030 IF L < -180 THEN L = L + 360
5040 IF L >  180 THEN L = L - 360
5050 IF L < 0 THEN X = 0 ELSE X = 1
5060 REM CONVERT L AND B TO RADIANS
5070 A1 = A * P8
5080 B1 = B * P8
5090 L  = L * P8
5100 REM COMPUTE DISTANCE ANGLE
5110 D = COS(L)*COS(A1)*COS(B1) + SIN(A1)*SIN(B1)
5120 IF D > .99998 THEN D = 0 ELSE  D = ATN(SQRT(1-D*D)/D)
5130 D1 = D / P8
5140 REM   ENSURE POSITIVE VALUE
5150 IF D1 < 0 THEN D1 = D1 + 180
5160 REM COMPUTE DISTANCE
5170 D1 = INT(D1 * 60 * 1.1512 + .5)
5180 D2 =  INT (D1*1.6093 + .5)
5190 REM COMPUTE BEARING ANGLE
5200 IF D = 0 THEN R = 0 ELSE R = COS(B1) * SIN(L) / SIN(D)
5210 IF R = 0 THEN R1 = 0 ELSE  R1 = ATN(R/SQRT(1 - R * R))
5220 REM CONVERT TO DEGREES
5230 R2 = INT ( ( R1 / P8 ) + 0.5 )
5240 REM PUT IN RIGHT QUADRANT
5250 IF ABS(R) > .99998 THEN 5360
5260 IF ABS(R) < .00174 THEN 5380
5270 B2 = ( B + 0.1) * P8
5280 R3 = COS(L) * COS(A1) * COS (B2) + SIN(B2) * SIN(A1)
5290 IF R3 >.999998 THEN R4 = 0 ELSE R4 = ATN(SQRT(1 - R3 * R3) / R3)
5300 IF R4 = 0 THEN R6 = 90 * P8 ELSE R6 = COS(B2) * SIN(L) / SIN(R4)
5310 IF X=1 THEN 5340
5320 IF ABS(R6) > ABS(R) THEN R2=180+ABS(R2) ELSE R2=360-ABS(R2)
5330 GOTO 5420
5340 IF ABS(R6) < ABS(R) THEN R2 = ABS(R2) ELSE R2 = 180 - ABS(R2)
5350 GOTO 5420
5360 IF X=1 THEN R2 = 90 ELSE R2 = 270
5370 GOTO 5420
5380 IF ABS(L) > 178 THEN 5410
5390 IF B < A THEN R2 = 180 ELSE R2 = 0
5400 GOTO 5420
5410 IF B > A THEN R2 = 180 ELSE R2 = 0
5420 RETURN
```

Fig. 6-2. QTHLIB program.

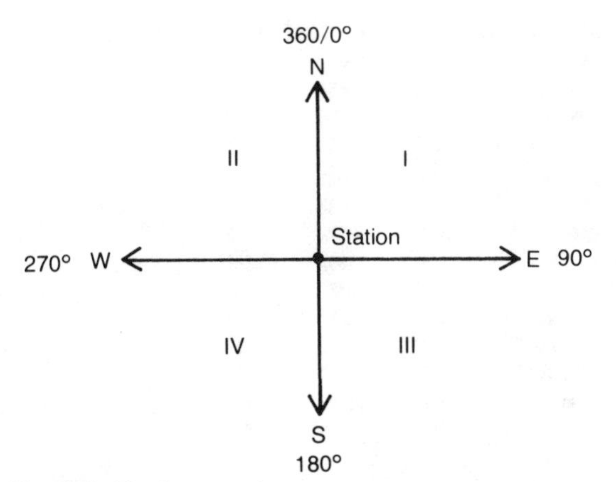

Fig. 6-3. The four quadrants.

the sign flag (x) is set (line 5050). The angle co-ordinates are converted from degrees to radians (lines 5070 to 5090).

The value of D is computed (line 5110). The inverse cosine function is figured to obtain the value of D using one of the trig identities described above (line 5120). If the angle is close to zero it is set to zero and the computation is not performed, thereby trapping a potential divide by zero error in the interpretor. The angle D is converted to degrees (line 5130) and set to a positive value (line 5150). The value of D is converted to statute miles and kilometers (lines 5170 and 5180).

The bearing angle (R2) is computed (lines 5200 and 5230) using the equation and trig identities given above. The value of the bearing at zero distance is detected before the computations take place to avoid having to perform operations on zero that could cause compiler errors (or interpreter errors). The remainder of the subroutine determines the quadrant and sets up the correct value of range bearing between 0 and 360 degrees. Note that the values are rounded off to the nearest integer by the INT(value + .5) function.

There are a number of different ways of determining the quadrant of the bearing. The four quadrants are shown in Fig. 6-3. Each quadrant contains 90 degrees. The sines and cosines of each quadrant have similar values (magnitudes) but different signs (positive or negative). The difference between the longitudes determines if the angle lies in quadrants I and II or in quadrants III and IV. For angles close to 0 or 90, (i.e., close to the boundaries of the quadrants) the following conditions are used, where 0 and 1 represent angles close to 0 or 1.

Sine	Angle
+0	0
+1	90
−0	180
−1	270.

For other values, the actual value is incremented by 0.1 and the difference in the value of the sine function is computed and tested, and the bearing placed in the correct quadrant depending on the result.

There are other techniques that can be used to determine the correct quadrant for the bearing but this one was handy, and it works well.

QTHTABLE

If a program is written to supply values of longitude and latitude to the subroutine just described, the distances and range bearing will be returned. Most radio amateurs can use such a program to print out a table of call areas, distances, and directions, such as that shown in Fig. 6-4. This listing was generated by the program shown in Fig. 6-5.

QTHTABLE reads the geographical co-ordinates stored in the STNDATA disk based file and generates the printout from a set of co-ordinates for different places stored in another disk file called QTHDATA in the following manner.

First, the program is identified (lines 10 to 20). The STNDATA file (covered earlier) is located and opened (lines 30 to 80). The same routine is used as in the logging package to search each of the floppy disk drives in turn

DIRECTION DISTANCE TABLE for G3ZCZ/4X LOCATED AT 31.8 35.2

Call Area	Miles	Kilometers	Bearing
ACAPULCO	7944	12784	312
ACCRA	2917	4694	239
ADDIS ABBABA	1587	2554	171
ALGIERS	1860	2993	289
ARCHANGEL	2271	3655	4
ATHENS	779	1254	307
ATLANTA	6439	10362	314
AUCKLAND	10120	16286	110
AUSTIN,TX	7154	11513	319
AZORES	3530	5681	296
B. ARIES	7611	12248	241
BAGHDAD	545	877	77
BAKER IS.	9403	15132	49
BERLIN	1799	2895	329
BERMUDA	5593	9001	303
BOGOTA	7165	11531	284
BOMBAY	2501	4025	102
BONN	1937	3117	321
BOSTON	5507	8862	314
BRASILIA	6221	10011	251
BRISBANE	8738	14062	103
BUDAPEST	1376	2214	327
CAIRO	259	417	244
CANARY IS.	2992	4815	279
CARACAS	6526	10502	285
CASABLANCA	2474	3981	284
CHRISTCHURCH	10002	16096	122
COLOMBO	3347	5386	111
CONCEPTION	8424	13557	244
COOK IS.	8856	14252	93
COSTA RICA	7498	12067	295
DENVER	6876	11066	330
DETROIT	5988	9636	319
FAIRBANKS	5759	9268	1
GILBERT IS	9050	14564	57
GUATEMALA	7614	12253	303
HAVANA	6834	10998	304
HELSINKI	2010	3235	350
HONG KONG	4856	7815	76
HONOLULU	8666	13946	15
ISTANBUL	722	1162	333
JAN MAYEN	3136	5047	341
JERUSALEM	0	0	0
JO,BURG	4034	6492	188
JUNEAU	6180	9945	355
KHARTOUM	1131	1820	189
KINGSTON	6772	10898	297
LAGOS	2688	4326	236
LENINGRAD	1953	3143	355
LIMA	7957	12805	271

DIRECTION DISTANCE TABLE for G3ZCZ/4X LOCATED AT 31.8 35.2

Call Area	Miles	Kilometers	Bearing
LISBON	2519	4054	294
LONDON	2228	3586	318
LOS ANGELES	7572	12186	337
LUANDA	3157	5081	211
MADRAS	3127	5032	104
MADRID	2232	3592	297
MALTA	1247	2007	289
MANILA	5451	8772	80
MARSHALLS	8281	13327	54
MELBOURNE	8523	13716	117
MEXICO CITY	7784	12527	313
MIAMI	6613	10642	305
MONTEVIDEO	7469	12020	241
MOSCOW	1658	2668	3
NAGASAKI	5311	8547	59
NAIROBI	2289	3684	177
NEW YORK	5685	9149	314
PARIS	2071	3333	314
PEKING	4423	7118	57
PERTH	6859	11038	122
PITCAIRN IS	11423	18383	294
PRAGUE	1654	2662	326
PT MORESBY	7117	11453	70
PUERTO RICO	6174	9936	292
QUITO	7605	12239	283
RANGOON	3927	6320	90
REYKJAVIK	3285	5287	330
RIYADH	859	1382	122
SAIGON	4744	7635	90
SAMOA	9297	14962	76
SAN FRANCISCO	7415	11933	342
SEATTLE	6774	10901	345
SEOUL	5012	8066	56
SINGAPORE	4931	7935	100
ST. JOHN'S	4547	7317	312
ST. LOUIS	6437	10359	320
SYDNEY	8779	14128	111
TAIPEI	4782	7696	68
TASHKENT	1983	3191	61
TEHRAN	967	1556	69
THULE	4287	6899	346
TOKYO	5686	9150	52
TORONTO	5787	9313	319
VANCOUVER	6679	10749	346
VLADIVOSTOCK	5061	8145	49
VOLGOGRAD	1270	2044	20
WARSAW	1586	2552	337
WASHINGTON	5896	9488	314
WINNEPEG	6082	9788	332

Fig. 6-4. Direction distance pointing data.

DIRECTION DISTANCE TABLE for G3ZCZ/W3 LOCATED AT 38.9 283

Call Area	Miles	Kilometers	Bearing
ACAPULCO	2053	3304	229
ACCRA	5262	8468	94
ADDIS ABBABA	7153	11511	66
ALGIERS	4225	6799	65
ARCHANGEL	4540	7306	25
ATHENS	5123	8244	54
ATLANTA	544	875	231
AUCKLAND	8624	13879	248
AUSTIN,TX	1319	2123	250
AZORES	2609	4199	75
B. ARIES	5219	8399	164
BAGHDAD	6193	9966	46
BAKER IS.	6716	10808	276
BERLIN	4168	6708	44
BERMUDA	822	1323	120
BOGOTA	2374	3820	175
BOMBAY	7991	12860	32
BONN	3975	6397	48
BOSTON	388	624	51
BRASILIA	4347	6996	145
BRISBANE	9476	15250	270
BUDAPEST	4557	7334	47
CAIRO	5806	9344	56
CANARY IS.	3556	5723	82
CARACAS	2052	3302	160
CASABLANCA	3797	6111	73
CHRISTCHURCH	8904	14329	241
COLOMBO	8939	14386	30
CONCEPTION	5235	8425	177
COOK IS.	8974	14442	274
COSTA RICA	2045	3291	194
DENVER	1492	2401	281
DETROIT	392	631	309
FAIRBANKS	3272	5266	327
GILBERT IS	7114	11449	281
GUATEMALA	1866	3003	210
HAVANA	1137	1830	198
HELSINKI	4308	6933	33
HONG KONG	8155	13124	347
HONOLULU	4827	7768	282
ISTANBUL	5216	8394	49
JAN MAYEN	3205	5158	26
JERUSALEM	5896	9488	52
JO,BURG	8099	13034	103
JUNEAU	2829	4553	318
KHARTOUM	6544	10531	66
KINGSTON	1444	2324	179
LAGOS	5425	8730	91
LENINGRAD	4472	7197	32
LIMA	3523	5670	180

DIRECTION DISTANCE TABLE for G3ZCZ/W3 LOCATED AT 38.9 283

Call Area	Miles	Kilometers	Bearing
LISBON	3564	5736	67
LONDON	3675	5914	49
LOS ANGELFS	2300	3701	275
LUANDA	6615	10646	97
MADRAS	8554	13766	27
MADRID	3780	6083	63
MALTA	4767	7672	60
MANILA	8566	13785	339
MARSHALLS	6975	11225	293
MELBOURNE	10177	16378	258
MEXICO CITY	1889	3040	231
MIAMI	925	1489	193
MONTEVIDEO	5202	8372	162
MOSCOW	4857	7816	33
NAGASAKI	7189	11569	337
NAIROBI	7540	12134	76
NEW YORK	210	338	53
PARIS	3825	6156	52
PEKING	6925	11144	350
PERTH	11571	18621	300
PITCAIRN IS	5590	8996	227
PRAGUE	4282	6891	47
PT MORESBY	8053	12960	310
PUERTO RICO	1556	2504	152
QUITO	2704	4352	182
RANGOON	8561	13777	8
REYKJAVIK	2802	4509	33
RIYADH	6736	10840	50
SAIGON	8992	14471	355
SAMOA	8009	12889	274
SAN FRANCISCO	2440	3927	283
SEATTLE	2321	3735	300
SEOUL	6940	11169	341
SINGAPORE	9656	15539	358
ST. JOHN'S	1352	2176	56
ST. LOUIS	715	1151	273
SYDNEY	9762	15710	262
TAIPEI	7643	12300	346
TASHKENT	6499	10459	25
TEHRAN	6320	10171	40
THULE	2676	4306	3
TOKYO	6771	10897	331
TORONTO	354	570	340
VANCOUVER	2351	3783	303
VLADIVOSTOCK	6496	10454	339
VOLGOGRAD	5401	8692	35
WARSAW	4455	7169	42
WASHINGTON	0	0	0
WINNEPEG	1247	2007	314

Fig. 6-4. Direction distance pointing data. (continued)

```
10 REM QTHTABLE
20 REM By Joe Kasser G3ZCZ June 1979
30 PRINT \ PRINT\ X$="STNDATA,"
40 FOR I=1 TO 4 \ I$=STR$(I) \ I$=I$(2,2)
50 IF FILE(X$+I$)=3 THEN EXIT 80
60 NEXT
70 PRINT"file STNDATA - station data file missing"\ GOTO 210
80 OPEN#0,X$+STR$(I)
90 READ#0,L$,C$,A,L1
100  REM L1 IS IN DEG (E)
110 PRINT#Z1 "DIRECTION DISTANCE TABLE for ",C$,"  ",
120 PRINT#Z1 "LOCATED AT ",A," ",L1
130 PRINT#Z1
140 OPEN#1,"QTHDATA"
150 READ#1,C$,B,L2
160 IFC$(1,1)= "*" THEN 210
170 GOSUB 5000
180 C$=C$+"          "
190 PRINT#Z1  C$,%10I,D1,%10I,D2,%10I,R2
200 GOTO 150
210 CLOSE #0 \ CLOSE #1
220 STOP
5000 REM SUBROUTINE TO PERFORM CALCULATIONS
```

Fig. 6-5. QTHTABLE program.

for the file. That data is read (line 90) for the station. The log data (L$) is then ignored. The value of longitude was stored as degrees EAST of Greenwhich in the STNDATA file by the STNINFO program (also covered earlier).

The heading for the table is printed (lines 110 to 130). Note the use of Z1 as the device 1 is the printer. Contrast this with Changing the value of Z1 changes the physical device that is assigned the output is routed to. For example, in Northstar BASIC, device 0 is usually the console while device 1 is the printed. Contrast this with the PRINT and LPRINT statements in other dialects of BASIC. In the logging package, the value of Z1 can be changed interactivelly; here it must be set by the user each time if a change is required. There is no reason why the routine cannot be modified by the user to do this. It has not been done here to keep the program flow simple, since the technique has already been described.

The QTHDATA file is then opened (line 140). The computing and printing is performed (lines 150 to 200). Each set of data is read from the disk (line 150). The data is tested to determine if the end of file has been reached (line 160). (The end of file here is signified by a "*" entry.) If it has, the program branches to line 210, if it has not, line 170 calls the subroutine that computes the distances and directions. The results are formatted and printed out (lines 180 and 190). The program continues (line 200), the disk files are closed (line 210), and the program stops (line 220).

Error-trapping and testing, and the return-to-system routines extensively used in the logging package have been ommitted in this chapter to allow the program flow to be readily observed. It is recommended that you add them.

QTHANY

QTHANY, shown in Fig. 6-6, allows a table to be printed out for any QTH. It is almost the same as QTHTABLE, except that the callsign, latitude, and longitude of

```
10 REM QTHANY
20 REM By Joe Kasser G3ZCZ June 1979
30 INPUT "What is your call sign ",C$
40 INPUT "What is your latitude ",A
50 IF A < 0 OR A > 90 THEN 40
60 INPUT "Are you north of the equator ",L$
70 IF L$ = "" THEN 60
80 IF L$(1,1) = "N" THEN A = - A
90 INPUT "What is your EAST longitude ",L1
100 IF L < 0 OR L > 360 THEN 90
110 PRINT#Z1 "DIRECTION DISTANCE TABLE for ",C$,"  ",
120 PRINT#Z1 "LOCATED AT ",A," ",L1
130 PRINT#Z1
140 OPEN#1,"QTHDATA"
150 READ#1,C$,B,L2
160 IFC$(1,1)= "*" THEN 210
170 GOSUB 5000
180 C$=C$+"          "
190 PRINT#Z1  C$,%10I,D1,%10I,D2,%10I,R2
200 GOTO 150
210 CLOSE #1
220 STOP
5000 REM SUBROUTINE TO PERFORM CALCULATIONS
```

Fig. 6-6. QTHANY program.

CITY	LATITUDE	LONGITUDE (E)
ACAPULCO	16.90 N	260.10
ACCRA	5.60 N	359.75
ADDIS ABBABA	9.05 N	38.70
ALGIERS	36.50 N	3.00
ARCHANGEL	64.50 N	40.70
ATHENS	38.00 N	23.70
ATLANTA	33.75 N	275.60
AUCKLAND	36.90 S	174.80
AUSTIN,TX	30.30 N	262.20
AZORES	38.50 N	332.00
B. ARIES	34.70 S	301.50
BAGHDAD	33.30 N	44.40
BAKER IS.	.23 N	183.50
BERLIN	52.50 N	13.40
BERMUDA	32.30 N	295.20
BOGOTA	4.63 N	285.90
BOMBAY	18.90 N	72.90
BONN	50.70 N	7.10
BOSTON	42.30 N	288.90
BRASILIA	16.20 S	315.50
BRISBANE	27.50 S	153.00
BUDAPEST	47.50 N	19.10
CAIRO	30.10 N	31.30
CANARY IS.	28.50 N	344.80
CARACAS	10.60 N	293.10
CASABLANCA	33.40 N	352.42
CHRISTCHURCH	43.50 S	172.70
COLOMBO	6.92 N	79.90
CONCEPTION	36.80 S	286.90
COOK IS.	21.00 S	158.00
COSTA RICA	9.98 N	275.90
DENVER	39.70 N	255.00
DETROIT	42.30 N	277.00
FAIRBANKS	64.80 N	212.20
GILBERT IS	.25 N	176.00
GUATEMALA	14.60 N	269.60
HAVANA	23.10 N	277.60
HELSINKI	60.10 N	25.00
HONG KONG	22.00 N	115.00
HONOLULU	21.30 N	202.20
ISTANBUL	41.00 N	29.00
JAN MAYEN	70.20 N	351.00
JERUSALEM	31.80 N	35.20
JO,BURG	26.20 S	28.00
JUNEAU	58.30 N	225.70
KHARTOUM	15.60 N	32.60
KINGSTON	18.00 N	283.20
LAGOS	6.50 N	3.50
LENINGRAD	59.90 N	30.40
LIMA	12.10 S	282.90
LISBON	38.70 N	350.90
LONDON	51.50 N	.17
LOS ANGELFS	34.00 N	241.70
LUANDA	9.83 S	13.30
MADRAS	13.10 N	80.30
MADRID	40.40 N	356.28
MALTA	36.00 N	14.00
MANILA	14.50 N	121.00
MARSHALLS	10.00 N	170.00
MELBOURNE	37.80 S	145.00
MEXICO CITY	19.40 N	260.80
MIAMI	25.80 N	279.70
MONTEVIDEO	33.90 S	303.80
MOSCOW	55.75 N	37.58
NAGASAKI	32.80 N	129.90
NAIROBI	1.30 S	36.80
NEW YORK	40.70 N	286.20
PARIS	48.90 N	2.30
PEKING	39.90 N	116.40
PERTH	32.00 S	115.80
PITCAIRN IS	25.10 S	230.00
PRAGUE	50.10 N	14.40
PT MORESBY	9.50 N	147.10
PUERTO RICO	18.50 N	294.00
QUITO	.23 S	281.50
RANGOON	16.80 N	96.20
REYKJAVIK	64.20 N	338.00
RIYADH	24.70 N	46.80
SAIGON	10.80 N	106.70
SAMOA	13.00 S	170.00
SAN FRANCISCO	37.80 N	237.50
SEATTLE	47.60 N	237.70
SEOUL	37.50 N	127.00
SINGAPORE	1.30 N	104.00
ST. JOHN'S	47.50 N	307.30
ST. LOUIS	38.70 N	269.70
SYDNEY	33.90 S	151.20
TAIPEI	29.10 N	117.60
TASHKENT	41.30 N	69.20

Fig. 6-7. QTH data.

CITY	LATITUDE	LONGITUDE (E)
VLADIVOSTOCK	43.00 N	131.80
VOLGOGRAD	48.80 N	44.50
WARSAW	52.30 N	21.00
WASHINGTON	38.90 N	283.00
WINNEPEG	49.90 N	262.80
TEHRAN	35.70 N	51.40
THULE	77.50 N	290.70
TOKYO	35.70 N	139.80
TORONTO	43.70 N	280.60
VANCOUVER	49.20 N	237.00

Fig. 6-7. QTH data. (continued)

the station for which the table is being prepared is entered from the keyboard (lines 30 to 100) rather than from the STNDATA disk file. Again, note the lack of error-checking in the user dialog.

THE QTHDATA FILE

The data stored in the QTHDATA file is used to generate the tables printed by the QTHTABLE and QTHANY program. It is stored in the following format: Call AREA, latitude, longitude. (See Fig. 6-7.) Latitudes south of the equator are stored as negative numbers, while longitudes are stored in degrees EAST of Greenwhich. Longitudes can either be entered as values in the range 0 to 360 or as values between −180 and +180 degrees. The use of a disk file for storing the QTH data does away with the requirements for storing the data in DATA statements in the main program. It allows a number of programs to access the same data, and only one copy of the data need be kept. Other data files, such as callsign for vhf/uhf or call prefix for the world, can also be set up, and the main program operated on them by changing the name of the QTHdata file being worked on at the time. (Change the name of the file in the OPEN statement.) Thus, different tables can be generated for different applications.

The last data element is an "*" stored in the city/callsign position. It serves as the end of file mark for the other programs.

A second data file and various direction distance tables for about 400 different callsign prefixes are presented in Appendix B.

QTHGEN

QTHGEN, listed in Fig. 6-8, is used to generate a QTHDATA file. The overhead statements appear (lines 10 to 20). The QTHDATA file is opened for reading and writing (lines 30 and 40). The table heading is printed (line 50). Lines 60 to 120 form a loop that reads any existing data, displays it in a formatted manner, and advances the read and write file pointers to point to the next position for the new data.

The entry is read and the read pointer advanced (line 60). If the callsign or location element (C$) is a "*" then the loop terminates using the branch to line 130 (line 70). The write pointer is advanced past the entry currently being read (line 80). The location and geographical coordinates of the entry are printed (lines 90 to 100). Negative values of latitude are displayed as degrees South of

```
10 REM QTHGEN
20 REM by Joe Kasser  G3ZC7 1979
30 OPEN #1,'QTHDATA'
40 OPEN #0,'QTHDATA'
50 PRINT#71 '  CITY           LATITUDE  LONGITUDE (E)'
60 READ#1,C$,A,L1
70 IF C$(1,1) =  '*' THEN 130
80 READ#0,C$,A,L1
90 C$=C$+'              '
100 PRINT#Z1 C$,%10F2,ABS(A),\IF A<0 THEN PRINT#Z1, ' S', ELSE PRINT#
    Z1, ' N',
110 IF L1 < 0 THEN PRINT#Z1 %10F2, 360 + L1 ELSE PRINT#Z1 %10F2,L1
120 GOTO 60
130 INPUT ' LOCATION ',C$
140 IF C$='*' THEN 220
150 INPUT 'LAT ',A
160 INPUT 'LONG E ',L1
170 PRINT C$,%10F2,A,%10F2,L1
180 INPUT 'OK ',A$
190 IF A$(1,1) = 'N' THEN 130
200 WRITE #0,C$,A,L1
210 GOTO 130
220 WRITE #0,C$,A,L1
230 CLOSE #1
240 CLOSE #0
```

Fig. 6-8. QTHGEN program.

the equator, positive values as degrees North. The longitude is displayed in the range of 0 to 360 degrees (line 110). The loop continues to the next entry (line 120).

The new data is accepted and verified (lines 130 to 190). The user is asked to verify that the entry is correct (line 180) before it is written to the disk file (line 200). If the "*" input is entered in response to the "LOCATION" question (line 140), it is used to end the entry sequence and terminate the loop. Line 210 causes the program to branch back to line 130 and continue the loop.

Following the termination of the loop, the "end of file" or "*" entry is written to the disk (line 220) and the disk files are closed (lines 230 to 240).

Lines 50 to 120 should be bypassed or deleted the first time that the program is run because the file created by the Northstar Disk Operating System (DOS) is blank and does not contain an "*" entry in the first position. Also, the data is added to an existing file. If a mistake is made, both the old and new data can be damaged. Lastly, the program contains a minimal amount of error-checking. Error-checking should be added to any version of the program that may be installed on your system.

QTHSORT

Once a list of locations or callsigns is entered into a QTHDATA file or its equivalent, the data may or may not be in alphanumerical order, especially if later additions are made. QTHSORT is provided to sort a QTHDATA file into alphanumerical order. The program is listed in Fig. 6-9, and contains the following features.

The program is identified as usual (lines 10 to 20). The matrix which will store the data from the QTHDATA file is set up (lines 30 to 40). The number of entries in the file (N) is set to 100 in this version. It can be adjusted to any value provided that the computer contains enough memory to hold all the data.

```
10 REM QTHSORT
20 REM by Joe Kasser  G3ZCZ 1979
30 N = 100
40 DIM C1$((N*10)+10),L2(N),A2(N)
50 F$ = 'QTHDATA' \  OPEN#1,F$
60 PRINT#Z1 '  CITY          LATITUDE  LONGITUDE (E)'
70 FOR I =  1 TO N
80 READ#1,C$,A,L1
90 C1$(I*10,(I*10)+10)=C$
100 L2(I)=L1
110 A2(I)=A
120 IF C$(1,1) =  '*' THEN EXIT 170
130 C$=C$+'          '
140 PRINT#Z1 C$,%10F2,ABS(A),\IF A<0 THEN PRINT#Z1, ' S', ELSE PRINT#
    Z1, ' N',
150 IF L1 < 0 THEN PRINT#Z1 %10F2, 360 + L1 ELSE PRINT#Z1 %10F2,L1
160 NEXT
170 I = I - 2  \ REM IGNORE * ENTRY
180 X = 0
190 FOR J = 1 TO I
200 J$ = C1$(J*10,(J*10)+10)
210 K$ = C1$((J+1)*10,((J+1)*10)+10)
220 IF J$ =< K$ THEN 300
230 I$=K$
240       C1$((J+1)*10,((J+1)*10)+10) = J$
250 A5 = A2(J+1) \ A2(J+1) = A2(J) \ A2(J) = A5
260 A5 = L2(J+1) \ L2(J+1) = L2(J) \ L2(J) = A5
270       C1$(J*10,(J*10)+10) = I$
280 PRINT K$,J$
290 X = 1
300 NEXT
310 IF X = 1 THEN 180
320 OPEN #2,F$
330 FOR J = 1 TO I + 1
346 PRINT#Z1, C1$(J*10,(J*10)+9),%10F2,A2(J),%10F2,L2(J)
350 WRITE#2, C1$(J*10,(J*10)+10),A2(J),L2(J)
360 NEXT
370 WRITE#2,'*',0,0
380 CLOSE#2 \ CLOSE#1
```

Fig. 6-9. QTHSORT program.

The name of the QTHDATA file is set up as F$ (line 50), and then opened. This technique allows other files to be operated upon by changing the name assigned to F$ either through the use of an i/o dialog, or by modifying the program statement in line 50. Once the change is made in line 50, it holds for all references to F$ later in the program.

The heading for the data to be read from the file is displayed (line 60), and the loop that reads data from the file into the memory array is started (line 70). The data is read (line 80), and stored in the arrays (lines 90 to 110). The end of file entry is detected and the loop is terminated (line 120). The individual entries are formatted and displayed (lines

130 to 150). The loop then continues (line 160).

The sorting routine is a modified bubble sort, slow but simple. It begins in line 180 in which a flag (X) is reset. This flag is used to signify that a bubble in the array was moved. Two consecutive elements in the matrix are pointed to (lines 200 to 210) and compared (line 220). If the secondary entry has a lower alphanumerical value than the first, a move takes place and the positions of the data in the two entries are swapped using the intermediate variable I$ and A$ (lines 230 to 270). If the first entry in the matrix has a lower value than the second a move does not take place, and the program branches to line 300 to test the next two entries (i.e., the second entry now becomes the first, and the next in line becomes the second). To reassure the operator that the system is still working, the program displays which bubble was moved (line 280) and the value of X is set (line 290) showing that something bubbled up towards the top of the data. The rest of the entries are tested when the NEXT instruction (line 300) causes the loop to execute another time.

The value of X is then tested (line 310). If it is set, then something bubbled during the last iteration through the loop and the program branches back to line 180 to test the matrix of data again to see if something else can bubble up. If X is not set, then nothing bubbled up during the last iteration and the data is assumed to be sorted. The data file is opened (line 320) and the sorted data is written back to the disk file (lines 330 to 360) and displayed. The end of file entry is written on the disk (line 370) after which the data files are closed (line 380). There is no error-checking.

ANTENNA-POINTING

An antenna-pointing program consists of two parts. The first part computes the direction in which the antenna is to be positioned, and the second part actually does the pointing.

```
10 REM QTHPOINT
20 REM By Joe Kasser G3ZCZ June 1979
30  X$="STNDATA,"
40 FOR I=1 TO 4 \ I$=STR$(I) \ I$=I$(2,2)
50 IF FILE(X$+I$)=3 THEN EXIT 80
60 NEXT
70 PRINT"file STNDATA - station data file missing"\ GOTO 210
80 OPEN#0,X$+STR$(I)
82 INPUT "What call area ", W$
90 READ#0,L$,C$,A,L1
100  REM L1 IS IN DEG (E)
140 OPEN#1,"QTHDATA"
150 READ#1,C$,B,L2
160 IFC$(1,1)= "*" THEN 206
162 IF C$(1,LEN(W$)) <> W$ THEN 150
170 GOSUB 5000
180 C$=C$+"          "
190 PRINT#Z1  C$,%10I,D1,%10I,D2,%10I,R2
202 REM, CALL POINTING SUBROUTINE HERE
204 GOTO 210
206 PRINT W$, " is not in the data file "
210 CLOSE #0 \ CLOSE #1
220 STOP
5000 REM SUBROUTINE TO PERFORM CALCULATIONS
```

Fig. 6-10. QTHPOINT program.

The computation part of the program is a modified version of QTHTABLE known as QTHPOINT, and is listed in Fig. 6-10. The modifications are in the following lines; Line 82 asks the user to decide which call area the antenna is to be aimed at (W$). Lines 150, 160, and 162 then read each entry in the disk file and compare the desired entry to the one just read. Line 206 notifies the user that the entry was not found (i.e., the end of file entry has been read before the desired entry was found). If line 162 finds a match, the desired pointing data is computed by the subroutine beginning in line 5000. Line 202 can be used to call the subroutine that positions the antenna.

The actual antenna-pointing subroutine is hardware-dependant. The interface requires an analog to digital converter to input the actual position of the antenna to the computer and digital outputs to control the motor relays. An example of an S-100 card that is ideal for antenna control was described in *Microcomputers in Amateur*

Radio (TAB Books #1305, Volume 1 in this series). A flow chart for a typical antenna positioning subroutine is shown in Fig. 6-11. The approach is to send a move signal to the antenna control hardware. The position of the antenna is then read and compared to the desired position. If they are the same, the move signal is removed. If they are different, the current position is compared to the difference between the actual and desired positions of the antenna as they were an instant before. If the difference has increased, the antenna is moving the wrong way and the direction component of the move signal is changed.

EDITING THE QTHDATA

When entering data into a QTHDATA file when using the QTHGEN program, errors can be made that are only noticed later when the data is displayed. These errors can be corrected by appropriate software such as the QTHFUDGE program listed in Fig. 6-12. The editing process edits data from one file to a second so that an error in the programming of the correction does not wipe out all the data. The usual error is to use an = instead of a <> so that all the entries instead of just the desired on are fudged, a situation that can be very frustrating if the old data has been destroyed by the editing process.

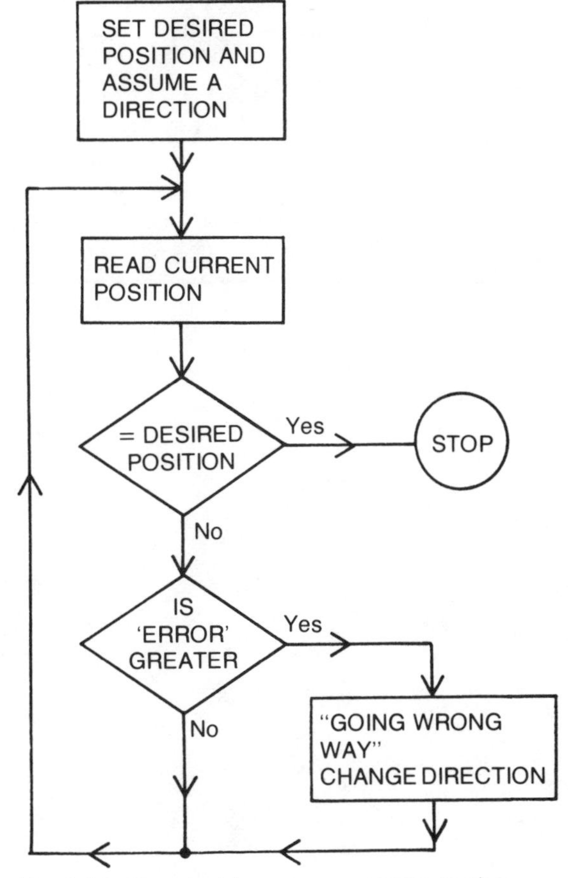

Fig. 6-11. Flowchart for antenna-pointing routine.

```
10 OPEN #1,'QTHDATA'
20 OPEN #0,'QTHNEW'
25 PRINT#7 '  CITY            LATITUDE  LONGITUDE (E)'
30 READ#1,C$,A,l1
40 IF C$(1,1) =  '*' THEN 190
51 IF L1=84.1 THEN L1=-84.1
52 IF L1=-29 THEN L1=29
60 C$=C$+'            '
65 IF C$(1,8)='NEW YOTK' THEN C$='NEW YORK'
72 PRINT C$,%10F2,ABS(A),\IF A<0 THEN PRINT ' S', ELSE PRINT ' N',
75 PRINT %10F2,L1,
80 IF L1 < 0 THEN PRINT %10F2, 360 + L1 ELSE PRINT %10F2,L1
170 WRITE #0,C$,A,L1
180 GOTO 30
190 WRITE #0,C$,A,L1
200 CLOSE #1
```

Fig. 6-12. QTHFUDGE program.

The program works in the following manner. The existing data file is opened (line 10), as is the new file, which will contain the adjusted data (line 20). A heading is displayed (line 25). The entry is read (line 30), and tested to see if it is the end of file entry (line 40). If it is, the program branches to line 190 to terminate the process; if not, the editing or fudging can take place (lines 51 to 65).

The actual editing is customised depending on which errors have to be corrected. In the example of Fig. 6-12, the sign of the value of the longitude (L1) associated with a call area/city is changed (lines 51-52), and spelling errors are corrected (lines 60 to 65).

The entries read from the disk file by

line 30 are displayed (lines 72 to 80). An entry that has undergone a correction will be displayed in its new values at this time. The data is written into the new file (line 170) and the program branches back to process the next entry (line 180). The program is closed out by writing the end of file entry into the new data file (lines 190 to 200). The contents of the new file can then be examined using QTHGEN to read and display its contents.

GENERATING GRAPHICS

The set of QTH programs presented in this chapter has been used to generate tabular listings. If a graphic output device such as a plotter is available, the output routines can be modified to plot points. If a large number of qth co-ordinates are available, maps can be plotted. For example, if the coordinates of the coastlines of the continents and the frontiers between countries are available, maps of the world can be plotted with any specified location at the center. These maps are easier to use than the tabular form used in this chapter, and are also ideal for plotting the tracks of the OSCAR satellites.

Satellites

The use of microcomputer to track satellites and process their telemetry is a natural. Keeping track of the position of a spacecraft in its orbit calls for a lot of number crunching. Processing the telemetry calls for the use of basic algebra at first, and can then lead to intense data analysis techniques.

The Radio Amateur Satellite Program, which celebrated its 20th anniversary in 1981, began with the launch of OSCAR 1 in December 1961. In the ensuing years, the satellites that followed demonstrated that radio amateurs could build, launch, and control spacecraft in orbit. Amateurs also pioneered new uses of spacecraft much as in the same manner they had pioneered the use of short waves fifty years previously. The early Phase 1 satellites were followed by the Phase 2 spacecraft that provided limited communications capabilities for years at a time, and at the time of writing, the Radio Amateur Satellite Corporation (AMSAT) is actively working towards the launch of the second Phase 3 satellite, the first having unfortunately been lost when the launch vehicle malfunctioned and had to be destroyed seconds after lift off.

SATELLITE TRACKING

Keeping track of the position of satellites is a monotonous, repetitive process. Satellites orbiting the Earth travel in predictable paths. This chapter describes two programs that can be used to keep track of satellites in low earth circular orbits. In order to best apply the computer to the OSCAR program, a little knowledge of the terminology of orbital physics is desirable.

Consider Fig. 7-1. The orbit of the satellite is shown edge on, or from the side as viewed by an observer out in space. As the

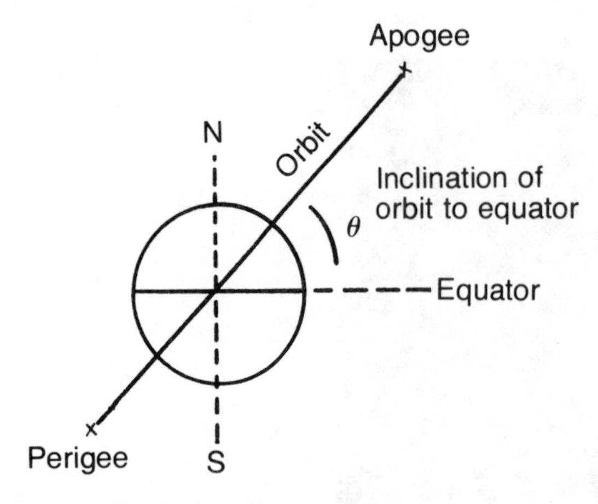

Fig. 7-1. The basic orbit.

spacecraft travels around the Earth on its orbital path it passes through a high point (apogee) and a low point (perigee). The angle between the plane of the orbit and the equator is known as the Inclination of the orbit (to the equator). Phase 2 satellites travel in orbits that have very small differences between their apogees and perigees; namely, they travel in almost circular orbits.

As the satellite travels along its orbit it returns to the same place over and over again. The time that the spacecraft takes to travel once around its orbit is called the period of the orbit. However, while the satellite is busy travelling along its orbit, the Earth is not standing idly by; it too is rotating along its own axis. (One revolution takes about 24 hours.) This means that if on one orbit the satellite crosses the equator due south of London on the 0 meridian, the next time that the satellite crosses the equator going in the same direction (one period later), the satellite will not be crossing the 0 meridian but will be at some longitude west of London since the Earth will have moved on its axis in that time.

This is shown graphically in Fig. 7-2 in which a section of the ground track of two successive orbits is shown. The ground track is the points on the surface of the Earth that the satellite passes over, ie the points on the Earth that are vertically below the satellite. In the figure it can be seen that the ground tracks at the equator show up as parallel lines separated by a fixed distance. This distance (measured in degrees longitude at the equator) is the longitude increment of the orbit, or the amount that the Earth has rotated due to its own motion while the satellite has travelled one orbit.

Conventional amateur satellite tracking information tags one point in the orbit. This position is the location where the satellite passes over the equator travelling from south to north. This is the point that is used to define the start of an orbit. It is known as the Ascending node since the satellite is presumed to be travelling up to the North Pole. The similar point on the other side of the globe where the spacecraft crosses the equator travelling towards the south is called the Descending node. The first orbit of the day is known as the Reference orbit (for the day) and data is published providing the time and location of the ascending node for the reference orbit. If the period of the orbit and the equatorial crossing increment are known, keeping track of the position of the satellite is merely a matter of repetitive additions.

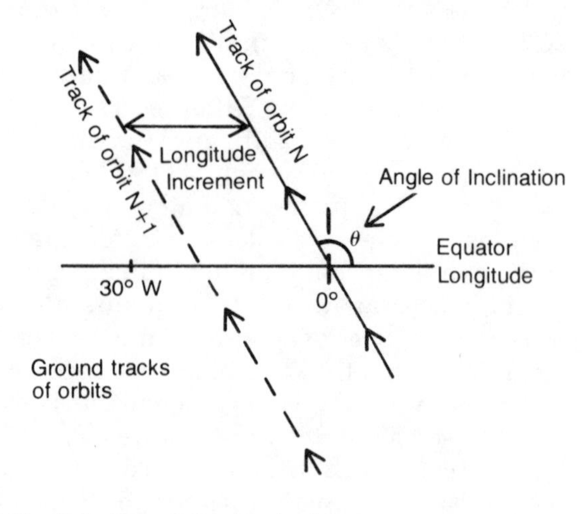

Fig. 7-2. Ground tracks of orbits.

The basic math involved is as follows. Starting with an orbit (N) crossing the equator (Ascending node assumed from now on unless otherwise stated) over longitude (L) at a time (T), then, orbit N + 1 crosses the equator at a time T + P, where P is the period of the orbit. The longitude at which orbit N + 1 crosses the equator is L + I, where I is the longitude increment per orbit. The position of the ascending node of successive orbits can be determined just by the use of simple addition.

A circle has 360 degrees, so when the value of the location of the crossing position becomes equal to or greater than 360, it should be converted to lie between 0 and 360. This can be expressed mathematically as

IF L >= 360 THEN L = L − 360

Similar techniques are used to keep the time and day information within the conventional 24 hours/day, 60 minutes/hour display.

A BASIC ORBIT PREDICTION PROGRAM

A basic satellite orbit prediction program is shown in Fig. 7-3. All so-called satellite tracking programs really predict the position of the spacecraft based on the parameters of the orbit; they do not actually track the position of the spacecraft.

The program begins with the usual overhead (lines 10 to 30). The arrays used in the program are set up (lines 40 to 50). The subroutines used by the rest of the program are next (lines 70 to 370). It is a little-known fact that most BASIC interpreters search the whole program starting from the top when searching for line number references in GOTO or GOSUB statements; thus placing the most commonly used subroutines near the start of a program speeds up the execution time of the program, because the search time for those line numbers is reduced.

The opening data is printed (lines 380 to 490). The CHR$(12) character (line 380) is the "clear screen and home cursor" character for many video displays. It may have to be changed for other CRT terminals. The user is then presented with a menu choice of nine different spacecraft. (The program can be expanded to include data for others.) Any other satellite in a Phase 2 type orbit (such as weather satellites) can be added as shown in the listing. The user is requested to choose one of the available spacecraft (line 500). Line 510 branches to lines 520–600 depending on which satellite was chosen so as to set the data pointer to the correct DATA statement for the desired satellite. The data is read (line 620), setting up the following variables:

I$ = the name of the spacecraft.
N0 = the number of the reference data orbit
D9 = the Julian date of the reference data orbit
H9 = the time (in hours) of the reference data orbit
M9 = the time (in minutes) of the reference data orbit
S9 = the time (in seconds) of the reference data orbit
W = the longitude of the Equator crossing point of the reference data orbit (in Degrees West of Greenwich)
P = the period of the orbit (in minutes
W2 = the longitude increment per orbit

The data is stored in lines 870 to 1160. The reference data orbit is the orbit for which the time/position data are stored in the program. It should not be confused with a reference orbit which was previously defined as the first orbit of the day. The Julian date is the number of the day of the year, where the 1st of January is Day 1 and the 31st of December is Day number 365 (or 366 in a leap year).

```
10 REM PHASE1 Version 5.0
20 REM Simple Orbit Tracking Program
30 REM By Joe Kasser G3ZCZ
40 DIM I$(32),M9$(36)
50 M9$ = 'JanFebMarAprMayJunJunAugSepOctNovDec'
60 GOTO 380
70 L9 = L9 - 1 \ IF L9 > 0 THEN 80 ELSE GOSUB 320 \ L9 = 60
80 D8 = D9
90 RESTORE 220 \ GOSUB 230
100 PRINT#P5,%5I,N9,%7I,D6,M8$,%10F4,T9,%10F2,W
110 RETURN
120 T9 = INT (S9 + M9*100 + H9*10000) / 1000000
130 RETURN
140 W = W + W2 \ IF W > 360 THEN W = W - 360
150 S9 = S9 + (P - INT(P9))/100*60
160 IF S9 < 60 THEN 170 ELSE S9 = S9 - 60 \ M9 = M9 + 1 \ GOTO 160
170 M9 = M9 + INT(P)
180 IF M9 < 60 THEN 190 ELSE M9 = M9 - 60 \ H9 = H9 + 1 \ GOTO 180
190 IF H9 < 24 THEN 200 ELSE H9 = H9 - 24 \ D9 = D9 + 1 \ GOTO 190
200 IF D9 > 365 THEN D9 = D9 - 365 \ REM NOT LEAP YEAR
210 RETURN
220 DATA 0,31,59,90,120,151,181,212,243,273,304,334
230 FOR I = 0 TO 11
240 READ Z1
250 IF Z1 >= D9 THEN EXIT 280
260 Z2 = Z1
270 NEXT
280 D6 = D9 - Z2
290 M8$ = ' '+M9$((I-1)*3+1,((I-1)*3)+3)
300 RETURN
310 REM
320 PRINT#P5
330 PRINT#P5, 'Equatorial crossing times for the ',I$,' spacecraft'
340 PRINT#P5,M$,' Orbits'
350 PRINT#P5
360 PRINT#P5,'ORBIT      DATE       TIME    XSING (W)'
370 RETURN
380 PRINTCHR$(12),'AMATEUR SATELLITE ORBIT PROGRAM'
390 PRINT
400 PRINT 'Satellite selection menu '
410 PRINT 'AMSAT - OSCAR 7         0'
420 PRINT 'AMSAT - OSCAR 8         1'
430 PRINT 'UoSAT - OSCAR 9         2'
440 PRINT 'RS 3                    3'
450 PRINT 'RS 4                    4'
460 PRINT 'RS 5                    5'
470 PRINT 'RS 6                    6'
480 PRINT 'RS 7                    7'
490 PRINT 'RS 8                    8'
500 INPUT 'Which satellite ? ', S \ IF S<0 OR S >8 THEN 400
510 ON S+1 GOTO 520,530,540,550,560,570,580,590,600
520 RESTORE 900 \ GOTO 620
530 RESTORE 930 \ GOTO 620
540 RESTORE 960 \ GOTO 620
550 RESTORE 990 \ GOTO 620
560 RESTORE 1020 \ GOTO 620
570 RESTORE 1050 \ GOTO 620
580 RESTORE 1080 \ GOTO 620
590 RESTORE 1110 \ GOTO 620
600 RESTORE 1140 \ GOTO 620
610 REM OTHER S/C CAN BE ADDED HERE
620 READ I$,N0,D9,H9,M9,S9,W,P,W2
630 RESTORE 220 \ GOSUB 230
640 PRINT\ PRINT I$,' Reference orbit is',N0,' on',D6,M8$,
650 GOSUB 120
660 PRINT ' at',%6F4,T9,
670 PRINT  ' crossing at',W,' W'
680 INPUT 'What is the first orbit ? ',N1  \ IF N1 >= N0 THEN 700
690 PRINT' Reference orbit is',N0 \ GOTO 680
700 INPUT 'What is the last orbit to be printed ? ',N2
710 IF N2<N1 THEN N2=N1
720 INPUT 'All orbits or just Reference orbits ? ',A$ \ IF A$='' THEN
    720
730 IFA$(1,1)='A'THENM$='all'ELSEIFA$(1,1)='R'THENM$='reference'ELSE
    720
740 INPUT 'Which output device (0-7) ? ',P5
750 IF N1 = N0 THEN 770
760 FOR N9 = N0 TO N1-1 \ GOSUB 140 \ NEXT
770 N9 = N1
780 GOSUB 120
790 GOSUB 70
800 REM MAIN LOOP
810 FOR N9 = N1+1 TO N2
820 GOSUB 140
830 GOSUB 120
840 IF A$(1,1) = 'R' THEN IF D8 = D9 THEN 860
850 GOSUB 70
860 NEXT
870 REM FORMAT OF DATA = SPACECRAFT NAME
880 REM ORBIT (#), DAY , EQR XSING TIME (HH,MM,SS), LONG ( DEG W)
890 REM PERIOD (MIN), LONGITUDE INCREMENT / ORBIT
900 DATA 'AMSAT OSCAR 7'
910 DATA 16804,200,00,36,52,67.5
920 DATA 114.945199,28.736297
930 DATA 'AMSAT OSCAR 8'
940 DATA 19370,357,01,08,10,83.77
950 DATA 103.179480,25.79703
960 DATA 'UoSAT OSCAR 9'
970 DATA 01171,357,00,16,58,138.2
980 DATA 95.31353,23.82.897
990 DATA ' R S 3 '
1000 DATA 0   ,018,01,05,60,231.9
1010 DATA 118.46293,29.756498
1020 DATA ' R S - 4'
1030 DATA 0   ,018,00,43,23,225.9
1040 DATA 119.33912,29.97578
1050 DATA 'R S - 5 '
1060 DATA 0   ,018,01,43,59,241.9
1070 DATA 119.49842,30.01562
1080 DATA ' R S - 6 '
1090 DATA 0   ,018,00,23,17,221.2
1100 DATA 118.66154,29.80622
1110 DATA ' R S - 7 '
1120 DATA 0   ,018,01,26,59,231.1
1130 DATA 119.14014,29.92596
1140 DATA ' R S - 8 '
1150 DATA 0   ,018,01,04,18,231.1
1160 DATA 119.70890,30.06830
```

Fig. 7-3. Simple phase 2 orbit-tracking program.

The data pointer is restored to line 220 (line 360) and the subroutine at line 230 which converts the Julian day number to day and month format is called. The Julian date equivalent of the first day of each month is set up (line 220), and in a leap year all values following the 31 have to be incremented. The program determines in which month the Julian day falls (lines 230 to 270), the actual day in the month (line 280), and M8$, which consists of the three letters for the month extracted from the year/month string (M9$), is created. This technique was also used in the logging package. The subroutine terminates in line 300.

The reference orbit data is displayed (line 640). Then, the subroutine which begins at line 120 is called. This subroutine formats the time in hours and minutes into a four digit number (T9) which is then displayed. The equatorial crossing location is also displayed (line 670). All the above data appears on a single line. The user is then asked to select the first orbit to be printed (N1). A test is performed to determine if the reference data orbit is valid for the desired starting orbit, i.e., if the desired orbit has a higher or equal number than that of the reference data orbit (N1 < N0). If it is, the program continues (line 700). If not, the user is reminded which reference data orbit is stored in the program (line 690), and the user is then prompted once again by the branch back to line 680. The user is asked to enter the desired last orbit for the printout (N2) (line 700). The value of that orbit is checked to make sure that it is always greater or equal to the number of the initial orbit (line 710).

The user is asked to set the "all orbit" or "reference orbit" flag, and the response is tested to ensure that it is at least one character long. The response is validated (line 730). It has to begin with an "A" (all) or an "R" (reference). If it does not begin with either of these letters, the question is asked again. M$ is then set up to identify the listing. Finally, the user is asked which output device should be used.

The desired starting orbit (N1) is then compared to the reference data orbit (N0). If they are equal, the program flow branches to line 770, if they are not, the orbits are advanced up to one orbit before the desired one by using the subroutine beginning in line 140 (line 760).

The orbit longitude increment (W2) is added to the value of the equator crossing location on the previous orbit, and then tests the result to see if it is greater than 360 (line 140). If it is greater than 360, it is converted to be less than 360. The time update (line 150) by updating the seconds by adding to them the fraction of the minute part of the period. A test is then performed to determine if the result is greater than 60. If it is, 60 is then subtracted from the number of seconds and the value of the minutes is incremented; the test is then repeated until the number of the seconds is less than 60. Similar operations occur on the values of the minutes and hours associated with the new equator crossing time (lines 180 to 190). So that the updated values of hours and minutes lie between the limits of 0 to 24 and 0 to 60 respectively. The end of the year is checked (line 200). (In a leap year, this will have to be modified.) The subroutine terminates (line 210) following the computation of the new time.

The main loop counter (N9) is set up (line 770). The time of the first orbit to be displayed is set up (line 780) by calling the subroutine beginning at line 120 for the print routine (which is itself called by line 790).

The print subroutine (beginning with line 70) first decrements the line counter (L9). If the line counter has counted down to zero (or is zero the first time around), the subroutine which starts at line 320 is called to print out the page heading. (This subroutine consists of the print statements in lines 320 to 360.) The value of the line counter is then reset to 60 after the return from the subroutine. Note that as this version, and many other versions, of BASIC automatically initializes the value of all variables to zero, the first line printed out will always be the page heading. The last day printed flag (D8) is set (line 80) equal to the current day being printed (D9). The program goes to the subroutine starting at line 200 to convert the Julian day number to day and month data (line 90) so that line 100 can print the orbit data in a formatted manner. The subroutine then terminates in the next line.

Now that the orbital data for the fist orbit requested has been printed out, the main program loop can begin (line 800). The DO LOOP is set up (line 810) and the program enters a subroutine starting at line 140 to update the time and equator crossing location by one orbit (line 820). Next, time information is formatted by a subroutine starting at line 120 (line 830). If all the orbits are to be printed, the test (line 840) fails at once and the program goes on to print the data (line 850). If only the reference orbits are to be printed, and the day associated with the last orbit printed (D8) is the same as the day associated with the current orbit, the current orbit is obviously not a reference orbit and so it is not printed. The main loop terminates (line 860) when data for all the orbits within the original limits have been computed.

The reference orbit data associated with each spacecraft are stored in lines 870 to 1160. This data is available from AMSAT and is published in most amateur radio magazines in one form or another. An

	-- AMSAT-OSCAR 7 --			-- AMSAT-OSCAR 8 --			
U.T.C. DATE	TIME (HMM:SS)	DEG.	ORBIT NUMBER	TIME (HMM:SS)	DEG.	ORBIT NUMBER	SCHED. MODE
WED. 1 APR. (91)	31:40	85.5	29169	117:15	78.4	15658	X
THU. 2 APR. (92)	125:54	99.1	29182	121:58	79.6	15672	A
FRI. 3 APR. (93)	25:12	83.9	29194	126:41	80.8	15686	A+J
SAT. 4 APR. (94)	119:27	97.5	29207	131:25	82.0	15700	J
SUN. 5 APR. (95)	18:45	82.4	29219	136:08	83.2	15714	J
MON. 6 APR. (96)	112:59	96.0	29232	140:51	84.4	15728	A
TUE. 7 APR. (97)	12:17	80.8	29244	2:22	59.8	15741	A+J
WED. 8 APR. (98)	106:32	94.4	29257	7:05	61.0	15755	X
THU. 9 APR. (99)	5:50	79.2	29269	11:48	62.2	15769	A
FRI. 10 APR. (100)	100:04	92.8	29282	16:31	63.5	15783	A+J
SAT. 11 APR. (101)	154:19	106.4	29295	21:14	64.7	15797	J
SUN. 12 APR. (102)	53:37	91.3	29307	25:56	65.9	15811	J
MON. 13 APR. (103)	147:51	104.8	29320	30:39	67.1	15825	A
TUE. 14 APR. (104)	47:09	89.7	29332	35:22	68.3	15839	A+J
WED. 15 APR. (105)	141:24	103.3	29345	40:05	69.5	15853	X
THU. 16 APR. (106)	40:42	88.1	29357	44:47	70.7	15867	A
FRI. 17 APR. (107)	134:56	101.7	29370	49:30	71.9	15881	A+J
SAT. 18 APR. (108)	34:14	86.6	29382	54:12	73.1	15895	J
SUN. 19 APR. (109)	128:29	100.1	29395	58:55	74.3	15909	J
MON. 20 APR. (110)	27:47	85.0	29407	103:38	75.5	15923	A
TUE. 21 APR. (111)	122:01	98.6	29420	108:20	76.7	15937	A+J
WED. 22 APR. (112)	21:19	83.4	29432	113:02	77.9	15951	X
THU. 23 APR. (113)	115:33	97.0	29445	117:45	79.1	15965	A
FRI. 24 APR. (114)	14:51	81.8	29457	122:27	80.3	15979	A+J
SAT. 25 APR. (115)	109:06	95.4	29470	127:09	81.5	15993	J
SUN. 26 APR. (116)	8:24	80.3	29482	131:52	82.7	16007	J
MON. 27 APR. (117)	102:38	93.9	29495	136:34	83.9	16021	A
TUE. 28 APR. (118)	1:56	78.7	29507	141:16	85.1	16035	A+J
WED. 29 APR. (119)	56:11	92.3	29520	2:47	60.5	16048	X
THU. 30 APR. (120)	150:25	105.9	29533	7:29	61.7	16062	A

Fig. 7-4(A). Sample satellite-tracking data parameters.

Satellite Reference Orbit Data for 8 Feb 1982 (Day 39)
Courtesy AMSAT

Satellite	Time(CUT) (CUT)	Equator Crossing (Deg W)	Period (Minutes)	Long Incr. (Deg/Orbit)
AMSAT-OSCAR 8	01:14:16	86.6	103.179	25.8
UoSAT-OSCAR 9	01:09:32	151.5	95.244	23.811
RADIO 3	00:09:32	259.7	118.519	29.76
RADIO 4	00:10:12	249.7	119.395	29.98
RADIO 5	01:51:27	275.0	119.555	30.02
RADIO 6	00:56:05	261.7	118.718	29.81
RADIO 7	00:03:43	248.3	119.196	29.93
RADIO 8	00:05:13	248.3	119.765	30.07

Fig. 7-4(B). Sample satellite-tracking data parameters.

example of two such sets is shown in Fig. 7-4. Figure 7-4A contains a set of data for the AMSAT-OSCAR 7 and 8 spacecraft. It lists the reference orbits for each day in a month. This type of data sheet is intended for people who use graphic plotters to locate the spacecraft. The data presented in Fig. 7-4B is intended to be entered into computer or calculator programs.

The reference orbit numbers for the Russian satellites are unknown; therefore, they're given as zero in the data statements of the program, while those of the OSCAR spacecraft have actual numbers.

Samples of the printouts obtained from the program listed in Fig. 7-3 are shown in Figs. 7-5 to 7-7. Figure 7-5 shows that the satellite made some 15 orbits during that one day (23 December 1981). Figure 7-6 shows a printout of the reference orbits for UoSAT-OSCAR 9, which shows that the satellite reference orbit crosses the equator at almost the same time and place each day. That means that if the satellite is heard at a particular time and place today, it will be heard at the same time and place tomorrow. The period of the orbit is about 1 hour 35 minutes, a fact that can be gathered from the data associated with orbits number 1531 and 1547. It is at this point that the gradual change in reference orbit time crosses the day boundary.

Figure 7-7 shows data for various Russian Radio Satellites which were placed into orbit by the same rocket on December 17,

Equatorial crossing times for the UoSAT OSCAR 9 spacecraft
all Orbits

ORBIT	DATE	TIME	XSING (W)
1171	23 Dec	.0017	138.20
1172	23 Dec	.0153	162.02
1173	23 Dec	.0329	185.84
1174	23 Dec	.0504	209.66
1175	23 Dec	.0640	233.48
1176	23 Dec	.0816	257.30
1177	23 Dec	.0952	281.12
1178	23 Dec	.1128	304.94
1179	23 Dec	.1304	328.76
1180	23 Dec	.1440	352.58
1181	23 Dec	.1616	16.40
1182	23 Dec	.1752	40.22
1183	23 Dec	.1928	64.04
1184	23 Dec	.2104	87.86
1185	23 Dec	.2240	111.68
1186	24 Dec	.0016	135.50
1187	24 Dec	.0152	159.32
1188	24 Dec	.0328	183.14
1189	24 Dec	.0504	206.96
1190	24 Dec	.0640	230.78
1191	24 Dec	.0816	254.60
1192	24 Dec	.0952	278.42
1193	24 Dec	.1128	302.24
1194	24 Dec	.1304	326.06
1195	24 Dec	.1440	349.88
1196	24 Dec	.1615	13.70
1197	24 Dec	.1751	37.52
1198	24 Dec	.1927	61.34
1199	24 Dec	.2103	85.16
1200	24 Dec	.2239	108.98
1201	25 Dec	.0015	132.80
1202	25 Dec	.0151	156.62
1203	25 Dec	.0327	180.44
1204	25 Dec	.0503	204.26
1205	25 Dec	.0639	228.08
1206	25 Dec	.0815	251.90
1207	25 Dec	.0951	275.72
1208	25 Dec	.1127	299.54
1209	25 Dec	.1303	323.36
1210	25 Dec	.1439	347.18
1211	25 Dec	.1615	11.00
1212	25 Dec	.1751	34.82
1213	25 Dec	.1927	58.64
1214	25 Dec	.2103	82.46
1215	25 Dec	.2239	106.28
1216	26 Dec	.0015	130.10

Fig. 7-5. Equatorial crossing times for UoSAT-OSCAR 9.

Equatorial crossing times for the UoSAT OSCAR 9 spacecraft
reference Orbits

ORBIT	DATE	TIME	XSING (W)
1171	23 Dec	.0017	138.20
1186	24 Dec	.0016	135.50
1201	25 Dec	.0015	132.80
1216	26 Dec	.0015	130.10
1231	27 Dec	.0014	127.40
1246	28 Dec	.0013	124.70
1261	29 Dec	.0012	122.00
1276	30 Dec	.0012	119.30
1291	31 Dec	.0011	116.60
1306	1 Jan	.0010	113.90
1321	2 Jan	.0010	111.20
1336	3 Jan	.0009	108.50
1351	4 Jan	.0008	105.80
1366	5 Jan	.0007	103.10
1381	6 Jan	.0007	100.40
1396	7 Jan	.0006	97.70
1411	8 Jan	.0005	95.00
1426	9 Jan	.0005	92.30
1441	10 Jan	.0004	89.60
1456	11 Jan	.0003	86.90
1471	12 Jan	.0003	84.20
1486	13 Jan	.0002	81.50
1501	14 Jan	.0001	78.80
1516	15 Jan	.0000	76.10
1531	16 Jan	.0000	73.40
1547	17 Jan	.0135	94.52
1562	18 Jan	.0134	91.82
1577	19 Jan	.0134	89.12
1592	20 Jan	.0133	86.42
1607	21 Jan	.0132	83.72
1622	22 Jan	.0131	81.02
1637	23 Jan	.0131	78.32
1652	24 Jan	.0130	75.62
1667	25 Jan	.0129	72.92
1682	26 Jan	.0129	70.22
1697	27 Jan	.0128	67.52
1712	28 Jan	.0127	64.82
1727	29 Jan	.0127	62.12
1742	30 Jan	.0126	59.42
1757	31 Jan	.0125	56.72
1772	1 Feb	.0124	54.02
1787	2 Feb	.0124	51.32
1802	3 Feb	.0123	48.62
1817	4 Feb	.0122	45.92
1832	5 Feb	.0122	43.22
1847	6 Feb	.0121	40.52
1862	7 Feb	.0120	37.82
1877	8 Feb	.0120	35.12
1892	9 Feb	.0119	32.42

Fig. 7-6. Reference orbits for UoSAT-OSCAR 9.

1981. The differences between the orbits of these spacecraft is interesting. The changes in their reference orbit times are plotted in Fig. 7-8. It can readily be seen that something is very different between Radios 3 and 6 and the rest. The plot also shows that while on the 28th of December the order in which the spacecraft will come over the observer's horizon is Radio 8, 7, 4, 3, 5 and 6, it will have changed to Radio 3, 8, 6, 7, 4 and 5 by the 5th day of January. If each spacecraft is in range of the listener for about 20 minutes, that same listener can spend most of the two hours between successive orbits of Ratio 8 copying signals from the other spacecrafts.

AZIMUTH AND ELEVATION PREDICTIONS

Generating reference orbit data is fine, but it does not provide any information about when the satellite is within range of an observer. Various tracking devices are available so that if the reference orbit time is known the position of the satellite with respect to the observer can be reduced. These graphic plotters work well, but the computer can be made to provide that information.

A typical program for circular orbit satellite az-el predictions is listed in Fig. 7-9. It is not that much larger than the simple program shown in Fig. 7-3, but it provides a lot more data, as you can see from the examples shown in Fig. 7-10 to 7-12. Not only are the equatorial crossing times and locations given, but the azimuth and elevation antenna-pointing information as well as the range or distance of the satellite and Doppler shift on its signals are all presented for two minute intervals throughout a pass (the time that the satellite is above the horizon of the observer). This program also runs a little slower than the other one because it is doing more number crunching.

The program begins with the usual identification overhead statements (lines 10

Equatorial crossing times for the R S 3 spacecraft
all Orbits

ORBIT	DATE	TIME	XSING (W)
0	18 Jan	.0106	231.90
1	18 Jan	.0305	261.66
2	18 Jan	.0504	291.41
3	18 Jan	.0703	321.17
4	18 Jan	.0902	350.93
5	18 Jan	.1102	20.68
6	18 Jan	.1301	50.44
7	18 Jan	.1500	80.20
8	18 Jan	.1659	109.95
9	18 Jan	.1858	139.71
10	18 Jan	.2058	169.46

Equatorial crossing times for the R S - 5 spacecraft
reference Orbits

ORBIT	DATE	TIME	XSING (W)
0	18 Jan	.0144	241.90
12	19 Jan	.0146	242.09
24	20 Jan	.0148	242.27
36	21 Jan	.0151	242.46
48	22 Jan	.0153	242.65
60	23 Jan	.0155	242.84
72	24 Jan	.0158	243.02
83	25 Jan	.0000	213.20
95	26 Jan	.0002	213.38

Equatorial crossing times for the R S 3 spacecraft
reference Orbits

ORBIT	DATE	TIME	XSING (W)
0	18 Jan	.0106	231.90
12	19 Jan	.0056	228.98
24	20 Jan	.0046	226.06
36	21 Jan	.0036	223.13
48	22 Jan	.0027	220.21
60	23 Jan	.0017	217.29
72	24 Jan	.0007	214.37
85	25 Jan	.0156	241.20
97	26 Jan	.0147	238.28

Equatorial crossing times for the R S - 6 spacecraft
reference Orbits

ORBIT	DATE	TIME	XSING (W)
0	18 Jan	.0023	221.20
12	19 Jan	.0013	218.87
24	20 Jan	.0003	216.55
37	21 Jan	.0153	244.03
49	22 Jan	.0143	241.70
61	23 Jan	.0133	239.38
73	24 Jan	.0124	237.05
85	25 Jan	.0114	234.73
97	26 Jan	.0104	232.40

Equatorial crossing times for the R S - 4 spacecraft
reference Orbits

ORBIT	DATE	TIME	XSING (W)
0	18 Jan	.0043	225.90
12	19 Jan	.0045	225.61
24	20 Jan	.0048	225.32
36	21 Jan	.0050	225.03
48	22 Jan	.0052	224.74
60	23 Jan	.0055	224.45
72	24 Jan	.0057	224.16
84	25 Jan	.0059	223.87
96	26 Jan	.0102	223.57

Equatorial crossing times for the R S - 8 spacecraft
reference Orbits

ORBIT	DATE	TIME	XSING (W)
0	18 Jan	.0104	231.10
12	19 Jan	.0106	231.92
24	20 Jan	.0109	232.74
36	21 Jan	.0111	233.56
48	22 Jan	.0113	234.38
60	23 Jan	.0116	235.20
72	24 Jan	.0118	236.02
84	25 Jan	.0121	236.84
96	26 Jan	.0123	237.66

Fig. 7-7. Equatorial crossing times for the Radio Phase 2 satellites.

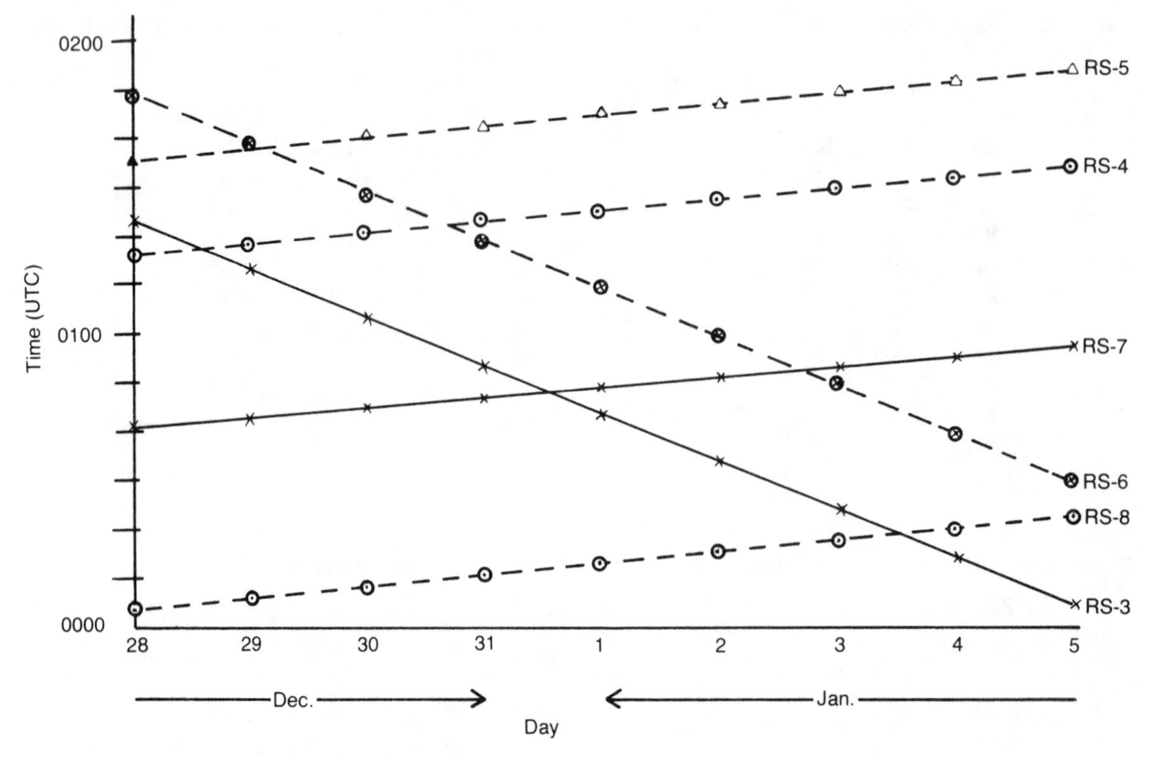

Fig. 7-8. Time changes for Radio Satellite reference orbits.

to 30). There is a note to indicate when the reference orbit data were last updated (line 30). The program allocates and defines the constant PI (3.1419) and the velocity of propagation of electromagnetic radiation (C) or light (line 40). The program then goes to line 420 to bypass the subroutines.

The menu is displayed and the data pointer is set to access the parameters associated with the chosen satellite (lines 420 to 650). This program uses spherical geometry to compute the desired data, and as such requires slightly different parameters to those used in the simpler program.

The data base for the desired spacecraft is set up (line 660) by reading the information associated with it from the DATA statements (lines 1050 to 1550). Remember that the pointer was already set up (lines 550 to 650). The variables read in line 660 are as follows:

I$ = the name of the satellite
N0 = the reference data orbit number
D = the Julian date and year of the reference data orbit
H9 = the hours part of the time of the equatorial crossing time of the reference data orbit
M9 = the minutes part of the reference data orbit crossing time
S9 = the seconds part of the time
W = the equatorial crossing location (Deg W) of the orbit
P = the period of the orbit of the satellite
I = the inclination of the orbit to the equator
H = the average height of the satellite orbit (kM) above the surface of the Earth.

The reference orbit data is displayed (lines 670 to 690). The user is asked to supply

```
10 REM Circular orbit satellite tracking program
20 REM originally by W3IWI, modified by G3ZCZ
30 REM  Ver 810120.4
40 DIM I$(30) \P1=3.141592654\P2=2*P1\P8=P1/180 \ C=2.998E5
50 GOTO 420
60 REM -- FNZ$ FILLS IN LEADING ZEROES IN STRING OF LEN=2
70 DEF FNZ$(V)
80 V1=INT(V/10) \ V2=INT(V-10*V1+.5)
90 RETURN CHR$(48+V1)+CHR$(48+V2)
100 FNEND
110 REM -- FNT CONVERTS FROM DD.DDD TO HH,MM,SS
120 DEF FNT(D9)
130 T7=INT(D9) \ T9=D9-T7 \ H9=INT(24*T9)
140 M9=INT(1440*(T9-H9/24))
150 S9=INT((T9-H9/24-M9/1440)*86400+.5)
160 IF S9<60 THEN 170 ELSE S9=S9-60 \ M9=M9+1 \ GOTO 160
170 IF M9<60 THEN 180 ELSE M9=M9-60 \ H9=H9+1 \ GOTO 170
180 IF H9<24 THEN 190 ELSE H9=H9-24 \ T7=T7+1 \ GOTO 180
190 RETURN T7
200 FNEND
210 REM -- GET SATELLITE COORD,AZ,EL,RANGE AT T=DD.DDD
220 DEF FNS(T9)
230 A0=(T9-T0)+W0 \ A0=(A0-INT(A0))*P2
240 N9=(T9-T0)/P0 \ A1=(N9-INT(N9))*P2
250 S0=SIN(A0) \ C0=COS(A0) \ S1=SIN(A1) \ C1=COS(A1)
260 X=R3*(+C0*C1+S0*S1*C2) \ Y=R3*(-S0*C1+C0*S1*C2)
270 Z=R3*S1*S2 \ S8=(X-X0)^2+(Y-Y0)^2+(Z-Z0)^2 \ S7=SQRT(S8)
280 S3=(R4-S8)/(2*R0*S7) \ E=-99 \ F7=(S7-L7)/(T5*C) \ L7=S7
290 IF S3<E3 THEN RETURN E ELSE E=ATN(S3/SQRT(1-S3^2))/P8
300 A1=R1*(Y*X0-X*Y0) \ A2=R7*7 - Z0*(X*X0+Y*Y0+Z*Z0)
310 A=ATN(A1/A2)/P8 \ IF A2<0 THEN A=A+180
320 IF (A1<0 AND A2>0) THEN A=A+360
330 A=INT(A+.5) \ S7=INT(S7+.5) \ RETURN INT(E+.5)
340 FNEND
350 REM -- HEADER PRINT ROUTINE
360 T6=T0+(N-N0)*P0 \ W6=T6-T0+W0 \ W6=(W6-INT(W6))*360
370 D7=FNT(T6)+D \ T$='/'+FNZ$(H9)+':'+FNZ$(M9)+':'+FNZ$(S9)
380 PRINT#P5
390 PRINT#P5,'ORBIT# ',%5I,N,' AT ',D7,T$,' UTC  AT W.LONG=',%6F
400 PRINT#P5,'    UTC        AZ       EL    RANGE     '+M$ \ RE
410 REM ---- PROGRAM BEGINS
420 PRINTCHR$(12),'AMATEUR SATELLITE AZ-EL PROGRAM'
430 PRINT
440 PRINT 'Satellite selection menu '
450 PRINT 'AMSAT - OSCAR 7          0'
460 PRINT 'AMSAT - OSCAR 8          1'
470 PRINT 'UoSAT - OSCAR 9          2'
480 PRINT 'RS 3                     3'
490 PRINT 'RS 4                     4'
500 PRINT 'RS 5                     5'
510 PRINT 'RS 6                     6'
520 PRINT 'RS 7                     7'
530 PRINT 'RS 8                     8'
540 INPUT 'Which satellite ? ', S \ IF S<0 OR S >8 THEN 440
550 ON S+1 GOTO 560,570,580,590,600,610,620,630,640
560 RESTORE 1100 \ GOTO 660
570 RESTORE 1170 \ GOTO 660
580 RESTORE 1230 \ GOTO 660
590 RESTORE 1280 \ GOTO 660
600 RESTORE 1320 \ GOTO 660
610 RESTORE 1360 \ GOTO 660
620 RESTORE 1410 \ GOTO 660
630 RESTORE 1460 \ GOTO 660
640 RESTORE 1510 \ GOTO 660
650 REM OTHER S/C CAN BE ADDED HERE
660 READ I$,N0,D,H9,M9,S9,W,P,I,H
670 PRINT\ PRINT I$,' Reference orbit is',N0,' on',D,
680 T$='   '+FNZ$(H9)+':'+FNZ$(M9)
690 PRINT ' at',T$, ' crossing at',W,' W'
691 INPUT 'Do you want to overide reference data ? ',A$ \ IF A$=''
    THEN 691
692 IF A$(1,1) = 'N' THEN 700 ELSE IF A$(1,1) = 'Y' THEN 693 ELSE 691
693 INPUT 'Reference Orbit number ? ',N0\INPUT 'Julian day number ? ',D
696 INPUT 'Reference orbit crossing time hours ? ',H9
697 INPUT 'Reference orbit crossing time minutes ? ',M9
698 INPUT 'Reference orbit crossing time seconds ? ',S9
699 INPUT 'Reference orbit crossing time location ( deg W ) ? ',W
700 INPUT'What Mode do you want ? ',M$
710 READ T$,F8,F9\IF T$=M$THEN720 ELSEIF T$='**'THEN740 ELSE710
720 IF F8=1 THEN M$='BEACON' ELSE M$='DOPPLER' \ F8=F8*F9
730 M$=T$(1,1)+'-'+M$ \ GOTO 760
740 PRINT 'Mode '+M$+' does not compute'
750 M$='  VELOCITY' \ F8=0 \ F9=C
760 RESTORE 1030 \ READ E3,T5,S$,W1,L1
770 P0=P/1440 \ W0=W/360 \ T0=H9/24+M9/1440+S9/86400
780 S2=SIN(I*P8) \ E3=SIN(E3*P8) \ C2=COS(I*P8) \ R1=6378.14
790 R3=R1+H \ R7=R1^2 \ R5=R3^2
800 S0=SIN(L1*P8)\C0=COS(L1*P8)\S1=SIN(W1*P8)\C1=COS(W1*P8)
810 F=1-1/298.22 \ R=R1/SQRT(C0*C0+F*S0*S0)
820 X0=R*C0*C1 \ Y0=-R*C0*S1 \ Z0=R*S0*F*F
830 R0=SQRT(X0*X0+Y0*Y0+Z0*Z0) \ R4=R5-R0*R0
840 INPUT 'What is the first orbit ? ',N1  \ IF N1 >= N0 THEN 860
850 PRINT' Reference orbit is',N0 \ GOTO 840
860 INPUT 'What is the last orbit to be printed ? ',N2
870 IF N2<N1 THEN N2=N1 \ I$=I$+' ORBITS '
880 INPUT 'Which output device (0-7) ? ',P5
890 PRINT#P5,CHR$(12),'AZ-EL POINTING FOR ',I$,N1,' TO ',N2
900 I$='FOR '+S$+' AT W.LONG '
910 PRINT#P5,I$,W1,' & LAT ',L1,
920 IFL1>=0 THEN X$=' N ' ELSE X$=' S '\ PRINT#P5, X$ +' MODE '+T$
930 PRINT#P5,'----------------------------------------------------'
940 T1=T0+(N1-N0)*P0 \ T1=INT(T1*1440)/1440
950 T2=T1+(N2-N1+1)*P0
960 T3=T5/1440 \ T5=T5*60 \ L7=9E9 \ N7=-99
970 FOR T=T1 TO T2 STEP T3 \ E = FNS(T) \ IF E<-90 THEN 1010
980 N=INT(N0+N9+.1) \ IF N<>N7 THEN GOSUB 360
990 N7=N \ D7=FNT(T) \ T$='   '+FNZ$(H9)+':'+FNZ$(M9)
1000 PRINT#P5, T$,%9I,A,E,S7,%14F3,F8-F7*F9
1010 REM
1020 NEXT \ PRINT#P5,TAB(20),'END'
1030 DATA -3,2 \ REM-MIN EL,TIME INTERVAL
1040 DATA 'G3ZCZ/4X', -31.8,35.2 \ REM CALL,LONG,LAT
1050 REM FORMAT OF DATA = SPACECRAFT NAME
1060 REM ORBIT (#), DATE (YYDDD),EQR XSING TIME (HH,MM,SS),LONG ( DEG W)
1070 REM PERIOD (MIN), INCL (DEG), HEIGHT (KM)
1080 REM FORMAT='MODE',DOP=0/FREQ=1,EQUIV FREQ,KHZ
1090 REM '**',1,1 FOR END OF DATA FOR SATELLITE
1100 DATA 'AMSAT OSCAR 7'
```

Fig. 7-9. Azimuth and elevation-pointing information for Phase 2 satellites.

```
1110 DATA 16804,78200,00,36,52,67.5
1120 DATA 114.945199,101.7,1449.6
1130 DATA 'A',0,175450.00,'AB',1,29502.00
1140 DATA 'B',0,-286250.00,'BB',1,145972.00
1150 DATA 'D',1,435100.00
1160 DATA '**',1,1
1170 DATA 'AMSAT OSCAR 8'
1180 DATA 19370,81357,01,08,10,83.77
1190 DATA 103.179480,98.7987,905
1200 DATA 'A',0,175450.00,'AB',1,29402.00
1210 DATA 'J',0,+289100.00,'JB',1,435100.00
1220 DATA '**',1,1
1230 DATA 'UoSAT OSCAR 9'
1240 DATA 01171,81357,00,16,58,138.2
1250 DATA 95.31353,97.462,536
1260 DATA 'AB',1,29.500,'BB',1,145825.00
1270 DATA '**',1,1
1280 DATA ' R S 3 '
1290 DATA 0   ,82018,01,05,50,231.9
1300 DATA 118.46293,82.9592,1632.7
1310 DATA 'AB',1,29321,'**',1,1
1320 DATA ' R S - 4'
1330 DATA 0   ,82018,00,43,23,225.9
1340 DATA 119.33912,82.9603,1666
1350 DATA 'AB',1,29360,'**',1,1
1360 DATA 'R S - 5 '
1370 DATA 0   ,82018,01,43,23,241.9
1380 DATA 119.49842,82.9629,1671.55
1390 DATA 'A',0,175.450,'AB',1,29.330
1400 DATA '**',1,1
1410 DATA ' R S - 6 '
1420 DATA 0  ,82018,00,23,17,221.2
1430 DATA 118.66154,82.9542,1641.7
1440 DATA 'A',0,175.450,'AB',1,29.453
1450 DATA '**',1,1
1460 DATA ' R S - 7 '
1470 DATA 0  ,82018,01,26,59,237.1
1480 DATA 119.14014,82.9629,1661.55
1490 DATA 'A',0,175.450,'AB',1,29.501
1500 DATA '**',1,1
1510 DATA ' R S - 8 '
1520 DATA 0  ,82018,01,04,18,231.1
1530 DATA 119.70890,82.9570,1675.25
1540 DATA 'A',0,175.450,'AB',1,29.502
1550 DATA '**',1,1
```

Fig. 7-9. Azimuth and elevation-pointing information for Phase 2 satellites. (continued)

information on which transponder or beacon the Doppler frequency is to be calculated for (line 700). Amateur satellite transponders have been built using frequencies in the following passbands;

Mode	Uplink	Downlink
A	2M	10M
B	70CM	2M
J	2M	70CM

The data associated with the mode and frequency is read (line 700.

T$ = the mode
F8 = the uplink/downlink flag
F9 = the equivalent frequency

```
AZ-EL POINTING FOR UoSAT OSCAR 9 ORBITS  1171 TO  1181
FOR G3ZCZ/4X AT W.LONG  -31.8 & LAT  35.2 N   MODE AB
--------------------------------------------------------
ORBIT#  1171 AT 81357/00:16:58 UTC  AT W.LONG=138.20
   UTC        AZ       EL    RANGE       A-BEACON
  00:50       20        2    2515          29.501
  00:52       27       12    1708          29.501
  00:54       47       29    1003          29.501
  00:56      121       41     795          29.500
  00:58      164       19    1342          29.500
  01:00      175        6    2121          29.499
  01:02      180       -2    2945          29.499

ORBIT#  1172 AT 81357/01:52:17 UTC  AT W.LONG=162.03
   UTC        AZ       EL    RANGE       A-BEACON
  02:24      349       -2    2928          29.501
  02:26      336        4    2283          29.501
  02:28      314       10    1832          29.500
  02:30      284       11    1743          29.500
  02:32      258        7    2064          29.500
  02:34      241        0    2646          29.500

ORBIT#  1178 AT 81357/11:24:10 UTC  AT W.LONG=305.00
   UTC        AZ       EL    RANGE       A-BEACON
  11:30      124        2    2481          29.501
  11:32      106        9    1867          29.501
  11:34       77       15    1530          29.500
  11:36       43       12    1656          29.500
  11:38       20        5    2166          29.500
  11:40        8       -1    2855          29.499

ORBIT#  1179 AT 81357/12:59:28 UTC  AT W.LONG=328.83
   UTC        AZ       EL    RANGE       A-BEACON
  13:02      186       -3    2998          29.501
  13:04      192        5    2194          29.501
  13:06      207       16    1456          29.501
  13:08      246       31     972          29.500
  13:10      304       24    1142          29.500
  13:12      327       10    1785          29.499
  13:14      337        1    2561          29.499

ORBIT#  1180 AT 81357/14:34:47 UTC  AT W.LONG=352.66
   UTC        AZ       EL    RANGE       A-BEACON
  14:44      278       -3    3048          29.500
                     END
```

Fig. 7-10. Azimuth and elevation-pointing data for UoSAT-OSCAR 9.

AZ-EL POINTING FOR R S - 8 ORBITS 0 TO 9
FOR G3ZCZ/4X AT W.LONG -31.8 & LAT 35.2 N MODE A

ORBIT# 2 AT 82018/05:03:43 UTC AT W.LONG=290.95

UTC	AZ	EL	RANGE	A-DOPPLER
05:08	114	-3	5258	.002
05:10	106	0	4887	.002
05:12	96	3	4586	.001
05:14	86	5	4375	.001
05:16	75	6	4270	.001
05:18	64	6	4282	-.000
05:20	53	5	4409	-.001
05:22	43	3	4642	-.001
05:24	34	0	4964	-.002

ORBIT# 3 AT 82018/07:03:26 UTC AT W.LONG=320.88

UTC	AZ	EL	RANGE	A-DOPPLER
07:02	169	-2	5163	.003
07:04	167	4	4515	.003
07:06	164	11	3877	.003
07:08	160	19	3266	.003
07:10	153	29	2708	.003
07:12	139	42	2253	.002
07:14	111	54	1979	.001
07:16	69	55	1968	.000
07:18	40	43	2223	-.001
07:20	27	30	2667	-.002
07:22	20	20	3220	-.003
07:24	16	11	3830	-.003
07:26	13	4	4468	-.003
07:28	11	-2	5118	-.003

ORBIT# 4 AT 82018/09:03:08 UTC AT W.LONG=350.81

UTC	AZ	EL	RANGE	A-DOPPLER
09:04	218	-1	5023	.003
09:06	223	5	4448	.003
09:08	230	10	3906	.003
09:10	239	17	3419	.002
09:12	251	23	3021	.002
09:14	268	28	2754	.001
09:16	288	31	2660	.000
09:18	308	28	2756	-.000
09:20	325	23	3023	-.001
09:22	337	17	3417	-.002
09:24	346	11	3897	-.002
09:26	353	5	4430	-.003
09:28	358	-1	4993	-.003

ORBIT# 5 AT 82018/11:02:51 UTC AT W.LONG= 20.74

UTC	AZ	EL	RANGE	A-DOPPLER
11:16	292	-2	5138	.001
11:18	301	-1	4997	.001
11:20	310	0	4932	.000
11:22	320	0	4947	-.000
11:24	330	-1	5039	-.000
11:26	339	-2	5204	-.001

ORBIT# 7 AT 82018/15:02:16 UTC AT W.LONG= 80.59

UTC	AZ	EL	RANGE	A-DOPPLER
15:40	26	-2	5114	.001
15:42	35	-1	4987	.001
15:44	45	0	4936	.000
15:46	55	0	4964	-.000
15:48	64	-1	5070	-.001
15:50	73	-3	5249	-.001

ORBIT# 8 AT 82018/17:01:58 UTC AT W.LONG=110.52

UTC	AZ	EL	RANGE	A-DOPPLER
17:36	360	-3	5270	.003
17:38	5	2	4698	.003
17:40	11	8	4150	.003
17:42	19	14	3642	.002
17:44	29	20	3204	.002
17:46	44	26	2871	.002
17:48	62	30	2690	.001
17:50	83	30	2694	-.000
17:52	101	26	2883	-.001
17:54	116	20	3223	-.002
17:56	126	13	3671	-.002
17:58	134	7	4189	-.003
18:00	140	2	4750	-.003

ORBIT# 9 AT 82018/19:01:41 UTC AT W.LONG=140.45

UTC	AZ	EL	RANGE	A-DOPPLER
19:38	348	1	4777	.003
19:40	346	8	4131	.003
19:42	343	16	3505	.003
19:44	337	25	2919	.003
19:46	328	37	2414	.002
19:48	308	50	2056	.002
19:50	269	57	1933	.001
19:52	231	49	2088	-.001
19:54	212	35	2469	-.002
19:56	203	24	2986	-.003
19:58	198	15	3576	-.003
20:00	194	7	4204	-.003
20:02	192	1	4849	-.003

END

Fig. 7-11. Azimuth and elevation-pointing data for Radio 8.

Modes can be A, B, or J for communications or AB, BB, or BJ for the beacons associated with the modes. The equivalent frequency is the sum of the uplink and downlink frequencies for the transponder if F8 = 0, or is just the downlink frequency if

```
AZ-EL POINTING FOR AMSAT OSCAR 8 ORBITS  19370 TO  19378
FOR G3ZC7/4X AT W.LONG  -31.8 & LAT  35.2 N   MODE JB
------------------------------------------------------------
ORBIT# 19371 AT 81357/02:51:21 UTC  AT W.LONG=109.56
   UTC        AZ       EL     RANGE      J-BEACON
  03:30       60       -1      3663    435104.040
  03:32       74        1      3466    435102.390
  03:34       89        1      3433    435100.400
  03:36      103        0      3571    435098.330
  03:38      116       -3      3862    435096.480
ORBIT# 19372 AT 81357/04:34:32 UTC  AT W.LONG=135.36
   UTC        AZ       EL     RANGE      J-BEACON
  05:08       20       -1      3648    435109.150
  05:10       25        6      2900    435109.050
  05:12       32       16      2187    435108.630
  05:14       48       30      1576    435107.390
  05:16       88       44      1244    435104.010
  05:18      139       36      1414    435097.950
  05:20      161       20      1957    435093.430
  05:22      171        9      2649    435091.630
  05:24      177        1      3394    435091.000
ORBIT# 19373 AT 81357/06:17:42 UTC  AT W.LONG=161.15
   UTC        AZ       EL     RANGE      J-BEACON
  06:50      359       -2      3725    435108.740
  06:52      352        5      3035    435108.340
  06:54      340       12      2422    435107.410
  06:56      321       20      1971    435105.450
  06:58      292       23      1816    435101.870
  07:00      264       19      2028    435097.440
  07:02      246       11      2512    435094.150
  07:04      235        4      3139    435092.420
  07:06      228       -3      3832    435091.620
ORBIT# 19374 AT 81357/08:00:53 UTC  AT W.LONG=186.95
   UTC        AZ       EL     RANGE      J-BEACON
  08:36      324       -2      3800    435102.630
  08:38      310       -2      3745    435100.660
  08:40      297       -3      3860    435098.610
ORBIT# 19378 AT 81357/14:53:36 UTC  AT W.LONG=290.13
   UTC        AZ       EL     RANGE      J-BEACON
  15:00      105       -2      3806    435107.100
  15:02       94        2      3301    435106.100
  15:04       79        6      2934    435104.440
  15:06       61        8      2766    435102.030
  15:08       43        7      2837    435099.140
  15:10       27        4      3130    435096.450
  15:12       15        0      3589    435094.450
                     END
```

Fig. 7-12. Azimuth and elevation-pointing data for AMSAT-OSCAR 8.

F8 = 1. If the communications passband is inverting (Mode B), then the equivalent frequency is negative (line 1140). The program sets up the mode information to be printed out (M$) (lines 720 to 750), and converts the value of F8 to correspond to the equivalent downlink frequency. The user is informed (line 740) when the choice of mode is not valid for the satellite (i.e., no data for that mode is present in the DATA statements). In that case, the printout displays the relative difference between C and the velocity of the spacecraft as seen by the observer.

The observer groundstation data stored in lines 1030 and 1040 is set up (line 760).

- E3 = the minimum angle of elevation to the satellite for the printout
- T5 = the time interval between successive lines in the printout
- S$ = the callsign of the observer
- W1 = the longitude of the location of the observer (Deg W)
- L1 = the latitude of the location of the observer (+ if N, − if S)

The data is converted to the format required by the subroutines used later in the program (lines 770 to 830). The user is asked (lines 840 to 870) to determine the range of orbits to be scanned for acquisition (i.e., in range of the observer/listener). If an orbit number equal to or less than that of the reference data orbit is chosen, the user is asked to choose an alternate (line 880). Also, the user is asked to decide which device is to receive the printout (line 880).

The heading is printed (lines 890 to 930). The CHR$(12) (line 890) is a form feed character used to set the printer to the start of a new sheet of paper. It may have to be changed for some output devices. The orbit data is converted to equivalent times (lines 940 to 960). N7 is used as an "orbit number changed" flag later.

The main program loop is comprised of lines 970 to 1020. While the orbit number (or equivalent time) is within the range to be printed out, the FNS function is performed. This function is used as a subroutine (lines 210 to 340). It computes the satellite's co-ordinates, azimuth, elevation, and range at a given time (T). The variable E is used either as the actual elevation of the satellite as seen by the ground station, or as a flag value to signal that the spacecraft is below the minimum angle of elevation to be printed (−99). Thus, the position data of the satellite is only printed if it is within range (equal to or above the minimum angle of elevation). First, however, there is a test to see if the satellite has passed the ascending node point and has begun a new orbit (line 880). If it has, a subroutine (starting at line 360) is invoked to print a new heading.

A formatted data line is set up and printed (lines 990 to 1000). When the main loop is completed, an END statement is printed (line 1020).

The two programs described in this chapter rely on the assumption that the orbit of the satellite is circular to simplify the calculations, an assumption that is not exactly true. While the data provided for orbits following close to the reference data orbit will be accurate, small errors due to the inaccurate assumptions made will accumulate and thus the further into the future the orbit predictions are projected, the less accurate they will be. As an example, the data shown in Fig. 7-4B is obtained from a later source than that contained in the data statements of the program listed in Fig. 7-3. If the program is used to project orbital information forward to the 8th of February the differences between the two sets of data will become apparent. This can be seen by comparing the data in Fig. 7-4B, with that projected for the same date by the computer, which is shown in Fig. 7-6. The lower the altitude of the orbit, the greater the error in the predictions. Thus, data for UoSAT-OSCAR 9 becomes inaccurate after a week or so, while that of AMSAT-OSCAR 8 tends to get inaccurate after a month or so. It is advisable to update the reference orbit data statements at a minimum of once a month using data published in various amateur radio magazines or obtained directly from AMSAT.

SPACECRAFT TELEMETRY

All of the satellites in the OSCAR series (Orbiting Satellite Carrying Amateur Radio) send status signals to the AMSAT Telemetry Command and Control Station Network. These signals, transmitted on beacons adjacent to the communications passband, are known as telemetry signals. By using a computer and published conversion equations, it is possible to decode the spacecraft telemetry and discover just what is happening up at the spacecraft.

Spacecraft telemetry is usually transmitted in blocks or frames. The telemetry decoding equations for the AMSAT-OSCAR 7 and 8 spacecraft are shown in Fig. 8-13 and Fig. 8-14. The AMSAT-OSCAR 7 Morse Code telemetry comes in a frame of 24 words, while the AMSAT-OSCAR 8 Morse Code telemetry comes in a six-word frame.

The telemetry is transmitted as a string of numbers in Morse Code. After listening to the signals for a few minutes it becomes obvious that the spacecraft is sending back a repetitive string of numbers. Each number contains three digits. The first digit is always the frame line number, and the remaining digits are the data associated with the channel. Thus, a typical AMSAT-OSCAR 8 telemetry frame could be HI 111 222 333 444 555 666 HI. The HI identifies the signals as coming from a spacecraft and also separates the frames. In this example, the data associated with each channel is:

Channel	1	2	3	4	5	6
Data	11	22	33	44	55	66

In the AMSAT-OSCAR 7 telemetry frame, which is divided into six rows of four channels, the first digit in the three figure group is the row number. Its telemetry could contain 355 444 456 467 478, 555, etc., where

Channel	Measured Parameter	Measurement Range	Preliminatry Calibration Equation
1A	Total Solar Array Current	0 to 3000 ma.	$I_T = 29.5$ (ma.)
1B	+X Solar Panel Current	0 to 2000 ma.	$I_{+x} = 1970 - 20\ N$ (ma.)
1C	−X Solar Panel Current	0 to 2000 ma.	$I_{-x} = 1970 - 20\ N$ (ma.)
1D	+Y Solar Panel Current	0 to 2000 ma.	$I_{+y} = 1970 - 20\ N$ (ma.)
2A	−Y Solar Panel Current	0 to 2000 ma.	$I_{-y} = 1970 - 20\ N$ (ma.)
2B	RF Pwr. Out–70/2 Rptr.	0 to 8 watts	$P_{70/2} = 8\ (1 - 0.01\ N)^2$ (watts)
2C	24-hour Clock Time	0 to 1440 minutes	$t = 14.4\ N$ (min.) or $0.24\ N$ (hrs.)
2D	Bat. Charge-Discharge Current	−2000 to +2000 ma.	$I_B = 40\ (N - 50)$ (ma.)
3A	Battery Voltage	6.4 to 16.4 volts	$V_B = 0.1\ N + 6.4$ (volts)
3B	Half-Battery Voltage	0 to 10 volts	$V_{1/2B} = 0.10\ N$ (volts)
3C	Bat. Charge Reg. #1 Vtge.	0 to 15 volts	$V_{cr1} = 0.15\ N$ (volts)
3D	Battery Temperature	−30° to +50° C.	$T_{Bat} = 95.8 - 1.48\ N$ (°C.)
4A	Baseplate Temperature	−30° to +50° C.	$T_{bp} = 95.8 - 1.48\ N$ (°C.)
4B	PA Temp. – 2/10 Rptr.	−30° to +50° C.	$T_{10} = 95.8 - 1.48\ N$ (°C.)
4C	+X Facet Temperature	−30° to +50° C.	$T_{+x} = 95.8 - 1.48\ N$ (°C.)
4D	+Z Facet Temperature	−30° to +50° C.	$T_{+z} = 95.8 - 1.48\ N$ (°C.)
5A	PA. Temp. – 70/2 Rptr.	−30° to +50° C.	$T_2 = 95.8 - 1.48\ N$ (°C.)
5B	PA. Emit. Curr. – 2/10 Rptr.	0 to 1167 ma.	$I_{10} = 11.67\ N$ (ma.)
5C	Modulator Temp. – 70/2 Rptr.	−30° to +50° C.	$T_m = 95.8 - 1.48\ N$ (°C.)
5D	Instr. Sw. Reg. Input Curr. (@14.3 V)	0 to 93 ma.	$I_{isr} = 11 + 0.82\ N$ (ma.)
6A	RF Power Out – 2/10 Rptr.	0 to 10,000 mw.	$P_{2/10} = N^2$ (milliwatts)
6B	RF Power Out – 435 Beacon	0 to 1,000 mw.	$P_{435} = 0.1\ N^2$ (milliwatts)
6C	RF Power Out – 2304 Beacon	0 to 100 mw.	$P_{2304} = 0.01\ N^2$ (milliwatts)
6D	Midrange Telemetry Calibration	0.500 volts	$N = 50 \pm 1$ counts

Note the wealth of detail that is available. Studies can be, and have been, made of all sorts of aspects of the spacecraft on board activity.

Fig. 7-13. Telemetry decoding equations for the AMSAT-OSCAR 7 spacecraft.

the channels associated with the numbers are 355 (3D) 444 (4A) 456 (4B) 467 (4C) 478 (4D) 555 (5A), etc.

Once the frame concept is understood, the received data can be identified according to channel number. It is then a matter of manipulating a few equations to find out what is happening in the spacecraft.

TELEMETRY DECODING

Telemetry decoding is an example of converting published equations into a form that the computer can process. The equation for the battery voltage of the AMSAT-OSCAR 8 spacecraft, for example, is:

$$Vb = 0.1N + 3.25 \text{ Volts}$$

Vb is the battery voltage and N is the number transmitted by the spacecraft in Morse code in the channel 4 position of the telemetry frame. A short routine could be written in BASIC to compute the voltage as follows:

```
10 INPUT "Channel 4 Counts"; N
20 PRINT (0.1 * N + 3.25); "Volts"
```

These two lines of code convert a raw number to an engineering unit. The routine is not very useful in its present state. If it was expanded to include the equations for all six channels, a more usable result would be obtained.

The telemetry decoding equations for the AMSAT-OSCAR 7 and 8 spacecraft are shown in Fig. 7-13 and Fig. 7-14. The AMSAT-OSCAR 8 spacecraft equations, for example, can be converted to BASIC as follows:

```
200  I1 = 7.15 * (101 - N1)
210  I2 = 57 * (N2 - 50)
```

Figure 8.14 Telemetry Decoding Equations for the AMSAT-OSCAR 8 Spacecraft.

Channel	Measured Parameter	Equation	
1	Solar Array Current	7.15 X (101-N)	mA
2	Battery Current	57 X (N-50)	mA
3	Battery Voltage	0.1 X N + 3.25	Volts
4	Baseplate Temperature	95.8 - 1.48 X N	Deg C
5	Battery Temperature	95.8 - 1.48 X N	Deg C
6	Mode J RF Output	23 X N	mW

Note that N is the morse code value for that channel

Fig. 7-14. Telemetry decoding equations for the AMSAT-OSCAR 8 spacecraft.

```
220  V1 = 0.1 * N3 + 3.25
230  T1 = 95.8 − 1.48 * N4
240  T2 = 95.8 − 1.48 * N5
250  P1 = 23 * N6
260  RETURN
```

This subroutine requires that two others be added to it to complete the program. The first is the input subroutine, and the second is the display routine. The input subroutine could look something like this;

```
100  For I = 1 to 6
110  PRINT "Channel" + 1;:N(I) IN-
     PUT N(I)
120  NEXT
130  RETURN
```

This is a simple FOR/NEXT loop that stores the input numbers in a matrix (N) having six positions. The output routine could be like this:

```
300  PRINT "AMSAT-OSCAR 8 Space-
     craft Status"
310  PRINT
320  PRINT "Solar Array Current", I1,
     "mA"
330  PRINT "Battery Current", I2,
     "mA"
340  PRINT "Battery Voltage", V1,
     "Volts"
350  PRINT "Baseplate Temperature",
     T1, "deg C"
360  PRINT "Battery Temperature",
     T2, "deg C"
370  PRINT "Mode J RF Output", P1,
     "mW"
380  RETURN
```

The three subroutines can be linked together as follows:

```
20  GO■SUB 100:REM INPUT DATA
30  GO■SUB 200:REM Computer En-
    gineering Units
40  GO■SUB 300:REM Printout Re-
    sults
50  STOP
```

If this program is typed in and executed as is, all results (i.e., values of I1, I2, etc.) will be zero. The reason is that the input data was stored in a matrix N(I) while the conversion subroutine (lines 200 to 260) uses six different variables (N1, N2, etc). The computation algorithm must thus be changed so that NI becomes N(1), N2 becomes N(2), etc., or a routine must be written and placed between lines 20 and 30 to perform the conversion. This routine could take the form;

```
25  N1 = N(1):N2 = N(2):N3 = N(3),
```

Obviously, any program should contain only one technique for storing the data. Either the matrix or the separate variables should be used. Both have been used in this example only to illustrate that there is more than one way of performing the task.

Now, in order to make the printout more useful, the orbit number can be added. A time can be added to show when the data was actually taken and, if the required hardware and software are available, there is no reason why the computer cannot convert the Morse code directly without the operator having to copy it down and then enter it into the computer.

Additional routines can be added to test each of the data channels to see if anything critical is happening. For example, the battery voltage can be monitored to ensure that it is charged correctly. The telemetry data display can be formatted on the screen or printout to be as complex and sophisticated as you want. The channels whose values have crossed limit values can be displayed in red, while regular channels can be displayed in other colors. The other spacecraft can also be displayed in different colors. Here there is no limit to what you can do; you can come up with your own version of a spacecraft control center.

If you add storage capability to the basic display capacity, you can save either the raw data or data averaged over an orbit and perform studies and analysis on the way that the spacecraft behaves over periods of time.

TELEMETRY ANALYSIS

The Russian Amateur Radio Satellites are in a Phase 2 type of orbit. Their telemetry (the signals transmitted by the spacecraft itself) are downlinked by means of Morse code in a slightly different manner to that of the AMSAT spacecraft. Each "word" is identified by a letter and each line of words is identified by a prefix letter. The basic channel words are K, D, O, G, U, S, and W. The satellite transmits the identification as RS (number) followed by a line of letters and numbers. The letters uniquely identify each word in the frame, whereas the AMSAT spacecraft just identified the line number and let the position in the line identify the word. The Radio Satellite word prefixes may be either none, E, I, S, N, R, A, U, M, or W. The data could appear as:

RS 8 K00 D84 O88 G01 U34 S49 W23
RS 3 EK00 ED84 EO82 EG01 EU01 ES29 EW26
RS 6 NK00 ND16 NO00 NG48 NU25 NS29 NW45

The provisional telemetry decoding equations for the Radio Satellites are shown in Fig. 7-15. Six of them (Radio Satellites 3-8) were launched by a single rocket in December 1981. Analysing their telemetry over a period of time shows similarities and differences in the spacecraft. A program to present the accumulated telemetry data can be written. It can store the received frames in DATA statements or in disk files and allow printouts of all the data or just some of it. Examples of such printouts are shown in Fig. 7-16. The data is shown in a format which consists of year, month, and day as a single number that increments daily. The remaining data is the decoded telemetry of the K, D, O, S, and W words of the first (no prefix) channel (line) which give information about the transponder power, battery, voltages, currents, and temperatures. Several examples of the printouts are shown, but at the time of writing there is not enough data available for any significant analysis other than to state that a resolution of one degree per morse code character is too coarse.

The program that was used to generate the data displays of Fig. 7-16 is shown in Fig. 7-17. It begins with the usual overhead (lines 10 to 40). The user is asked to choose one of the satellites (line 60). No error-checking is performed here. The heading is printed (lines 70 to 80) and the main loop starts (line 90).

The first set of data elements is read (line 100). The data is stored in the DATA statements (lines 270 to 440). The variables used are:

- S1 = the satellite number
- Y1 = the year of the orbit
- M1 = the month of the orbit
- D1 = the day of the orbit
- T1 = the time of the orbit (in hours and minutes)
- K = the telemetry data value of the K channel
- D = the telemetry data value of the D channel
- O = the telemetry data value of the O channel
- S = the telemetry data value of the S channel
- W = the telemetry data value of the W channel.

The data is tested to see if the last element is being read (line 110). If it is, the program terminates at line 450. If not, the telemetry is read from the data area.

The program then determines if the data is to be printed (lines 130 to 140). The data is only printed if the user asked for all the data to be printed (test in line 130) or if the data is from the satellite that the user wants printed (test in line 140). If the data is

RS Satellite First Channel: Prefix, none or (active 'E' e.g., 'K' or 'EK' etc.

Letter	Content	Calculation
K	Output power	$0.2 \times N^2$ = op in mW of transponder
D	Voltage of source	$n \times 0.2$ = power source in volts
O	Charge current	$20 \times (100-n)$ = charge in mA
G	Believed to be TLM calibration constant test level	
U	Not given	
S	Temp. Regulator	T = n = Temp. of Voltage Regulator in C
W	Temp. 10m TX cooling fins	T = n = Temp. of 10-meter output stage in C

Second Channel: Prefix: 'I' or (active) 'S'; e.g., 'IK' or 'SK', 'ID' or 'SD' etc.

Letter	Content	Calculation
K	Output pwr transp.	As previous
D	Zero adj. of TLM	Figure given
O	Beacon output pwr	$0.2 \times N^2$ = Beacon output in mW
G	Sensitivity transp.	N = −dB (regulated)
U	'S' meter 1st RX	$0.1 \times (N-10)$ = 'S' units
S	'S' meter ROBOT RX	as above
W	'S' meter 2nd service RX	as above

Third Channel: Prefix 'N' (quiet) or 'R' (active); e.g., 'NK' or 'RK' etc.

K	As previous two 'K' channels
D-W	Regret no further information yet to hand

Fourth Channel: Prefix 'A' (inactive) or 'U' (active); e.g., 'AK' or 'RK' etc.

K	Output power of transponder	as previous
D	9 V transponder line	$0.1 \times N$ = transponder supply 'V'' in volts
O	7.5 V transponder line	as above
G	9 V 1st stabilizer	as above
U	7.5 V 1st stabilizer	as above
S	9 V 2nd stabilizer	as above
W	7.5 V 2nd stabilizer	as above

Fifth Channel: Prefix 'M' (inactive) or 'W' (active) e.g., 'MK' or 'WK' etc.

K	Output power of transponder	as previous
D	On board log	N = no. of QSO's ±1 (assumed on ROBOT)
O	Heater radiation control	$N \times 0.1$ = watts, power of heating system
G	ROBOT input power	$n \times 20$ = power in mW
U	Power of service channel	$n \times 20$ = nW (assumed to be transponder)
S	Sensitivity pad of ROBOT	N = −dB of ROBOT RX
W	Sensitivity of service RX	N = −dB

Fig. 7-15. Provisional telemetry decoding equations for the Radio Satellites.

RADIO SATELLITE TELEMFTRY TABLF

DATA FOR RADIO 0

R-S	DATE	TIME	TX PWR	BATT V	CHARGE I	REG T	TX T
3	820117	2005	0	15	340	27	27
8	820122	821	0	17	380	21	20
5	820128	1928	0	16	340	30	26
3	820206	1733	0	16	320	27	15
6	820206	1752	1960	16	540	19	30
8	820213	1630	1378	17	340	29	26
8	820213	1840	1378	17	300	39	25

RADIO SATELLITE TELEMETRY TABLE

DATA FOR RADIO 3

R-S	DATE	TIME	TX PWR	BATT V	CHARGE I	REG T	TX T
3	820117	2005	0	15	340	27	27
3	820206	1733	0	16	320	27	15

RADIO SATELLITE TELEMETRY TABLE

DATA FOR RADIO 8

R-S	DATE	TIME	TX PWR	BATT V	CHARGE I	REG T	TX T
8	820122	821	0	17	380	21	20
8	820213	1630	1378	17	340	29	26
8	820213	1840	1378	17	300	39	25

Fig. 7-16. Printouts of accumulated Radio Satellite telemetry data.

not going to be printed, then the program goes back to read the next set of data (line 150).

The raw telemetry values are converted to engineering units in accordance with the telemetry decoding equations (lines 160 to 180) which are then printed in a formatted manner (lines 190 to 250). The program then jumps back to process the next set of data. Data is stored starting at line 270. New data can be added between lines 430 and 440, renumbering the program after each addition.

The program listed in Fig. 7-17 can be modified for any satellite(s). With the right changes it can be used to display data from AMSAT-OSCAR 8, UoSAT-OSCAR 9, or any other spacecraft. The interesting point about the Radio Satellites is that there are six of them in orbit at this time and data from one can be compared directly to that on the corresponding telemetry channel of another one, while the AMSAT spacecraft are solitary "birds."

After data has been collected for a period of weeks or months, relationships

```
10 LPRINT 'RADIO SATELLITE TELEMETRY TABLE'
20 REM UPDATED 13 FEB 1982
30 REM BY JOE KASSER 1982
40 REM ADD NEW DATA AT END SEE TEXT FOR DETAILS
50 LPRINT
60 INPUT 'Which Satellite (0 FOR ALL) ';X
70 LPRINT ,,'DATA FOR RADIO';X
80 LPRINT 'R-S  DATE    TIME   TX PWR   BATT V  CHARGE I  REG T   TX T'
90 REM MAIN LOOP STARTS HERE
100 READ S1,Y1,M1,D1,T1
110 IF S1 = -999 THEN 460
120 READ K,D,O,S,W
130 IF X = 0 THEN 160
140 IF X = S1 THEN 160
150 GOTO 100
160 K = .2 * K * K
170 D = D * .2
180 O = 20 * ( 100 - O )
190 LPRINT USING '###';S1;
210 LPRINT USING '########';Y1*10000+M1*100+D1,
230 LPRINT USING '######';T1,
250 LPRINT USING '########'; K,D,O,S,W
260 GOTO 100
270 REM DATA STATEMENTS
280 REM   LINE 1 S/C,YR,MNTH,DAY,TIME
290 REM   LINE 2 K  ,D , O  , S , W
300 DATA 3,82,1,17,2005
310 DATA 00,73,83,27,27
320 DATA 8,82,1,22,0821
330 DATA 00,84,81,21,20
340 DATA 5,82,1,28,1928
350 DATA 00,82,83,30,26
360 DATA 3,82,2,06,1733
370 DATA 00,82,84,27,15
380 DATA 6,82,2,06,1752
390 DATA 99,82,73,19,30
400 DATA 8,82,02,13,1630
410 DATA 83,83,83,29,26
420 DATA 8,82,02,13,1840
430 DATA 83,83,85,39,25
440 DATA -999,0,0,0,0
450 REM DUMMY EOF ENTRY
460 END
```

Fig. 7-17. RADIOTLM program.

between the battery, charge current, temperatures, transmitter power, and outside happenings will become apparent. Much has been learned and much more has yet be learned from analyzing spacecraft telemetry.

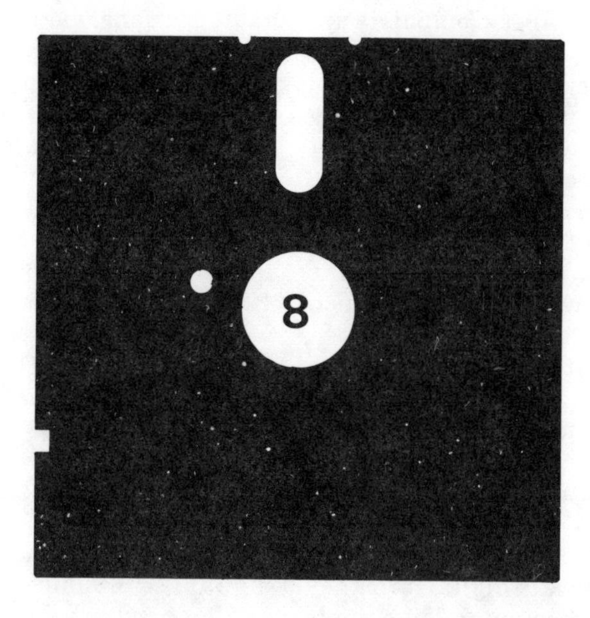

Hardware-Dependent Software

The previous chapters of this book have dealt with software written in BASIC that can be readily adapted to run on most microcomputers. They dealt mainly with data processing (logging, contests, or record keeping), "number crunching" (satellite orbits), and simulations (contest game). This chapter describes some hardware-dependent software in which the microcomputer is directly interfaced to the rest of the amateur radio station.

The software concepts described in this chapter require specific hardware to interface the radio equipment to the computer. Preferably, they should also be implemented in assembly language so that they can run at a higher speed than they would if they were written in BASIC. Assembly language is also more flexible in the sense that the interface driving subroutines are easier to write in assembly language than in BASIC. Since the actual microprocessor inside a microcomputer and the input/output (i/o ports) circuitry are different from computer to computer, and the interface circuitry between the radio equipment and the computer (i/o) ports varies so much, the software in this chapter will be described conceptually in terms of its performance rather than in terms of the actual code in order to give you ideas about the potential of the microcomputer when tightly coupled to the amateur radio station hardware.

INTERFACES

An interface is a direct connection that connects or *interfaces* the computer to a peripheral. The standard CRT/keyboard that all general-purpose computers contain is an interface to the user.

Interfaces come in two forms, analog and digital, and these two forms contain two varieties, serial and parallel. Computers are digital machines in which things are very clear. An electronic circuit has two states, "on" and "off." The outside world is usually an analog world in which things are not carefully defined. Special circuits, now usually contained within an integrated circuit, have been developed to convert the analog outside world to the digital world of the computer. There are two such circuits which provide two-way conversions.

An analog-to-digital converter provides the conversion of signals from the outside world to the computer, and a digital-to-analog converter performs the reverse function, converting digital signals generated by the computer to analog signals the outside world uses.

Once the signals are in digital form they have to be fed into the computer. The signals can be fed into the computer in a serial or a parallel manner. The serial interface feeds one piece of information at a time to the computer. The most common serial interface is the UART or USART that interfaces a serial data line, as you might find between a CRT terminal and a computer.

The parallel interface feeds a number of pieces of information to the computer at once. It usually feeds eight bits, commonly called one byte, of information to the machine at a time.

RTTY

RTTY is a digital mode of communication between radio stations. It began with surplus military equipment and has today progressed to sophisticated and complex computer-controlled communications circuits both at hf and at vhf.

RTTY today is transmitted using two different and incompatable codes, ASCII and the BAUDOT. RTTY started with surplus equipment using the BAUDOT code and only a year or so ago, when microcomputers were rapidly appearing in amateur radio stations, did the use of ASCII become legal. After all, why bother to convert from ASCII (in the computer) to BAUDOT (on the air) and back to ASCII again (in the computer at the receiving station), when both sides can happily communicate using ASCII? The BAUDOT code shown in Fig. 8-1 contains 64 possible characters, while the ASCII code shown in Fig. 8-2 contains 128 different characters. It is not very convenient to exchange computer programs on the air using BAUDOT.

RTTY is a serial mode of communication. Characters are transmitted one at a time at a number of different speeds or *baud rates*. The usual ASCII speeds are 110 or 300 baud, while Baudot is transmitted at the slower rates of 45.5 or 50 baud. Being a digital communications mode using an analog medium (radio), the serial digital signals entering and leaving the computer are interfaced to the radio via a terminal unit or modem that converts the digital signals to two tones, one representing the mark (high) frequency, the other representing the space (low) frequency. Thus RTTY is transmitted on the air as a pair of tones.

Simple RTTY Transmit/Receive Techniques

The basic RTTY interface is shown in Fig. 8-3. It contains an analog part and a digital part. The analog section converts the on-the-air tones to digital levels, while the digital section interfaces the serial data to the computer. The serial port in the computer must be able to lock up on data at the various baud rates used by amateur RTTY. Most commercial serial interfaces are not suitable for amateur RTTY because they cannot work at 45.5 bauds which is at present the most commonly used RTTY communications rate.

Inside the computer, the main piece of software required is code converter that converts Baudot to ASCII and back again. This software usually uses a look-up table in

Bit Numbers	Letters Case	Figures Case				
			US Alphabets			
5 4 3 2 1	International Alphabet #2	International Alphabet #2	Military Std	Weather	TWX	Telex
0 0 0 0 0	Blank*	Blank*	Blank*	—	Blank*	Blank*
0 0 0 0 1	E	3	3	3	3	3
0 0 0 1 0	Line Feed	Line Feed	Line Feed	Line Feed	Line Feed	Line Feed
0 0 0 1 1	A	—	—	↑	—	—
0 0 1 0 0	Space	Space	Space	Space	Space	Space
0 0 1 0 1	S	(Apos)	Bell	Bell	Bell	(Apos)'
0 0 1 1 0	I	8	8	8	8	8
0 0 1 1 1	U	7	7	7	7	7
0 1 0 0 0	Car. Ret	Car. Ret	Car. Ret	Car. Ret	Car. Ret	Car. Ret
0 1 0 0 1	D	WRU	$	↗	$	WRU
0 1 0 1 0	R	4	4	4	4	4
0 1 0 1 1	J	Aud Sig	(Apos)'	↙	(Comma),	Bell
0 1 1 0 0	N	(Comma),	(Comma),	O		(Comma),
0 1 1 0 1	F	†	!	→	1/4	$
0 1 1 1 0	C	:	:	O	WRU	:
0 1 1 1 1	K	(	(	←	1/2	(
1 0 0 0 0	T	5	5	5	5	5
1 0 0 0 1	Z	+	"	+	"	"
1 0 0 1 0	L	)	)	↖	3/4	)
1 0 0 1 1	W	2	2	2	2	2
1 0 1 0 0	H	†	Stop	↓		#
1 0 1 0 1	Y	6	6	6	6	6
1 0 1 1 0	P	0	0	0	0	0
1 0 1 1 1	Q	1	1	1	1	1
1 1 0 0 0	O	9	9	9	9	9
1 1 0 0 1	B	?	?	·ƀ	5/8	?
1 1 0 1 0	G	†	&	↘	&	&
1 1 0 1 1	Figures	Figures	Figures	Figures	Figures	Figures
1 1 1 0 0	M	.	.	.	.	.
1 1 1 0 1	X	/	/	/	/	/
1 1 1 1 0	V	=	;	Ⓣ	3/8	;
1 1 1 1 1	Letters	Letters	Letters	Letters	Letters	Letters

Notes. Transmission Order Bit 1→Bit 5.
*"Blank" in US; "No Action" in International Alphabet #2.
† Unassigned (domestic variation, not used internationally).

Fig. 8-1. Baudot code.

which the characters are stored in the Baudot order, as shown in Fig. 8-4. The flowcharts for the code conversion subroutines are shown in Fig. 8-5. Once the code conversion is performed, the characters may be sent to the radio or to the CRT as desired.

Semi-Smart RTTY Stations

With computer power at our beck and call, the properties of the computer can be used to smarten up the RTTY station. Standard items of information describing the station (commonly known as brag tapes, because they were originally stored on paper tapes) can be stored in the computer and transmitted as desired. The computer can also be used in a split screen/buffer mode, in which incoming messages are displayed on the top section of the screen, and messages being typed in for transmission are displayed on the lower half. The operator can thus type a reply as the incoming message is being received, and the computer will send the whole message at the full speed when instructed to do so. The computer will also

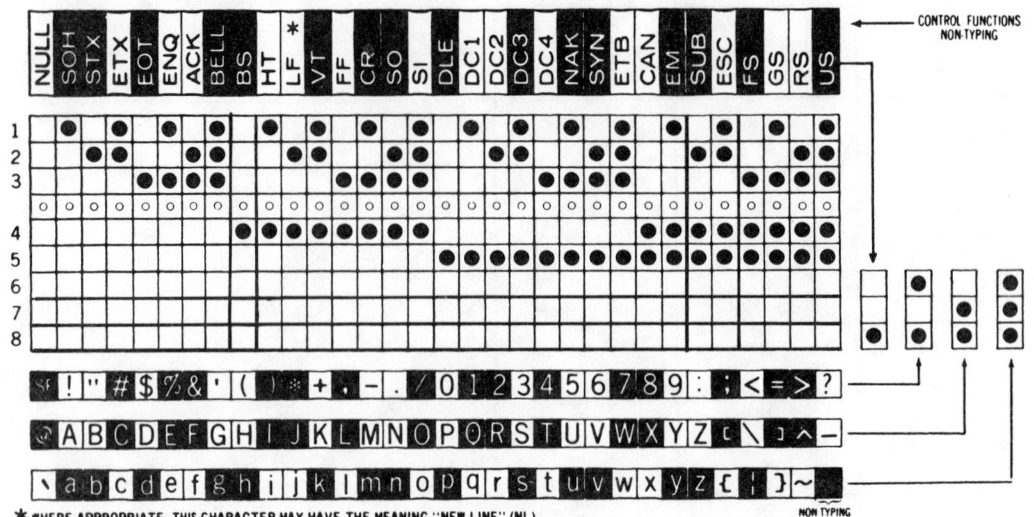

Fig. 8-2. ASCII code.

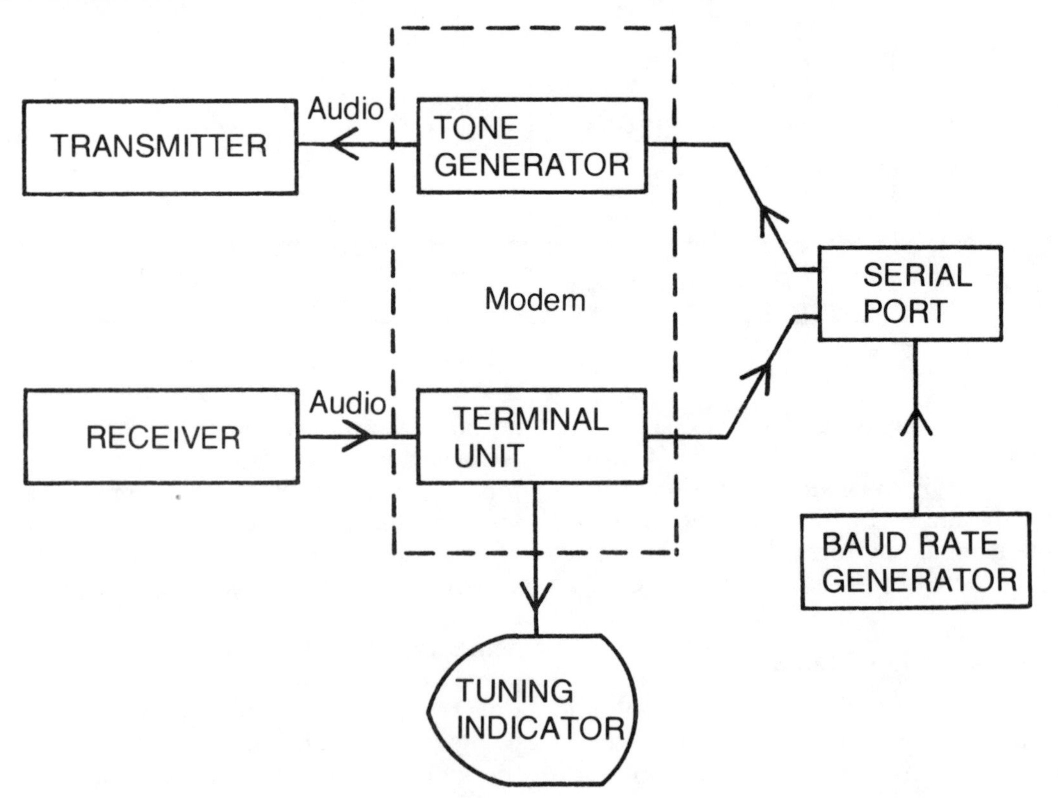

Fig. 8-3. Basic RTTY Interface.

```
                              ;LOOK UP TABLE FOR CODE CONVERSION
FB04 00450A41    BCN1:  DB    0,'E',0AH,'A'
FB08 20534955           DB    ' SIU'
FB0C 0D                 DB    0DH      ;CR
FB0D 44524A4E46         DB    'DRJNF'
FB12 434B545A4C         DB    'CKTZL'
FB17 5748595051         DB    'WHYPQ'
FB1C 4F4247             DB    'OBG'
FB1F 00                 DB    0        ;FIGS
FB20 4D5856             DB    'MXV'
FB23 00                 DB    0        ;LETS
FB24 00                 DB    0        ;BLANK
FB25 33                 DB    '3'
FB26 0A                 DB    0AH      ;LF
FB27 2D2007             DB    '- ',7   ;7= BELL
FB2A 3837               DB    '87'
FB2C 0D                 DB    0DH      ;CR
FB2D 2434272C21         DB    '$4',27H,',!:'
FB33 2835222932         DB    '(5")2'
FB38 1B                 DB    1BH      ;ESC KEY = ^H
FB39 363031393F         DB    '6019?'
FB3E 26                 DB    '&'
FB3F FF                 DB    0FFH     ;FIGS
FB40 2E2F3B             DB    './;'
FB43 FF          BCN2:  DB    0FFH     ;LETS
```

Fig. 8-4. Look up table for ASCII/Baudot conversions.

take care of the Morse code identification requirement. If a time of day clock is available, it can be used to *time tag* the transmissions. A frequency counter can be used to log the frequency. If floppy disks are available, the whole QSO can be logged automatically.

RTTY messages are pretty standard. The computer can be programmed to recognize a CQ call or your own callsign. This selective call (SELCAL) technique has been used for years. Initially it used hard-wired digital logic, but it has now graduated to microcomputer-based designs.

An Automatic RTTY Contest Station

The software and hardware assembled for use in the semi-smart RTTY station can be integrated to provide a fully automatic contest station. In a contest the messages are standard. The computer is not very smart but can be readily programmed to recognize and generate standard contest messages. If the performance of a contest station is analyzed, a flowchart can be written describing what the operator does during the contest. This flowchart, shown in Fig. 8-6 can then be coded to provide an automatic RTTY contest station.

SLOW SCAN TELEVISION

Slow Scan Television (SSTV) is a medium for transmitting still pictures via amateur radio. It is almost identical to facsimile in concept. The standards commonly used in amateur SSTV are listed in Fig. 8-7. In the early days of SSTV, the picture was displayed on a cathode ray tube having a long persistence. Once the novelty of receiving pictures had worn off, the display was seen to be crude and inconvenient. The tube had to be viewed in a darkened room. The picture updated slowly (every eight seconds) and faded rapidly. Most people, in fact, photographed the picture and viewed the photograph rather than the screen. See Fig. 8-8. The second generation of SSTV equipment incorporated scan converters (Fig. 8-9). These are relatively expensive, but convert the SSTV pictures into a form displayable on a standard fast scan TV monitor.

The advent of the microcomputer introduces the third generation of SSTV equipment, which improves the capabilities of the medium by at least an order of magnitude (Fig. 8-10). It allows image processing

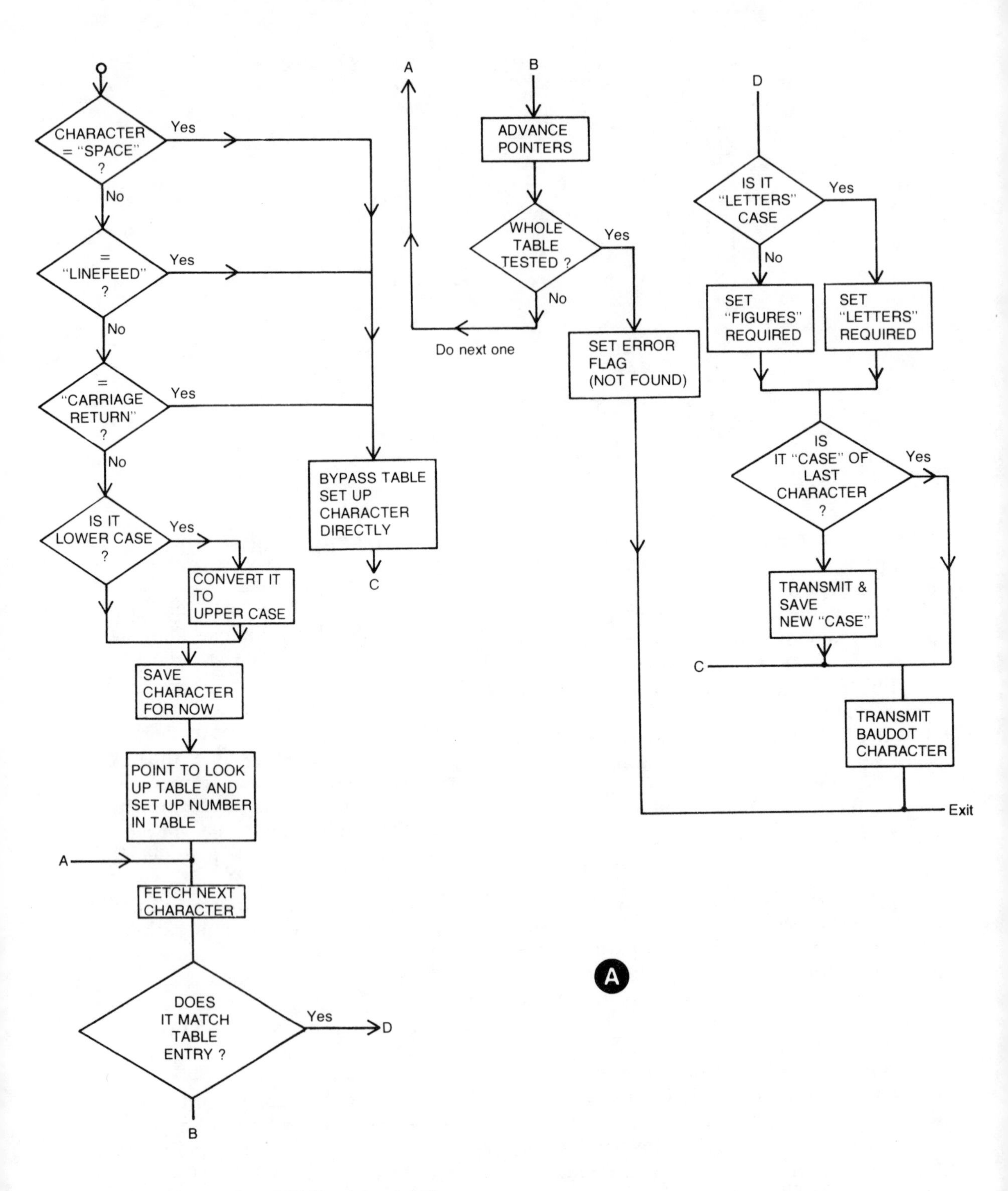

Fig. 8-5. Flowcharts for ASCII/Baudot conversions.

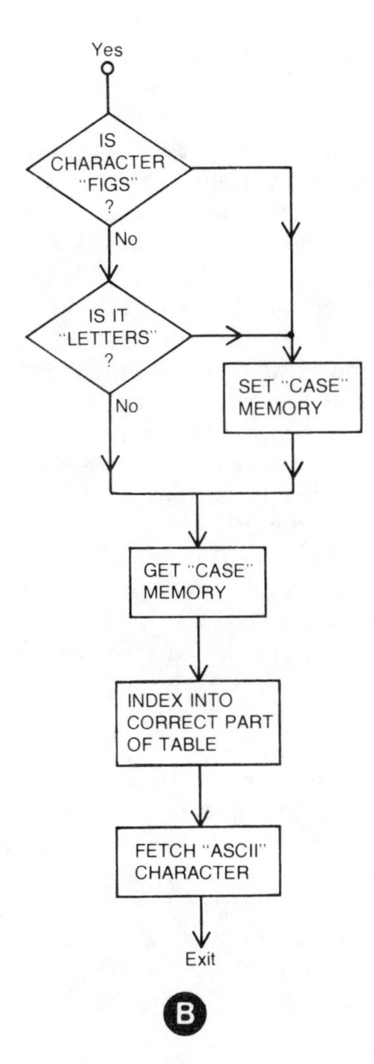

techniques to be used to cut down the effects of interference, and it allows real-time color displays using a field sequential color transmitting scheme.

Basic SSTV

Basic SSTV equipment is a picture source and display. Many amateurs, when first entering the mode, purchase a display and generate a tape recording on a friend's system to use as a video source. Later, as funds permit, they add cameras, electronic pattern generators, and other signal sources.

The microcomputer can be added to the SSTV station in stages. It can be used to generate patterns or characters. For example, it can display any characters typed in at the console. It can digitize and scan

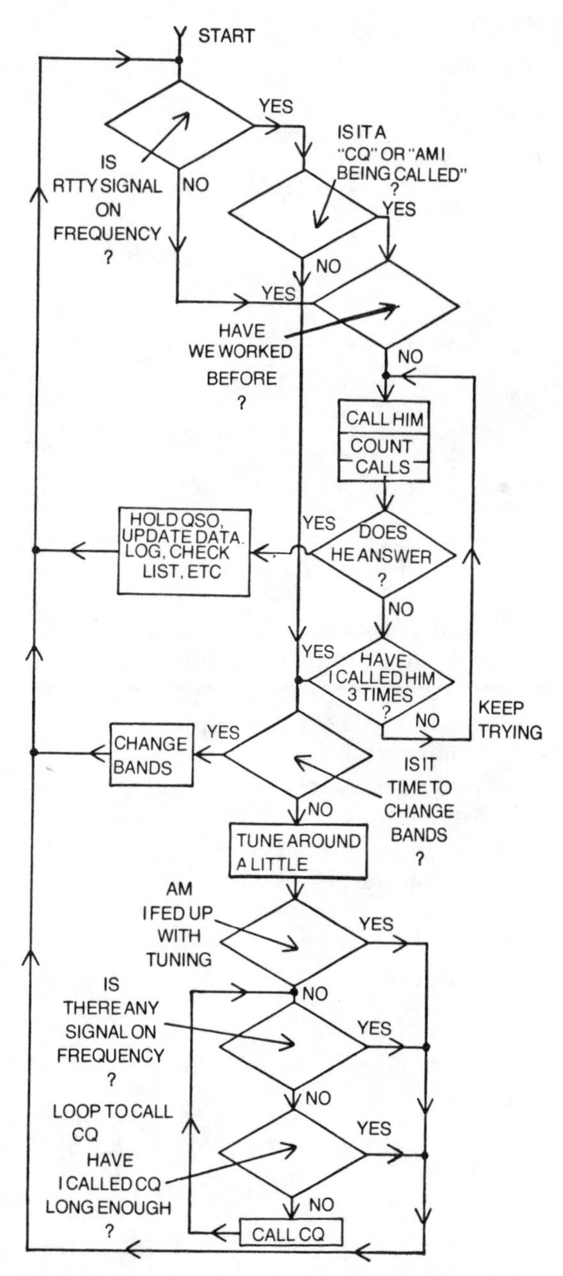

Fig. 8-6. Flowchart for automatic RTTY contest station.

Sweep Rates	60 Hz	50 Hz
HORIZONTAL	15 HZ	16.667 HZ
(Division	60/4	50/3)
Sync Pulse	5 mS	5 mS
VERTICAL	8 Seconds	7.2 Seconds
Sync Pulse	30 mS	
LINES	120	
ASPECT RATIO	1:1	
DIRECTION OF SCAN		
Horizontal	Left to Right	
Vertical	Top to Bottom	
SUBCARRIER FREQUENCY		
Sync	1200 Hz	
Black	1500 Hz	
White	2300 Hz	
BANDWIDTH	1.0 TO 2.5 KHZ	

Fig. 8-7. Amateur Radio SSTV standards.

convert incoming pictures. A picture library can be built up on floppy discs. Then, as image processing software gets added, interesting things begin to happen.

Initially, the use is to compensate for interference on received signals. Simple techniques (algorithms), such as displaying the average picture received over a number of sequential frames, can be used. Grey scale adjustments can be made. A whole picture editing system can be developed using the console or a light pen or both as control inputs. Color can be added if three pictures (red, blue and green) can be stored separately and combined in the display.

A fast scan display at the system console makes the pictures much more viewable. Later additions to the software can include overlaying one picture upon another, merging pictures, and merging sections of different pictures. A picture of a person can be overlayed onto different backgrounds. Composites can be built up. Data can be added to annotate the picture. The narrow bandwidth of the transmission medium as well as the visual impact of the display when used with the microcomputer have the potential to make SSTV as popular as SSB voice in the amateur field.

Color

Slow scan color may be transmitted using a field sequential technique. The picture to be transmitted is separated into blue, red, and green components. Each component is transmitted as a separate picture frame, and assembled at the receiving station into a color picture. The transmission of one slow scan color picture thus requires three frames.

Transmission is quite simple because the regular station camera can be used.

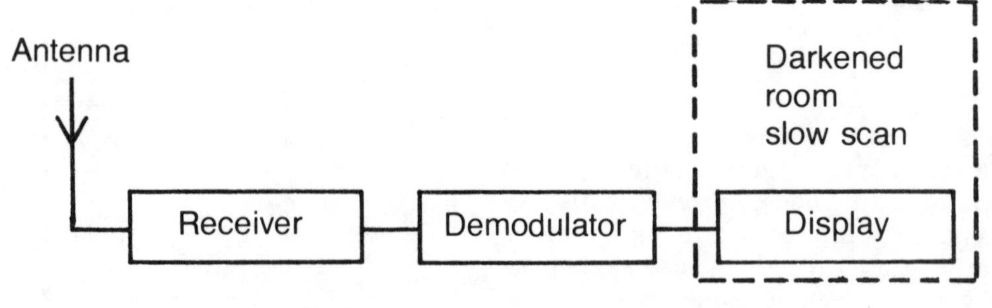

Fig. 8-8. First generation SSTV receiving setup.

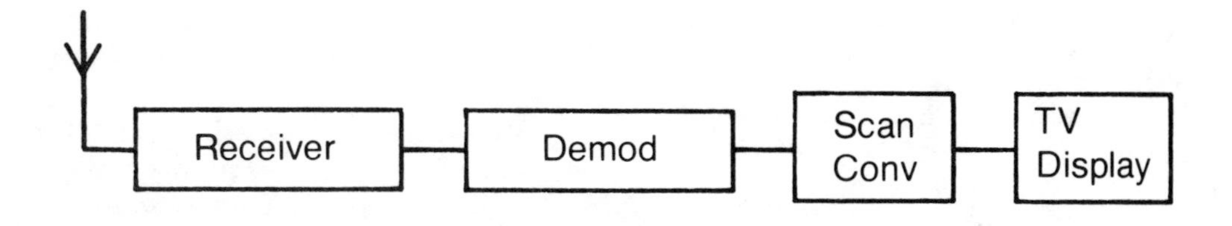

Fig. 8-9. Second generation SSTV receiving setup.

Color separation is performed by inserting a filter between the camera and the subject. Three filters are required; namely, green, red, and blue. The filters can be glued to a disc and rotated by a stepping motor so that the disc is advanced one filter band for each field sync pulse, as shown in Fig. 8-11. The more filter bands that are present on the disc, the fewer steps the motor has to sequence to switch colors. This simple technique allows the transmission of "live" color pictures, but ignores the problem of the receiving station deciding which frame is allocated to which color. This problem can be solved by always transmitting the colors in the same sequence and verbally announcing which color is coming first, or by transmitting some kind of color-reference synchronizing signal. Since color slow scan television is still in its infancy, these standards still have to be worked out.

Picture Generation

The microcomputer can be used to generate electronic test patterns and captions. A CQ message can be readily implemented using standard character generator circuitry. A video QSL card can be displayed and transmitted simply enough. The home station callsign is standard and the call of the other station is typed in for each contact.

Software can also be written to generate pictures. A light pen or even the arrow keys on the keyboard can be used to draw a picture on the CRT screen. Alphanumeric captions can be added, and the whole picture stored on a floppy disk or tape cassette. The picture can then be transmitted at will. Alternatively, an incoming picture can be stored and altered, then sent back to the originator. Your caption or callsign can be superimposed on a picture received from an-

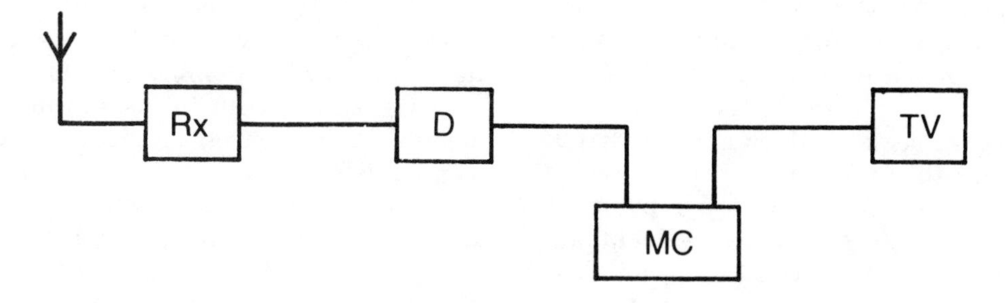

Fig. 8-10. Third generation SSTV receiving setup.

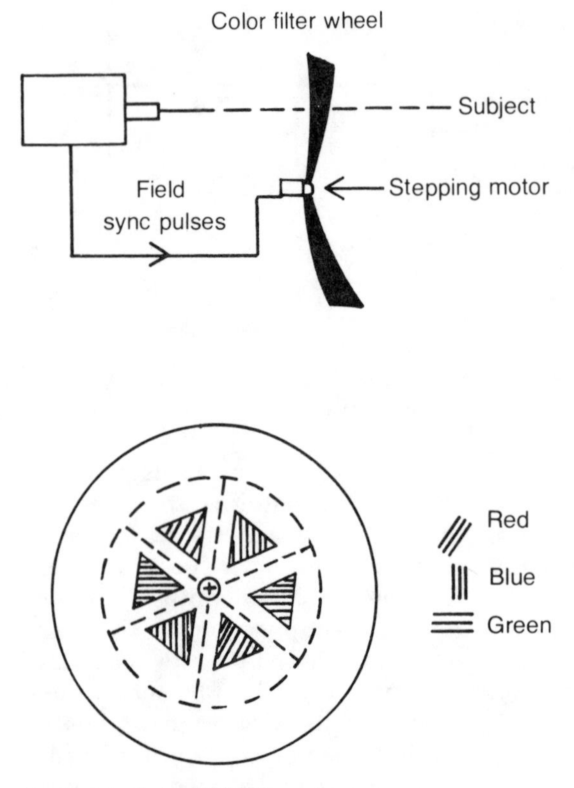

Fig. 8-11. Color SSTV filter wheel.

other station either on the air or via tape. If a TV camera is not available, this technique can provide an inexpensive means of generating pictures using the station computer.

Picture Processing

Once pictures are stored in memory they can be processed. For example, black and white pictures are usually transmitted as a series of identical frames. The computer can compare the sequential frames and display those parts that are identical. The bad parts can be blacked out or the computer can guess what they should be by looking at what came before them and what came after them. If a color display is available, the QRM'd parts can be shown in color while the rest of the picture would appear as black and white. Different frames can be displayed in different colors and the effects of interference on the different frames readily noted.

If the drawing capability described in the previous section is available, the incoming picture can be edited by hand to remove any interference effects and then saved.

All the real-world techniques of computer picture enhancement can be applied to SSTV. For example, if the 120 by 120 resolution is not good enough for you and you don't like the squares that the dots are blown up to, you can make the display 240 by 240 if you have the hardware, by interpolating one pixel of your own between each of the received pixels, and setting the value of this new pixel to the average value of the intensity of the two received pixels on each side of it. If you want to get fancier, put the pixels in different positions on each alternate line and average the intensities of the received pixels both from the current line, and from the lines above and below the inserted element. This will not really improve the resolution any but will make the picture more viewable.

Digital Slow Pictures

Videographic displays and digital communications can be merged to provide picture transmission capabilities. A videographic display causes the contents of a memory area in the computer to be displayed on a TV screen. The display may be color or black and white with grey scales. If the contents of a display memory in one computer can be transmitted to the display memory of another one, a picture will have been transferred.

If two radio amateur stations have the same video display hardware, the transfer of the contents of the video memory in one computer to the other one in effect transfers a picture from one station to the other. If the displays have 1024 by 1024 pixel resolution with color and luminance, then a very high-resolution picture can be transferred, albeit at a slow rate. The majority of videographic displays at this time have resolutions of 128 by 128 or 256 by 192. Some may be color, some may be black and white, and some may be color but viewable as black and white on a black and white monitor. Now it is desirable that anyone with a graphics video display capability in his computer should be able to transmit and receive pictures if a suitable modem is available.

The digital format of transmission means that any format could be used providing that suitable software exists. In order to avoid resolution problems, it is desirable that the formats be compatable with all kinds of displays, a situation that is hardly likely to occur in practice.

The digital format is really a digitized analog signal. Thus, a 0 level, or a sync pulse level, will always be a 00 signal and a very bright signal would always be a OFF Hex or a maximum. If signals are transmitted in consecutive lines with a header in the first line, format-independent pictures can be transmitted. The header would contain the data shown in Fig. 8-12.

Sync	4 bytes	(Binary)
Destination station callsign	10 byes	(ASCII)
Sending station callsign	10 bytes	(ASCII)
Horizontal resolution	2 bytes	(Binary)
Vertical resolution	2 bytes	(Binary)
Colors	1 byte	(Binary)
Luminance	1 byte	(Binary)
Spare	2 bytes	
Total	32 bytes	

Each byte contains 8 bits

Sync header	Pixels	Sync header

Fig. 8-12. Digital picture transmission information.

The sync vector of four bytes serves to synchronize the system and notify the receiver that a header is coming. The callsigns of the sender and recipient are then transmitted in the next 20 bytes. Ten bytes are sufficient for 99.9% of the amateur radio callsigns allocated today. For example, GW3ZCZ/KH6 is only ten bytes long.

The next two bytes contain in binary code the number of pixels per line, followed by two bytes that contain the number of lines in a frame also in binary. One byte each is allocated to color and luminance information. The color byte could be an ASCII "R," "G," or "B" to signify which frame of a field sequential color picture is being transmitted or any other information. The luminance byte contains information about how many grey levels are present in the picture. Two spare bytes are allocated at this time so as to bring the byte count to 32.

The picture information then follows. A sync header synchronizes the receiving station software and the pixels are then transmitted. The number of pixels is known from the header, and when the line is completed, the sync information is again transmitted to ensure that the receiver stays in sync with the transmitter. When a whole frame has been transmitted, the sequence may stop, or another frame may be transmitted.

The picture transmission has been defined in terms of a memory-to-memory transfer. If both stations have identical videographics hardware they can display the same picture. What happens, however, if the second station does not have the same hardware as the first?

If the receiving station has hardware with the same resolution as the transmitting station, the same picture can be displayed. If color or luminance are the same they will be lucky. If the receiving station does not have color, it will probably be able to display a black and white picture.

If the receiving station has a different resolution than the transmitting station, a number of choices are open. Assume that the picture being transmitted is 256 by 256 pixels, and the receiving station has a 128 by 128 display. The receiving station has a number of options. The first option is to perform some kind of signal processing on the picture data to combine two pixels in both horizontal and vertical directions to reduce the resolution of the picture. This allows the whole picture to be displayed, but at a lower resolution. The remaining options are shown in Fig. 8-13. Here a reduced area of the picture is displayed. A corner, as shown in Fig. 8-13A, or a middle section, as shown in Fig. 8-13B, may be chosen. This technique is sometimes used to zoom in on selected portions of a high-resolution picture.

As the display options are in software, the receiving station has a choice of how to display the picture. If the receiving system has better resolution than the sending station, similar techniques can be used to either display the picture in a part of the screen together with other desired information (such as callsigns) in the remaining portions of the screen or the picture information can be doubled both horizontally and vertically to fill the screen. If the receiving station has a 256 by 192 resolution, for example, the choices can be applied to the vertical portion only.

The picture has been transmitted as a memory-to-memory transfer. The software in the receiving station can thus decide if processing has to be performed for the display. The processing can be performed on the picture as it is received in systems with minimal amounts of memory, or the picture can be stored in a memory bank and moved to the video display area by the processing algorithm.

The mode of transmission of the digital data can be ASCII, RTTY, or packet, depending on what is available to the users. A checksum was not placed on the end of each line because packet transmissions contain their own error-detecting and correction techniques.

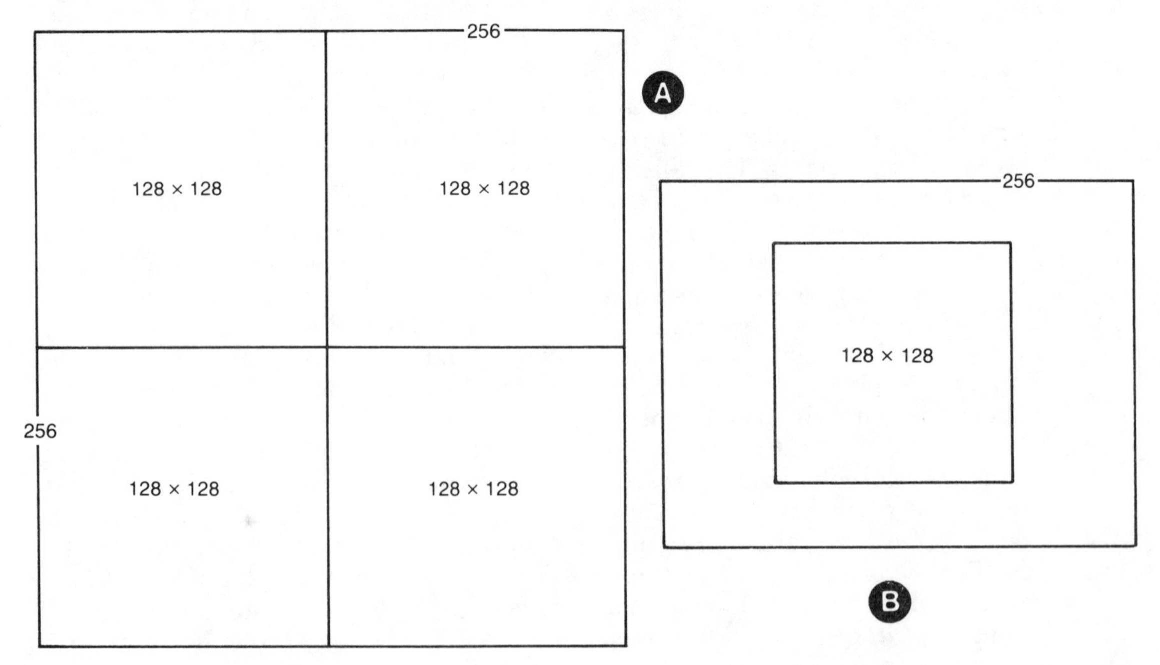

Fig. 8-13. Lower resolution receiving options.

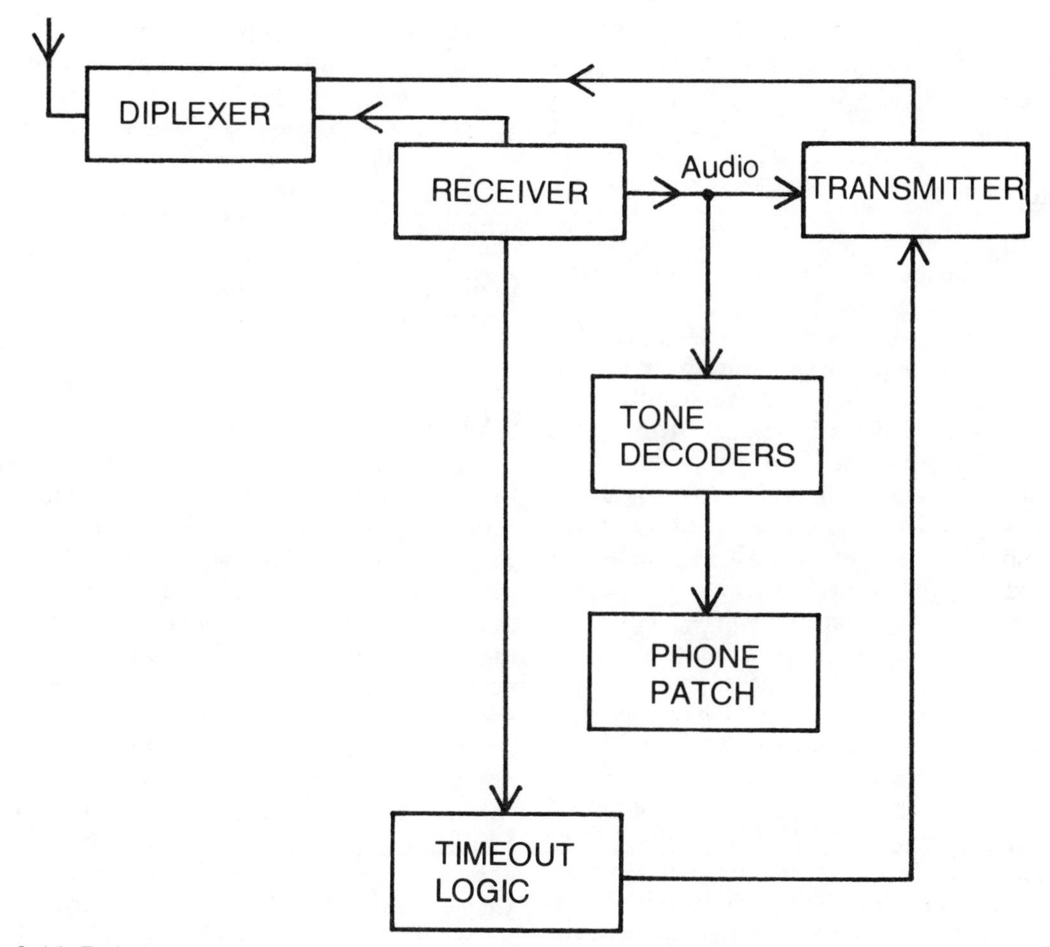

Fig. 8-14. Typical repeater configuration.

The use of digital transmission formats for slow scan television changes the meaning of slow, for it no longer applies to low-resolution pictures, but can now refer to the slow data rate for transferring a picture of any resolution.

REPEATERS

Although repeaters can be considered as a receiver and transmitter pair in which the audio output of the receiver is connected to the audio input of the transmitter and the signal detected or squelch output of the receiver connected to the transmitter transmit control line as shown in Fig. 8-14, the typical repeater contains a number of other circuits. These circuits include an identification device, usually in Morse code, a timer for calculating when to identify, a delay on the squelch to keep the transmitter on for a short period of time after the received signal disappears, and a phone patch.

The incorporation of a dedicated microprocessor in the repeater control circuitry can simplify the hardware, while at the same time increasing the performance capability.

Identification Function

The ID timing function can be performed in software. The code generation for the ID can be performed using the Morse code generation subroutines. The actual tones themselves may also be generated in software.

Autopatch

Many repeaters having autopatch facilities restrict them to club members by having an access code. This means that the code is changed once a year when dues time arrives. This requires that dues be paid on a calendar basis or there is the possibility that someone can freeload for part of the year, should their membership expire sometime before the access code is changed. By using a microprocessor to decode the access code, each individual club member can have a different access code, which can be flagged when the membership expires. Thus, attempts at accessing with an expired membership could result in a message reminding the member to send in his or her dues.

The first digit and the number of digits can be monitored to inhibit long distance calls. The microprocessor can log the time of a call, the number called, and the access code used. Logging requirements would then be met and each user could be presented with a summary of their calls in the event that the repeater club requests payment for calls made through the machine.

Emergency or other standard numbers could be made via two-digit access codes using # or * as the first digit.

Self Test

The frequency of the signal into the receiver can be monitored and indicated to the user by inserting K, L, or H signals following the end of each incoming transmission. The K means correct or on frequency; L means that the signals were low; H means that they were high. This helps people set crystal-controlled equipment on frequency.

By using particular touch tones it is possible to supply users with an indication of their received signal levels, as well as their center frequency. Users can then adjust transmitters or antennas during non-peak-use hours using the repeater as a remote IF monitor. If the repeater does not have a phone patch, a user ought to be able to dial in on the control line, and be able to listen to his audio.

Cross Banding or Linking

The basic repeater serves a limited area. In many areas it is desirable to increase the coverage for specific applications. In this way, amateurs with specific interests who have repeaters can communicate—for example, two cities close together which having RTTY repeaters. Each can operate independently, but they can also be linked. Any amateur on either machine could send a CQ on the local machine, and, if no response is obtained, could access via a touch tone code the link, and try again on the other machine. Clubs having DX alert channels for country chasers or contests could link machines if they were dispersed over wide geographical areas. VHF repeaters can be linked to HF. A 2 Meter repeater can have an auxiliary 10 Meter channel, accessable via tone.

If the repeater is run by a 6 Meter club, or other VHF/UHF DX activity club, a beacon receiver on the band of interest could be placed at the receiver site. The microcomputer could indicate, upon request, the level of activity on the beacon frequency. Thus, 6 Meter DXers need not rush home from work expecting sporadic E activity only to find that there was none, and people would not miss openings because they were not at their stations. A simple access of the repeater and the right touch tone signal would tell them at once how band conditions are.

Phoneline Access

Qualified club members should be able to dial up the repeater control line and talk over the air to locate someone by radio when all other techniques fail. If the repeater is equipped for RTTY, the microprocessor at the repeater can even convert telephone modem tones and speeds to Baudot tones and speeds and allow RTTY operation in that manner. The microprocessor would have to contain a first-in, first-out (FIFO) buffer to allow for the change in baud rates.

REMOTE VHF/HF OPERATION

The new processor-controlled HF transceivers make remote control techniques easy to implement. It is now possible to operate your one kilowatt HF station while out driving around in your car using the VHF/UHF bands as the link between you in your car and the HF transceiver in the shack. The hardware needed to perform the task is a VHF/UHF transceiver in the mobile or remote site, and both VHF/UHF and HF transceivers, as well as an interface or control unit, at the base station site.

Consider the requirements for such techniques and how they can be implemented. The basic requirements are the ability to tune the HF radio, ease of control from transmit to receive, and secure operations and tight control.

The control functions are performed in a number of ways. Many HF transceivers are now equipped with a microprocessor-based accessory frequency-controlling device that has the ability to scan, memorize, and enable remote up/down tuning of the band. The VHF/UHF transceiver used at the

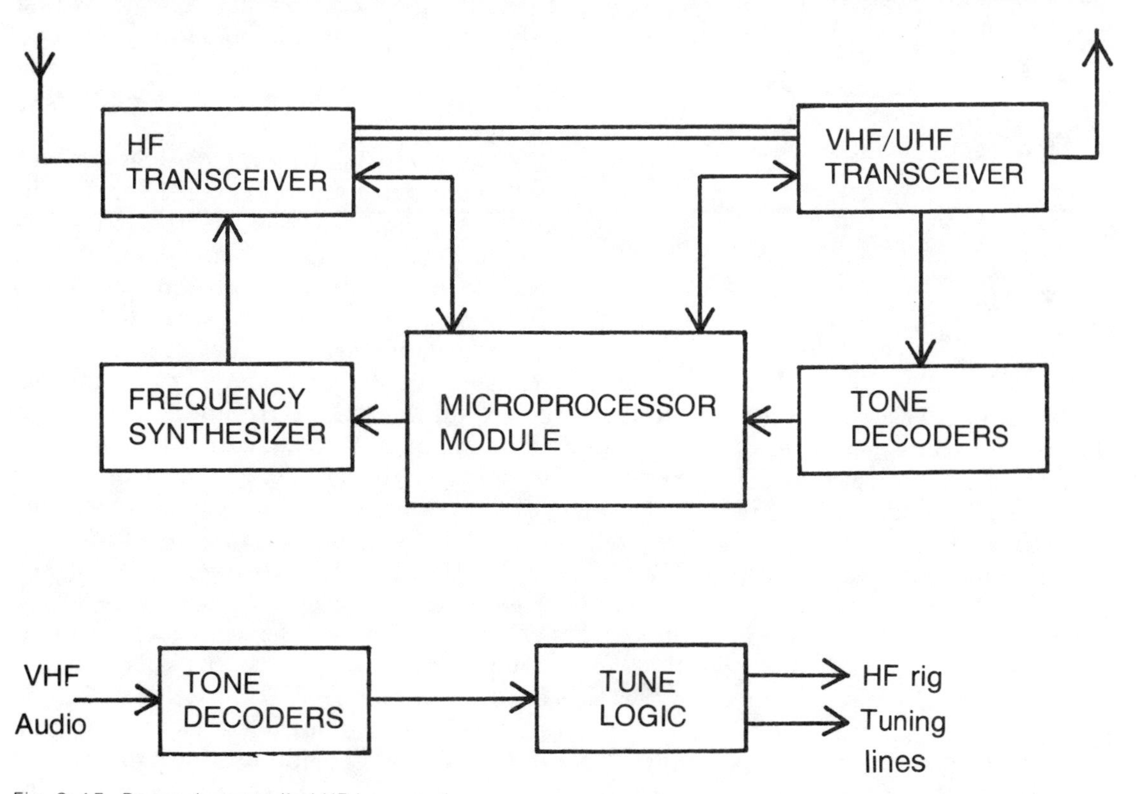

Fig. 8-15. Remotely controlled HF base station.

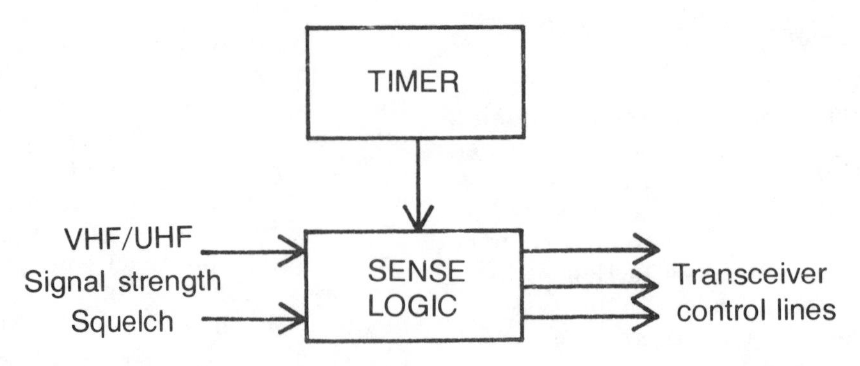

Fig. 8-16. Remote control interface unit.

remote location (portable or mobile) is equipped with a tone pad if a base station is set up as shown in Fig. 8-15. These tones can be used to perform the control operations as follows:

- 1—start tuning HF
- 2—control function #1
- 3—start tuning LF
- 4—tune HF one step
- 5—control function #2
- 6—tune LF one step
- 7—rotate base station antenna clockwise
- 8—control function #3
- 9—rotate base station antenna anticlockwise
- 0—stop tuning
- *—special function
- #—activate the link

There are two modes of tuning the base-station transceiver, slow and fast. In the slow mode, the base-station HF transceiver VFO can be tuned one increment higher or lower in frequency each time the respective digit (4 or 6) is received on the VHF/UHF link. Since the basic increment is very small (about 10 or 20 Hz), tuning across the band can take awhile. The second mode allows for faster tuning. The VFO begins to tune when the start tone is transmitted. The audio output of the HF transceiver is radiated on the VHF/UHF link and when the wanted frequency is reached, the stop tuning tone is activated. In case of overshoot, fine tuning can be implemented using the slow mode.

A typical block diagram for a simple interface unit is shown in Fig. 8-16. The tone decoders generate the control signals as and when tones are received. The tuning pulses for the HF transceiver VFO are generated in the tune logic module, while control of which transceiver is transmitting at any one time is governed by the sense-logic module. The sense-logic module also controls the system timeout.

The operation (the HF receiver output), is transmitted by the VHF/UHF transmitter, but every few seconds the carrier is dropped for about 400 ms. During this dropout time the sense logic monitors the S meter in the VHF/UHF receiver. If a signal is received, the sense logic goes into the HF transmit mode and relays anything received on the VHF/UHF link to HF, except IF tone signal. A subaudible tone or tone-burst detection circuit can also be placed into the sense logic to inhibit unwanted VHF/UHF signals from being retransmitted at HF. The dropout time has to be just greater than the stabilization time of the VHF/UHF transceiver and can be made small enough so that it is virtually indetectable by the VHF/UHF listener. The HF transmitter is controlled using its VOX circuitry. This technique allows the use of a single band VHF/UHF transceiver in the half-duplex mode. If two bands are available, then full duplex at VHF/UHF can be achieved. The sense logic also shuts down both transmitters if an input does not appear on the VHF/UHF link for a preset time interval.

The basic system can be made more sophisticated in a number of ways. If a synthesized VFO is used at either end, one of the special-function digits can be used to prefix a frequency input. It would then be possible to punch *320*, for example, and have the HF transceiver instantaneously tune to 14.320 MHz. A similar function could instruct the VHF/UHF transceiver to change its frequency. The remaining digits are available for other uses, such as antenna rotation, power control, identification, and invoking any memorized frequencies in the microprocessor-based frequency-controlling accessory.

MULTICHANNEL COMMUNICATIONS

In many instances, two or more radio amateurs wish to have a communications link using VHF fm channels. Conventionally, they would all tune to a single channel and receive all signals on that channel. The number of radio amateurs using VHF fm is constantly growing and, as such, no single group can hope to be the sole user of a single channel in any location for very long. As more people start to use that channel, the group may incorporate sub-audible carrier or tone burst techniques so that transmissions not directed at the group do not activate their receivers and are not heard. This overcomes the problem of how to ignore unwanted signals but does not overcome the problem of the channel being in use when one member of the group desires to initiate a transmission to another member of the group.

As more amateurs operate VHF fm, the common simplex channels become crowded, and there is a tendency to move into less-used channels. As synthesized transceivers become popular, any channel can be accessed as easily as any other, and semi-private channels become a thing of the past. Thus communications links tied to a single channel are subject to ever-increasing interference.

The probability of any named channel being occupied at any given time is much higher than the probability of any one channel out of a named set being unused at any given time. The exact relationship is a function of many variables, including the number of channels, the number of amateurs having equipment capable of operating on those frequencies, and the time of day. If the group monitors one channel out of a set which is occupied by a lengthy conversation so that they cannot use it, communications cannot be established even if all the other channels are not in use, since none of the group is monitoring those channels. The fact that they are unused is of no practical use to the group.

The microprocessor can be used to enable multichannel communications capability in this situation if each station in the group uses the same algorithm in both the transmit and receive modes. They can communicate using any free channel in the band, without even knowing which channel they are using. The algorithms are described below.

Receive

Each receiver scans the channels in sequence. If a signal is not detected on any channel, the next channel is scanned. If a signal is detected, then a test is made for the coded tone. If it is not received, the next channel is scanned. If the tone is detected, scanning stops and the audio is passed through to the speaker. When the transmission terminates, scanning continues with the next channel in the sequence. When all channels have been scanned, the sequence starts again with the first channel. The flowchart is shown in Fig. 8-17.

Once a signal tone has been received, the technique to test for the end of the transmission depends on whether a tone burst or sub-audible tone was used to encode the transmission. If a tone burst was used, the test is performed on the signal level

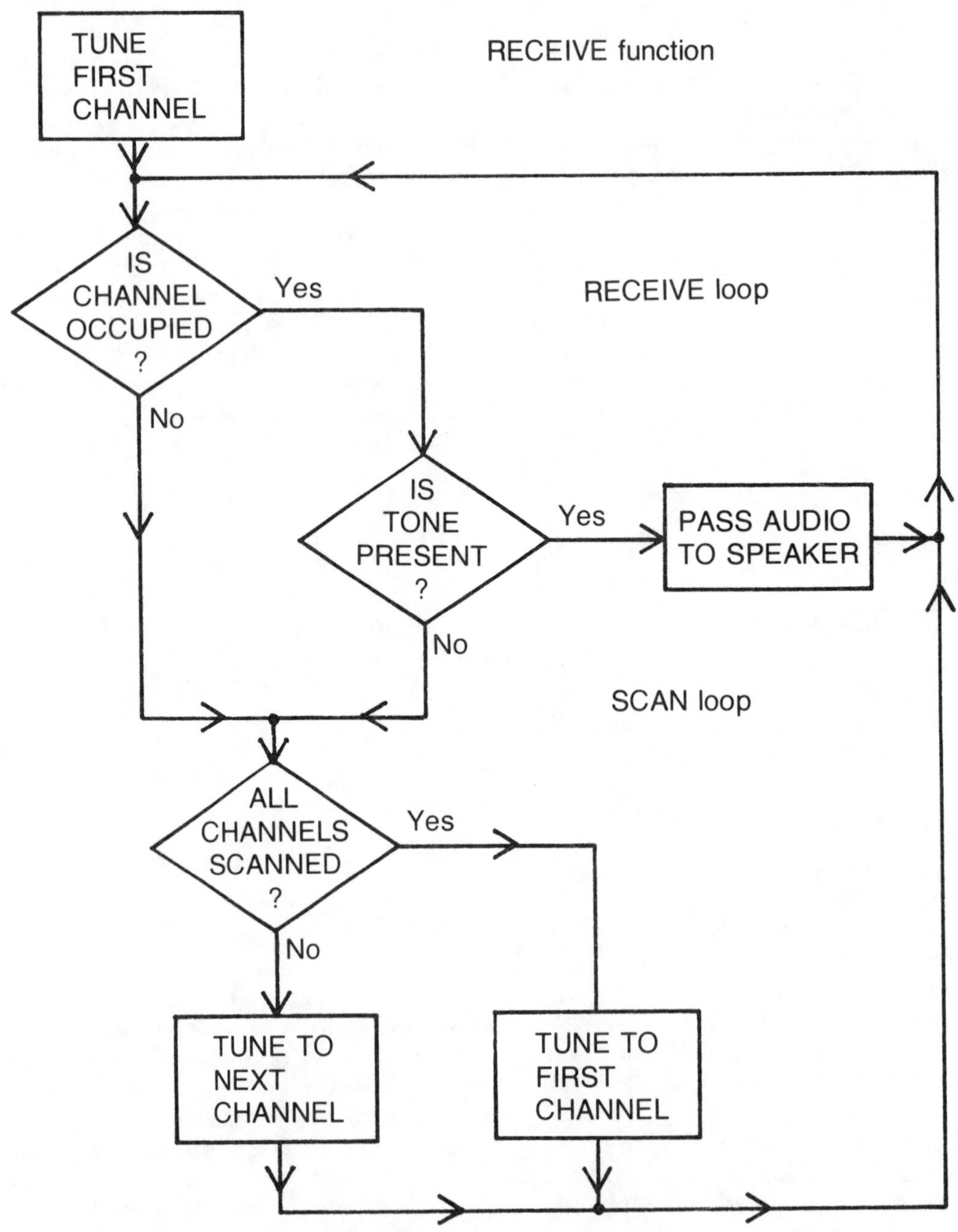

Fig. 8-17. Flowchart for multichannel communications receive function.

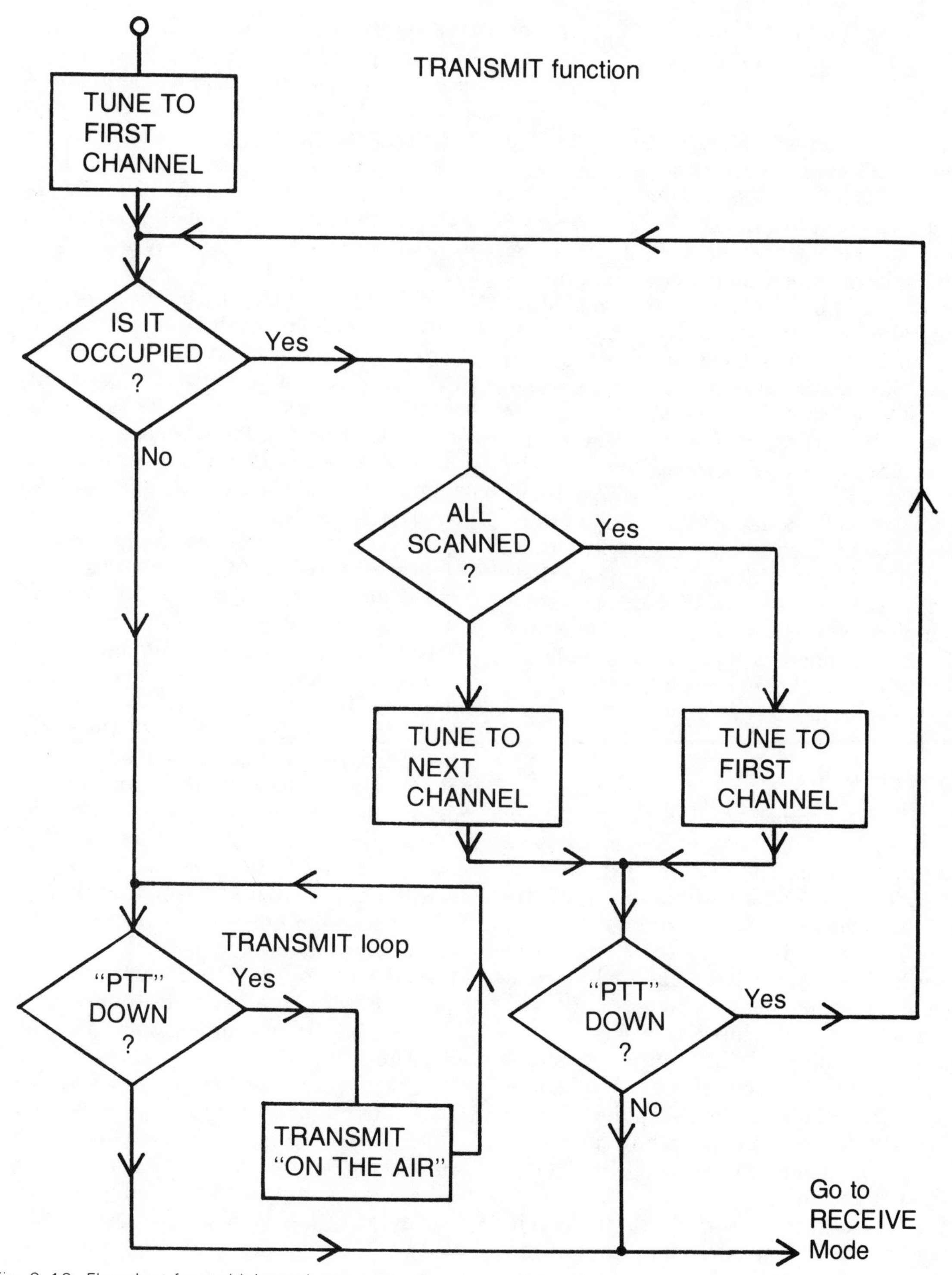

Fig. 8-18. Flowchart for multichannel communications transmit function.

(squelch); if a sub-audible tone is used, the test is performed on the tone.

Transmit

The flowchart is shown in Fig. 8-18. The first loop locates a clear channel. Then that channel is held as long as the push to talk circuit is activated. In use, the operator may depress the push to talk button on the microphone, and when a clear channel is located, a light can turn on (or some other indication can occur) showing that a clear channel has been located and transmission has commenced. The operator then transmits his message and reverts to the receive mode. The transmit mode would be an interrupt to the receive mode.

Since the scanning times are fast (in the msec range), none of the transmission would be lost at the receive end, and there would only be significant delay in establishing a channel if the level of band occupancy is high. That delay would be much less than if only a single channel was available, for the group would then have had to wait up to several minutes for another contact to terminate.

Repeater Utilization

If the members of the group cannot always establish contact using simplex channels, the concept can be expanded to include repeaters. Additional advantages can be obtained in this manner.

If a group is distributed over a wide geographical area, it is likely that some members will be out of simplex range with others, and thus not have direct communications capability. It was to provide such a capability that repeaters were established in the first place. In the event that the one repeater does not provide that capability, the scanning concept may be utilized on the repeater channels.

Consider the following example. A city has a number of repeaters that can be heard over the entire metro area. Each member of a group cannot access all of them, but may access some that the other members cannot. It is possible that some members cannot communicate directly because they cannot mutually access even one repeater. A truth table of such a situation is shown in Fig. 8-19. In this example, there are 10 repeaters (1–10) in the city and 7 members in the group (A–G).

A table showing which repeaters must be used to communicate between individuals is shown in Fig. 8-20. It can be seen that there is no direct communications capability between members A and E or C and G. In many cases, redundancy exists whereby more than one repeater channel can provide communications between two or more members at any particular time.

If each member of the group uses the simplex channel scanning algorithm described above, they can copy all repeater channels and receive any transmission aimed at them. If their transmit algorithm is modified, so the transmit mode only tests the channels that they can access, they will be able to communicate with all other members of the group because each station will not care which channel is carrying a transmission at any particular time. The transmit algorithm must also include a "blip" routine to test for the repeater being non-operational (down) and ignore that channel under those circumstances. Regular communications using the repeaters will not be affected, although other listeners not equipped with the scan technique may be intrigued by hearing only parts of conversations.

The members of the group would thus have direct communications capability under these conditions. This technique that shares frequencies in a particular environment is an example of frequency division multiplexing on a demand assignment basis.

Group Members Accessibility	Repeaters									
	1	2	3	4	5	6	7	8	9	10
A	1	0	0	1	0	1	1	0	0	1
B	0	1	1	0	0	1	0	0	0	0
C	1	1	1	0	0	0	0	1	1	0
D	1	1	1	1	1	1	1	1	1	1
E	0	1	1	0	1	0	0	1	1	0
F	1	1	1	0	0	0	1	1	0	0
G	0	0	0	0	1	1	1	0	0	0
Total members having access to repeater	4	5	5	2	3	4	4	4	3	2

Transmit Capability
1 = Can Access
0 = Cannot Access

Fig. 8-19. Repeater utilization.

SPREAD SPECTRUM TECHNIQUES

The multichannel communications technique described above showed that if a microprocessor was connected to a frequency synthesizer, the channel used by the transmitter could be changed according to the occupancy of the channel.

Consider regular HF communications. In any sky wave QSO (that is, a contact between two stations in which the signals have been bounced off the ionosphere), there is usually some interference or QRM. The amount of this interference varies as time goes by. The interference may be caused by atmospheric noise (QRN) or other stations on the same frequency, one or more of which who may not be able to hear your contact, but which also having contacts on that frequency. The occupancy of each frequency in the band is different, so that if your contact is being interfered with, changing frequency may allow you to make a QRM-free QSO. Changing frequency is a hassle, because one station has to locate a clear frequency, then return to the original frequency, and then try and get a message through the QRM telling the other station which frequency to QSY to.

TRANSMIT

RECEIVE

	A	B	C	D	E	F	G
A		6	1	1, 4, 6, 7 10	NONE	1, 7	7
B	6		2, 3	2, 3 6	2, 3	2, 3	6
C	1	2, 3		1, 2, 3, 8, 9	2, 3, 8, 9	1, 2, 3, 8	NONE
D	1, 4, 6, 7, 10	2, 3, 6	1, 2, 3, 8, 9		2, 3, 5, 8, 9	1, 2, 3, 7, 8	5, 6, 7
E	NONE	2, 3	2, 3, 8, 9	2, 3, 5, 8, 9		2, 3, 8	5
F	1, 7	2, 3	1, 2, 3, 8	1, 2, 3, 7, 8	2, 3, 8		7
G	7	6	NONE	5, 6, 7	5	7	

Fig. 8-20. Communications channels available between stations.

Suppose the microcomputer was set to change frequency at certain intervals. These intervals could be as small as one second or so. If both stations knew when the changes would occur and what frequencies would be used, the receiving station would follow the transmitted station as it wandered about the band, and no information would be lost. The background QRM level would change as a function of the occupancy of the frequencies being used. The ear of the operator at the receiving station would probably ignore the background noise because it would appear to be random.

If the stations used a predetermined sequence of frequency hops, locking up a receiver to a transmitter would not be very difficult. The frequency changes would take place according to a long sequence so that most of the band would be used before any one frequency is again sampled. There are a large number of such patterns that can be used. The computers would probably use what is known as a pseudo-random pattern. That is a fixed pattern that is very long so that it appears to be random.

There are many such patterns, and if each station is using one, any one station could have a considerable problem in finding another one to talk to. The problem is not as bad as one would think, because the existing bands already have a large number of frequencies and yet contacts are made. SSB uses one part of the band, and RTTY, SSTV, and CW use others. Nets use certain frequencies so that the problem of how stations find each other is not that bad in real life.

A number of calling patterns would be known to all amateurs and programmed into their rigs. Once a QSO is established, they would QSY to a different pattern sequence. The concept of calling channels and working channels is well known on VHF fm communications, especially in countries where repeaters are not used. Special pattern sequences could also be used for nets or group QSOs.

Spread spectrum communications may be the next amateur communications mode to become popular. Several groups are already experimenting with it to find out if it offers the average amateur any improvement in communication capability. Much time will be spent in writing software to implement these features before spread spectrum rigs show up on the market.

PROPAGATION RESEARCH

Microprocessor control of a frequency synthesizer allows extensive propagation research to be performed. There are many HF/VHF propagation beacons installed around the world. Fig. 8-21 shows a sample of 6M and 10M beacons. Many serious operators monitor beacon frequencies so they know when propagation exists between the area that a beacon is located in and themselves. When they are receiving signals from the beacon, they know that the band is open and they stand a good chance of making a contact. This is of interest particularly to stations operating in the 50 MHz or 6 Meter band.

Given that the microprocessor can control the frequency of the receiver, the receiver/computer can be programmed to monitor a sequence of beacon frequencies. If the receiver has a builtin microprocessor, then one is limited at first to the program in the rig. Frequencies can be scanned and an audio output or squelch output signal obtained when a signal is found.

If a general-purpose microcomputer is connected to the frequency synthesizer as shown in Fig. 8-22, then a more flexible

COUNTRY	CALL SIGN	FREQUENCY
Australia	VK2WI	28.2625 MHz
Bermuda	VP9A	28.2350 MHz
Canada	VE2TEN	28.2175 MHZ
Cyprus	5B4CY	28.2200 MHz
England	GB3SX	28.2150 MHz
Mauritius	3B8MS	28.2100 MHz
New Zealand	ZL2MHF	28.2300 MHz
Norway	LA5TEN	28.2375 MHz

This list is not a complete list of current 28MHz beacons.

Fig. 8-21. Sample list of HF propagation beacons.

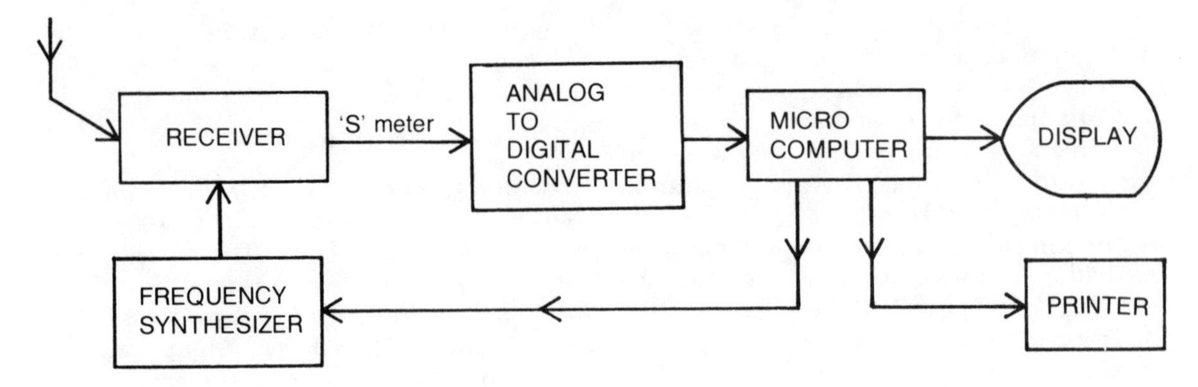

Fig. 8-22. Hardware configuration for propagation research.

10 Meter Beacon reception data

DATE . . . (simulated)

TIME (GMT)	BEACON							
	VK2WI	ZL2MHF	3B8MS	5B4CY	GB3SX	LA5TEN	VP9A	VE2TEN
0500	0	0	0	0	5	0	0	0
0530	0	0	0	0	5	0	0	0
0600	0	0	0	0	5	0	0	0
0630	0	0	0	0	5	0	0	0
0700	0	0	0	2	5	0	0	0
0730	0	0	0	5	5	0	0	0
0800	2	2	3	9	5	0	0	0
0830	4	5	6	9	5	4	0	0
0900	6	9	9	9	5	9	0	0
1000	8	8	9	9	5	6	0	0
1030	6	6	8	8	5	5	0	0
1100	0	0	0	0	5	0	0	0
1130	0	0	0	0	5	0	0	0
1200	0	0	0	0	5	0	0	0
1230	0	0	0	0	5	0	0	0
1300	0	0	0	0	5	0	2	0
1330	0	0	0	0	5	0	3	1
1400	0	0	0	0	5	0	4	1
1430	0	0	0	0	5	0	5	3

Fig. 8-23. Beacon monitoring data.

system results. The switches in the synthesizer are removed and replaced by output port bits. The squelch or receiver S meter signal level is digitized and sent to the computer. The computer can then set up a frequency and, after waiting for all the time constant delays to settle down, can determine if a signal is present on the frequency. The computer can, for example, be programmed to sequentially listen on all the beacon frequencies in the band listed in Fig. 8-21. The received signal strength reading can be noted each time. If the sequence is set up to take place at periodic intervals of 1, 5, or 10 minutes, then a table can be plotted of the results. An example of such a table is shown in Fig. 8-22, in which received signal strength is tabulated as a function of time of day in samples of 30 minutes. In this case, the data is assumed to be the average received signal strength at one minute intervals in the 30 minute time frame. This tends to reduce noise induced errors. The data shows that the receiving station is located within groundwave range of the GB3SX beacon. The band can be seen to open at about 0700 to the Middle East and after 0800 it is open to Australia and New Zealand. Short skip to Norway is also present. Something occurred at 1030 to kill propagation completely, and the band is dead for two hours or so until it opens up to the Carribean and North America.

This data is useful for predicting propagation and DX chasing. If the operator comes into the shack, he can see what part of the world the board is open to and decide if he wants to search for stations in new countries in that area. The data can be referenced to solar and weather information for study.

145 MHz and 432 MHz beacons can also be monitored to determine band openings. A similar setup can monitor VHF/UHF repeater channels and plot the occupancy. It would be interesting to determine which of the repeaters in a city is: most used, least used, most used during commuting times, and least used during commuting times.

The techniques are the same. Serious 6 Meter operators could even program their computers to telephone them at work announcing that the band is open.

Digital Communications

Digital communications include Morse code and RTTY communications. Morse code is a digital communications medium in which the presence or absence of a signal and the spacing of the signals define the content of the data. RTTY uses either Baudot or ASCII codes to relay written information which is displayed by a radio teletypwriter or a cathode ray tube terminal. This section deals with RTTY communications at VHF/UHF frequencies.

VHF/UHF RTTY is usually transmitted using Audio Frequency Shift Keying (AFSK) on fm equipment. Transmissions can be asynchronous and of random length using ASCII or Baudot codes, or fixed format packets of data using ASCII or some other 8 bit word code. Putting microcomputers into the communications link can allow anybody equipped with digital communications hardware to communicate with anybody else also equipped with digital communications hardware, even if their equipment cannot communicate directly. Thus G3ZCZ/W3 who has ASCII 300 Baud equipment could communicate with WA3LOS who is equipped with a Baudot Model 15 teletypewriter. This is an ideal way to provide low cost low speed communications at the entry level as well as high speed communications at advanced levels. This chapter examines various aspects of digital communications.

RTTY REPEATERS

The RTTY repeater usually provides coverage of a large metropolitan area. The frequencies of 146.10 MHz (input)-146.70 MHz (output) have been specifically assigned to such repeaters in the USA, although often other frequencies are used.

Simple repeaters receive the AFSK tones and re-radiate them directly just as if they were audio signals in a conventional repeater. More advanced repeaters demodulate and then regenerate the signals.

RTTY repeaters first came into operation for the same reasons that audio repeaters did; to provide an extended coverage area as shown in Fig. 9-1. Anyone in the coverage area with suitable equipment can copy signals on the frequency.

Repeater

A

B

Repeater

If there is a hill between two local stations, it is not possible for them to communicate by means of VHF. If a repeater station is positioned at the top of the hill as in b, communication becomes possible.

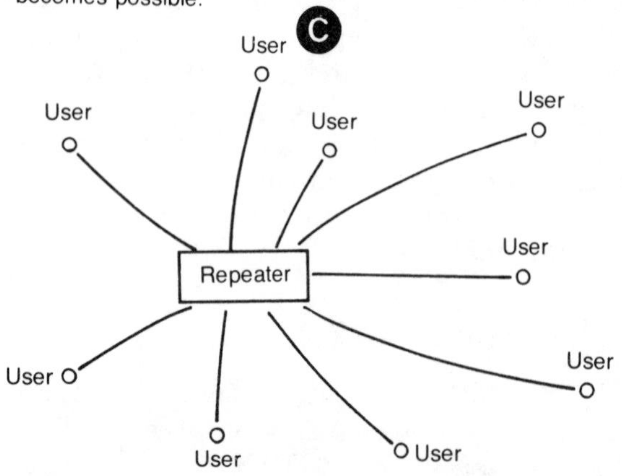

A city-wide radioteletypewriter VHF repeater link.

Fig. 9-1. Extending communications range by the use of a repeater.

However, RTTY is a slow mode of communicating. Even when sending pre-stored messages at full machine speed, messages still take a long time to send. A two-way RTTY contact can take an hour or so to pass information that can be passed in minutes by voice. RTTY does have one major advantage over voice communications; its unattended operation or autostart capabilities.

RTTY NETWORK

The RTTY network works as follows. All stations monitor the same frequency, either HF or VHF. Messages are sent blind; that is, when a message is originated into the network, the sender does not know for certain if the destination station is monitoring the frequency, unless contact is first established. If contact cannot be established directly, the message can still be sent, but there is the probability that the destination will not be available, and it will be lost.

If the message can be stored in a central computer by the sender, and then retrieved later by the receiver, the successful transmission of the message from sender to receiver becomes a certainty. The addition of a computer thus becomes an asset to the network.

If several stations in the network have computers capable of answerback, the utilization of the computer may be reduced. A sender can put out a direct call. If an answer is not received (indicating that the destination is not on line or monitoring at the time), the message can either be transmitted to the computer for storage or held and transmission retried at a later time. It is also possible for the network computer performing the store and forward operation to rotate among the various member station computers on an "as available" basis, as long as the network computer has a distinctive identification.

Some repeaters allow ASCII signals at data rates of 300-1200 baud to be carried. Stations with Baudut equipment cannot directly communicate with stations having ASCII equipment. The network computer can contain a conversion capability wherein messages received on one mode are converted to and re-transmitted using the other mode.

In use, ASCII messages are relayed directly. The "high speed" ASCII message being input can be stored in a memory buffer and an output program can transmit the contents of the buffer at 45.5 baud even while the buffer is filling up at the ASCII data rate. Incoming Baudot messages can either be transmitted directly in ASCII at the higher baud rate (but still at a real character spacing of 45.5 baud) or can be stored and then transmitted later as a single message at full speed. In the latter case, some sort of tone or signal has to be placed on the output frequency to notify all users that an incoming signal is present and is being stored but not transmitted. It would probably be better in this case to retransmit the Baudot message if it is received and then follow it with the ASCII message upon completion of the reception.

In use, the operator at his station types up and transmits a message. The message is transmitted directly to the target station or is stored in the computer. Some time later, the operator will check to see if a reply has been received. Depending on the degree of sophistication of the network, he may even be able to interrogate the network computer to see if the message has been forwarded. Thus the concept of "store and forward" in the network computer is really a logical extension of auto start techniques.

These techniques for Baudot/ASCII conversions allow amateurs equipped for different modes of operation to communicate. The scheme presented above suffers from the limitation that only one message can be transmitted at any one time.

The Baudot network can be classified as a dumb network. Users are usually operating in the manual mode, possibly using paper or audio tape to facilitate operations. Incoming messages are printed out and possibly punched on tape. Very little selectivity exists to separate messages addressed to a station from others on the frequency. (A few hard-wired Selective Calling units do exist). Error detection and correction techniques are minimal.

The ASCII network can be classified as a semi-smart network. Most users have some kind of microcomputer-based system. Communications are at 300-9600 baud, but again have a minimal amount of error-correction facilities. This network can be used to transfer files between computers and in fact is being used as such.

The packet network can be classified as a smart network since error-detecting and error-correcting techniques can be used. Thus if the receiving station detects an error in the message it can automatically request a retransmission of the bad message to ensure that the traffic is correct.

BURST MODE COMMUNICATIONS

Consider a digital repeater operating at 1200-9600 Baud ASCII. Each user has a small microprocessor-based terminal that contains a minimal amount of hardware and software to perform the following operations: store a few lines of text, remember who the message is going to, remember the callsign of the station, and display incoming and outgoing messages.

These capabilities are not too advanced on current smart, dedicated, microprocessor-based RTTY terminals.

In use, any amateur would start typing a message at the terminal. When a line of text has been entered, the microcomputer checks the frequency to see that it was clear and then transmit that line of text at the high speed rate (verifying it on the repeater output frequency to ensure that it was reradiated properly). The amateur typing away at the terminal need not even know when the transmission burst is sent. Since most people type slowly compared to 1200-9600 baud, the terminal will spend most of its time in the non-transmitting state. Thus a number of amateurs could be using it at the same time. Anyone monitoring the channel would pick up all the signals. Each line of text could belong to different messages and thus would appear to be garbled. If, however, each line of text was prefixed by the callsign of the target station (and suffixed by the callsign of the sending station) and the microprocessor in each terminal was programmed to respond to and display messages only addressed to its callsign, the traffic on frequency would become invisible and thus time sharing of the repeater would be unnoticed by the users. Incoming messages would be displayed a line at a time instead of a character at a time as in the conventional network. These lines of text that are transmitted in a burst mode can be called packets of data.

Once a microprocessor is put into use in storing the entered characters and then bursting them out as a packet, it can also be used to provide some error-detection capabilities.

PACKETS

A packet of data is a high-speed burst of information. The typical RTTY frequency can only be occupied by one QSO at a time, and data is sent at the rate that it is typed. Thus, although a Baudot network can pass data at 60 words per minute (wpm) using conventional mechanical teletypewriters, a real data throughput of 60 wpm is only achieved when running at machine speed. Since the actual typing speed in a contact varies as a function of the digital dexterity of the operator, the data throughput is slow. Computers can be used to speed up the flow of information and improve the channel occupancy.

Supposing the data being typed is buffered by the computer. The contents of the buffer can then be output at high speed (say 1200-9600 baud) as a burst. If the computer checks that the channel is unoccupied before transmitting, there will be a minimal amount of loss of data due to interference (two stations transmitting simultaneous bursts are the only practical cause of such interference). If each packet or burst was prefixed by the callsign of the destination station, it would be uniquely identified. The computer at the receiving station would ignore all bursts addressed to other stations. Thus many QSO's could take place time-sharing the channel. An example of such a scheme is shown in Fig. 9-2. Any station could display all information relayed or just the messages addressed to itself. Thus the addition of a minimal amount of software would improve the use of the basic RTTY repeater network.

Once computers are used for high-speed data burst communication links, advantage may be taken of the capabilities of the computer to provide error-checking and error-correcting capacity. Protocols can be defined and adopted with those ends in mind.

The main problem here is that new stations joining the network can bomb it if their equipment (hardware or software) is not working correctly. If an average of one new station per week joins the network and bombs it for two evenings each time, the network will suffer a lot of down time.

Several techniques can be used to minimize this problem. The station software can be tested out on a simplex or different channel, or a cheap special-purpose microprocessor-based circuit card could be de-

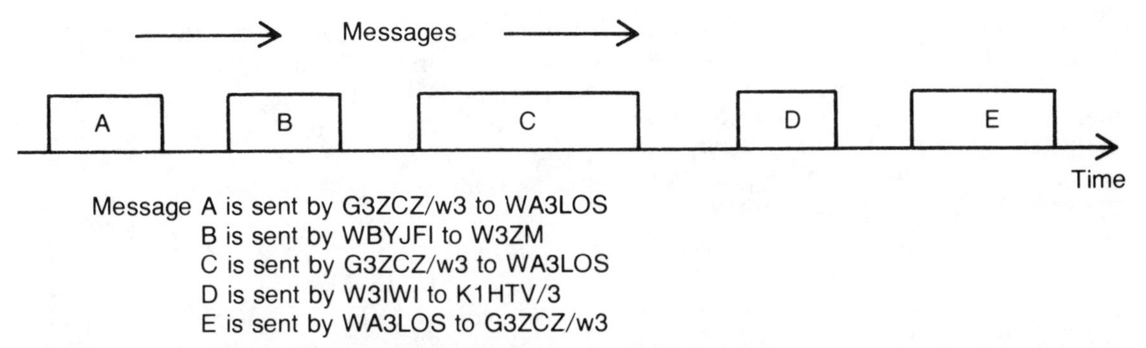

Fig. 9-2. Timesharing a communications channel.

veloped that would act as a front end processor fitting between the computer and terminal unit. It would contain the buffers and network communication algorithm. Anyone wishing to access the network would be required to obtain the unit in similarly to the way that a tone burst or subaudible tone is required for access to a large number of 2 Meter (audio) fm repeaters. The front end processor card could be mass produced at low cost once protocols are established. If designed properly, the protocols could be PROM based, and the same unit could be used for a number of different protocols by plugging in a different PROM for each protocol.

The actual protocol provides a means for ensuring error-free transmission of a message and is transparent as far as the message itself is concerned. The analogy in conventional amateur radio is that the sounds emerging from a speaker at the receiving station are the same as those entering the microphone at the transmitting station. In an interference-free situation it does not matter to those sounds if the modulation technique is am, fm, ssb, or dsb.

THE PACKET NETWORK

The packet network is set up for stations who can communicate directly using packet techniques. The advantages of packet communications are many and include timesharing of the channel, relatively high speeds, and error detection and correction.

The block diagram of a packet network would be identical to an RTTY Baudot or ASCII network, but packet transmissions offer one big advantage in that a packet repeater can operate in the simplex or single frequency mode. In this network, all stations monitor the same frequency. All stations can transmit to and receive from the central store and forward station (repeater). In use, a station would store the message as received. It would then transmit the message on the same frequency so that the intended recipient would be able to receive it. If the intended recipient was not able to copy the original sender, the message is delivered. If the recipient was able to copy the original message, it would be able to detect that it had received the same message twice because the repeater would have set a flag byte in the message header indicating that the packet was a relayed version.

The conventional repeater requirement for two frequencies (input and output) at one time has now been replaced by the requirement for two time frames, original (input) and retransmitted (output), on one frequency.

If you know the Q code, you will know that QUA means "send me all new messages," and QBM means "send me the message from "----." Other examples are: WR3ABU :QRT: DE G3ZCZ/W3 which signs G3ZCZ/W3 off the network, and WA3VXE :QRL: DE G3ZCZ/W3 which asks WA3VXE if he is busy. No response within a short period of time means that he is not there. If he is, the replay would be G3ZCZ/W3 :QRU: DE WA3VXE; i.e., an automatic answerback.

Note that WR3ABU would not respond to the QRU because its callsign was not recognized. G3ZCZ/W3 would then send his message as follows:

> WA3VXE :QSO: ALAN, IT LOOKS LIKE WR3ABU IS DOWN, SO I TRIED YOU DIRECT :QSL: DE G3ZCZ/W3.

The response would come in a flash (or at least at 60 wpm):

> G3ZCZ/W3 :QSL: DE WA3VXE

Hence, even if WR3ABU was monitoring the transmission and recognized its call in the text, since the callsign was not immediately followed by the :Q sequence it would forget that it had just recognized its call and go back to sleep.

This message:

> WR3ABU :QSP: GB3LO :QSP: G8BTB :QSO: PAT ARRIVING ON THURSDAY 22 JUNE :QSL: DE G3ZCZ/W3.

would instruct WR3ABU that a message is to be sent to GB3LO who will then forward it to G8BTB. This extension assumes that GB3LO is the store-and-forward computer in a second network in which G8BTB is operating.

NETWORK COMMUNICATIONS LANGUAGE

The Baudot and ASCII RTTY networks require some routing signals to ensure that messages are routed to their intended destinations. A suitable source for these signals is the Q code commonly used by radio amateurs. The use of slightly modified Q code signals makes the messages easily readable by both man and machine.

For example, a message such as WR3ABU :QSP: WA3VXE :QSO: ALAN PLEASE CALL ME ON THE TELEPHONE AFTER NINE TONIGHT :QSL: DE G3ZCZ/W3 is almost already understandable even without explaining that the Q codes used mean:

:QSP: (please) relay to callsign following
:QSO: the message following
:QSL: end of message/confirm reception

In other words, the store and forward computer at WR3ABU was asked to forward (QSP) a message to WA3VXE and confirm its reception by G3ZCZ/W3. Later on when WA3VXE signs in to the network, he would send WR3ABU :QRU:DE WA3VXE which means, "WR3ABU, do you have any messages for me?" WR3ABU would reply WA3VXE :QRU: G3ZCZ/W3, WB2YUX/3, DE WR3ABU, meaning that there are messages from G3ZCZ/W3 and WB2YUX/3. WA3VXE would then send either WR3ABU :QUA: DE WA3VXE or WR3ABU :QBM: G3ZCZ/W3 DE WA3VXE.

The : placed before and after the three letter group if the Q code makes recognition and decoding easier since all control language statements begin with :Q and a : is the fifth character in the sequence. An example of some of the Q codes suitable for use in the dumb and semi-smart networks are shown in Fig. 9-3.

Amateurs using Baudot equipment would have to type the control language statements in full. Those having microcomputers could type ASCII control characters which would be software-converted to the equivalent 5-letter control group.

The Network Control Language (NCL) provides the computers with information as

Code	Question	Answer or Advice
QRA	What is your identification or callsign?	My call sign is . . . .
QRL	Are you busy?	Yes, I am in use by . . . .
QRM	—	Your transmissions are being interfered with.
QRQ	Shall I speed up to . . . . bauds?	Yes.
QRR	Are you equipped for automatic operation?	Yes.
QRS	Shall I slow down to . . . . bauds?	Yes.
QRT	—	Signing off (log off).
QRU	Have you any messages for me?	Yes, messages are from . . . .
QRV	Are you ready?	Yes.
QRX	Will you wait?	Yes.
QRY	What is my turn?	Your turn is number . . .
QSG	—	Send . . . . messages.
QSK	Can you operate full duplex?	Yes.
QSL	Will you confirm?	Confirmed.
QSM	—	Repeat last mesage.
QSO	—	Message for . . . .
QSP	—	Relay via . . .
QTA	Cancel message to	Cancelled
QTC	—	The message is . . . .
QTH	What is your address?	My address is . . . .
QTR	What is the correct time?	It is . . . . UTC.
QTX	—	Log on.
QUA	Send me all new messages.	—
QUC	Who did the last message I sent go to?	It went to . . . .
QBM	Send me the message from . . .	—
QDB	—	The message to . . . . is forwarded.
QIC	May I call . . . direct?	Yes.
QJG	—	Revert to message mode (log off interactive mode).
QNO	—	Negative response or action.

Fig. 9-3. Extracts from The Q code.

to what is to be done with the data in a message. Numerous languages exist to provide computers with instructions, but few exist for communications purposes. NCL is written in some other language and is not a true language as such, but is an implementation of an NCL program in which the man-machine (or machine-machine) dialog is in specific format. Most radio amateurs are familiar with the Q code. Words such as QRM, QSL, or QSO are understood by them all. Others, such as QRA or QSP, may not be understood unless the radio amateur is used to traffic handling; but, since they already have some knowledge of the Q code and—better still—an idea of the concept behind it, the Q code is an ideal "language" for telling the computer how to route or process data.

The NCL based on the Q code can be used at all levels of digital networking, starting with a lowly Baudot circuit all the way up to a packet network carrying video as well as audio packets of data. Of course, the packet network with its fixed length packets can simplify the actual transfer of data by using positions in the packet to convey information. Thus, the callsign of the sending or receiving station could always be placed in a certain position in the packet, rather than—as in the random length RTTY message—use the Q code to specify originator or destination.

NCL is used to communicate with the communications software in the computer or in a stand-alone packet terminal interface. Apart from the use of the : as prefix and suffix for the control word (:QRM:) the Q code can be used in the conventional meanings. Thus, the Q code shown in Fig. 9-3 can be converted to NCL as shown in Fig. 9-4. The callsigns can also be expanded upon, drawing upon the usage of wild card characters used in several microcomputer operating systems. These wild card characters are known as general call characters and allow the sender to send a message to a category of stations.

The ? character may be used to match any single character or number; for example, G3Z?? matches any callsign in the G3ZAA-G3ZZZ series. W1??? matches any W1 call with a three letter suffix. The * character matches any section of a callsign (including a null character) as follows:

G3*	matches anybody with the G3 prefix
G**	matches anybody with the G prefix
GM3*	matches all calls with the GM3 prefix
3	matches any call with the three digit in it
***	matches any call in the world

The two general call characters can be mixed at will. For example, G*3A?? matches any call in the United Kingdom in the G3AAA-G3AZZ series, including those with the GM, GW, GC, GJ, GU and GB prefixes. Thus, GB3AAA, GM3ART, and GW3AAA would be matched. Note that G3AA would not match because three ? characters were used.

NETWORK IMPLEMENTATION

How do RTTY, packets, and ASCII interconnect? How does an amateur who only has a Model 15 RTTY machine communicate with an amateur who has a packet terminal? Should he? In the conventional communication modes an amateur who only has a Morse code (cw) station can communicate with another amateur who is using voice (ssb) but neither of these can communicate with someone else using the teletypewriter (RTTY). There are many Baudot stations in existence, newer amateurs may come on the air with ASCII using microcomputers, and advanced amateurs can use packet techniques. In general or

	Statement	Response (if any)
:QRA:	What is your call sign?	My call sign is . . . .
:QRG:	What is my exact frequency?	Your frequency is . . . . kHz
	What is the frequency of . . . . ?	His frequency is . . . kHz
:QRH:	Does my frequency vary?	Yes
:QRK:		Your bit error rate is . . . .
:QRL:		I am busy now, please call me later
:QRM:		Your signals were interfered with
:QRN:		There is noise on the frequency
:QRO:	Increase transmitter power	
:QRP:	Decrease transmitter power	
:QRQ:	Speed up to . . . . bauds	OK
:QRS:	Slow down to . . . . bauds	OK
:QRT:		Signing off from the network
:QRU:	Have you any messages from me?	Yes, messages are from . . . ., . . .
:QRV:		Signing on to the network (includes QRU by implication)
:QRX:		I am busy now, I will call you later
:QRY:		It is your turn to send a message to . . . .
:QRZ:	Who is calling me?	
:QSA:		Your report is signal strength . . .
:QSB:		Your signals are fading
:QSD:		Your signals were mutilated (negative acknowledgment or not received) try again
:QSG:		I can accept packets from . . . . stations concurrently
:QSK:	Can you operate full duplex?	I can operate full duplex
:QSL:		Acknowledging correct reception of packet
:QSM:		Repeat packet
:QSN:		I can copy you directly

Fig. 9-4. Basic NCL dictionary.

	Statement	Response (if any)
:QSO:		The message follows
:QSP:		Please relay to . . . .
:QSU:		Send your reply via . . . . (gateway or repeater station)
:QSW:	Which frequency/channel will you reply on?	I shall reply on . . . .
:QSX:	Can you copy . . . . direct?	I can copy . . . . direct
:QSY:	Change to channel/frequency . . . . ?	OK, I shall change to channel/frequency . . . .
:QSZ:		Repeat message or last packet
:QTA:		Cancel message
:QTB:		The character count is . . . . (used in RTTY messages only)
:QTC:	How many messages do you have for me?	I have . . . . messages for you?
:QTH:	What is your location?	My location is . . . .
:QTR:		Message was originated at (day, time)
:QUA:		Send me all the new messages
:QUC:	Has the last message I sent been forwarded	Yes, it has been forwarded to . . . .
:QBM:		Send me the message from . . . .
:QDB:		The message to . . . . has been forwarded
:QIC:		This is a direct call to . . . .
:QJG:		Reverting to automatic mode
:QNO:		Negative acknowledgment

:QSA: and :QRK: can form the basis for signal reports.
:QSM: could be used to flag a message that has been passed via a store and forward repeater.
:QSU: can be used for routing control, whereas :QSP: defines final destination.
:QTA: is used by the operator to delete received messages from his system.
:QUA: can be used transferring the function of network computer from one computer to another.

Fig. 9-4. Basic NCL dictionary. (continued)

:QDB: could form an intermediate acknowledgment when tracking the routing of a message through the network.

:QIC: can be used to find out if a station is logged on at any particular time.

:QNO: is the standard negative acknowledgment to state that the receiving station cannot perform the desired operation; i.e., it cannot QSY or QRS.

This figure contains an initial proposal for the NCL dictionary which will, of course, be changed as NCL comes into use.

NOTES

Fig. 9-4. Basic NCL dictionary. (continued)

random communications, in which one amateur calls CQ and wants to see who (what?) comes back, all modes usually work other stations equipped with the same mode. Thus, Baudot RTTY stations have established frequencies within the amateur bands where they have the greatest probability of finding others suitably equipped. It is conceivable that ASCII and packet stations could do the same. The big advantage of packets and computers in the RTTY area is that delivery of messages can be guaranteed, and by using a hierarchy of rf links, messages can be relayed between amateurs having different digital equipment. Thus, a Baudot station could send a message to an ASCII station.

Many local area nets exist using Baudot equipment. These nets may be on VHF or on HF. Each mode has its advantages and disadvantages. Figure 9-5 shows a potential situation for interconnecting such a network with a new ASCII network in the same local area. In its simplest implementation, two repeaters are co-sited. One is a conventional Baudot RTTY repeater, the second an ASCII repeater. In the normal mode the two are separate; ASCII stations talk to

ASCII stations at 300 baud or even at 1200 baud and Baudot stations talk to Baudot stations at 45.5 baud.

However, by using a translator, Baudot stations can communicate with ASCII stations, since the translator will perform the code/speed conversions from one to the other. The translator can in a sense be thought of as a third repeater.

Each network can operate independently. When somebody on one network wishes to send a message to someone on the

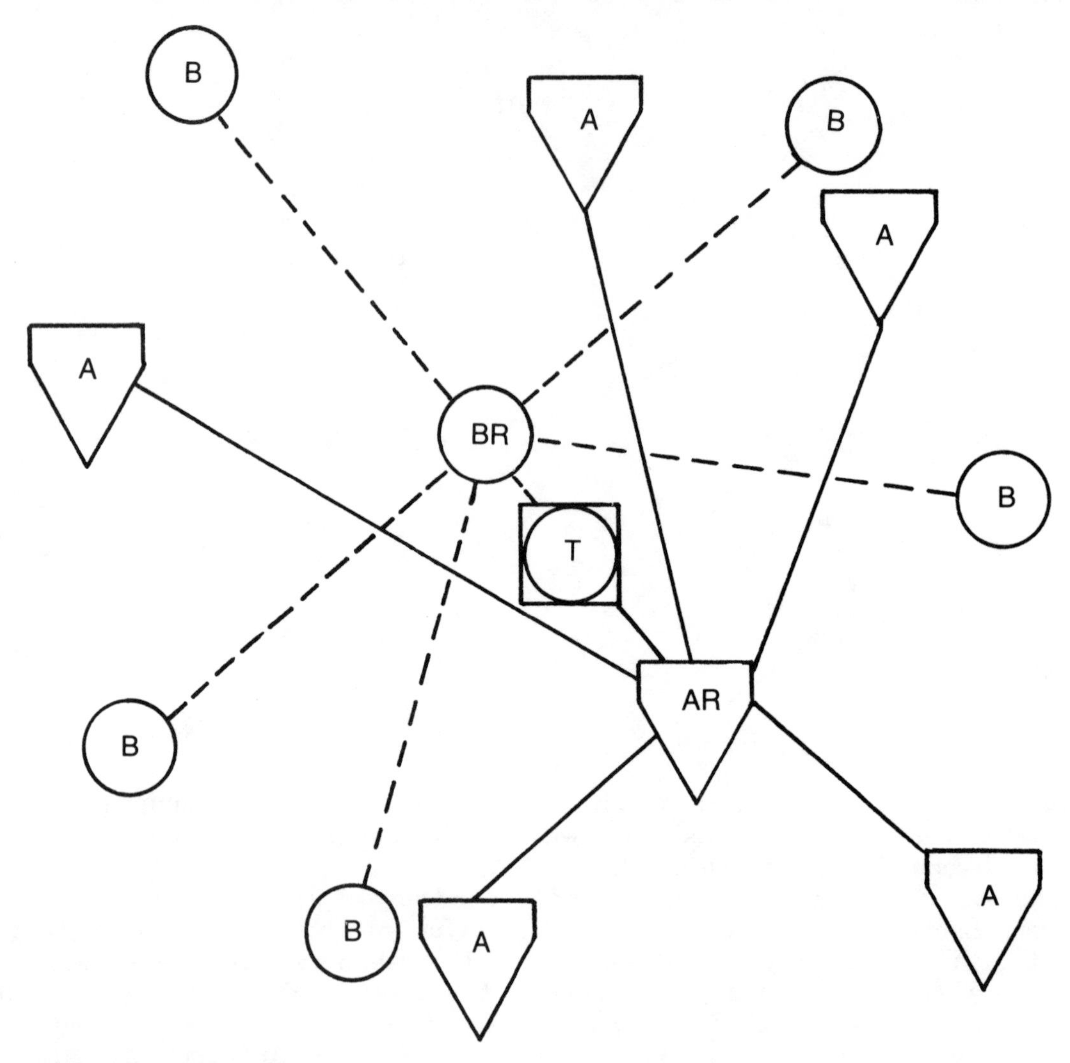

Fig. 9-5. Linking Baudot and ASCII networks.

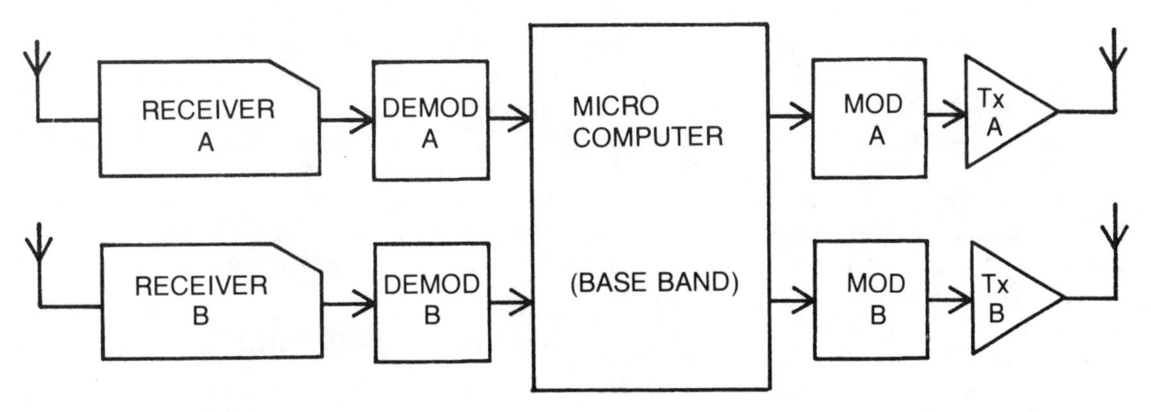

Fig. 9-6. A basic digital repeater.

other, he can use the network control language based on the Q code to instruct the translator accordingly. The translator should thus be able to store the message for later forwarding in case the other network is carrying traffic at the time that the message is originated. The translator will transmit the message later when the other network is free, or can hold it until the intended recipient signs in (on either network) and requests his message.

The translator can be used to perform the store and forward function for both networks at the same time. Once computers are put into the network, they can perform as many or as few tasks as their owners desire. The different functions can be split up between computers, or all the computers in the network can have the capability to perform all the tasks, thus providing a high degree of redundancy and reliability in operation.

NETWORK HIERARCHIES

The lowest level is the 45.5 Baud Baudot network. Baudot machines are usually available for less than $50 at local hamfests. They can be large and noisy, but they work and can easily be interfaced to a computer. They can thus be used in basic or conventional RTTY stations and then when a computer is incorporated in the station, can be used as a hard copy printout device, or even as a system console. Since thousands are already in use, they should not be obsoleted just because newer and better things such as 1200 Baud ASCII are now available. Their limitations will soon become apparent to the user, who will then upgrade to the newer and better devices, passing the Baudot equipment to someone else, allowing them to join in the fun. Thus Baudot operators will still be able to enjoy simplex contacts at HF and UHF.

The next level is the ASCII user. Here the baud rates can go as high as 9600 baud and yet remain within a 3KHz audio bandwidth. Common tone pairs used by amateurs in the U.S.A. are Bell 103 tones for 300/110 Baud contacts and Bell 202 tones for 1200 baud. It is thus possible to build a digital repeater that can monitor the incoming signal and perform conversion to a different code/tone pair for retransmission on the output. An example of such a device is shown in Fig. 9-6. The incoming signals are demodulated and converted to serial data by the different receive terminal units. These are fed to a microprocessor module via serial ports. The microprocessor is operating as a dedicated controller in this environment. Under normal conditions, the microprocessor retransmits the signals in the same format as they were received. Thus if 45.5 baud Baudot signals were received, that is what would be transmitted. If, however, the message is prefixed by a control code, the microprocessor would cause the signals to be transmitted using a different modulation technique. The microprocessor would also perform speed conversions as well as time conversions and provide a first-in, first-out (FIFO) buffer in the event that the retransmitted signals were at a slower baud rate than the input signals. With this kind of arrangement, amateurs with 1200 baud ASCII terminals will be able to communicate with other amateurs having Model 15's or similar Baudot equipment. The limitations of 45.5 baud, as compared to the high speeds, will soon cause a decrease in the number of stations using Baudot, and a corresponding increase in the number of ASCII operators. This approach, however, does allow the newcomer to join in with a minimal investment, encourages upgrading of equipment (by example), and does not penalize those with older equipment.

The basic data link allows errors to creep into the message. Errors are caused by noise or interference entering the communications path. In most of today's amateur digital communications the occurrence of errors is not too serious and can easily be detected by visually scanning or reading the received text. When computers are doing the communications, they also have to have a means to detect errors that have occurred in the transmission of the message.

LINKING OF NETWORKS

The discussion so far has limited itself to single area networks serving a common set of users. The requirement to interlink networks exist. For radio amateurs, the interlinking technology will of course be by radio. An example of interlinked area networks is shown in Fig. 9-7. Each network has at least two links to the outside world. The link may be through the central store and forward computer or it may be through one of the users. For example, if it is used to link two networks the link may be either via the central computer or via the user, but if vhf/uhf is used, the link would probably be

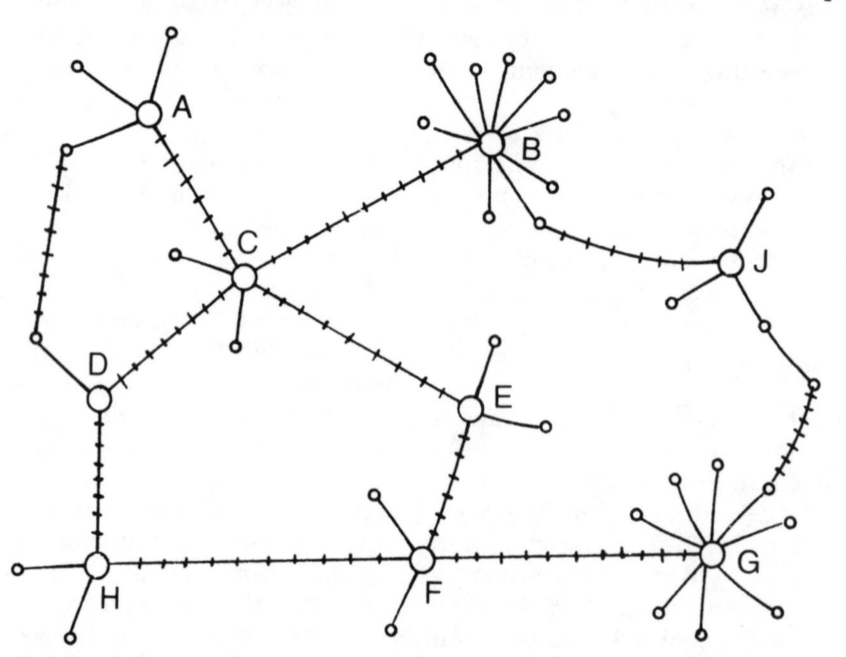

Fig. 9-7. Interlinking networks.

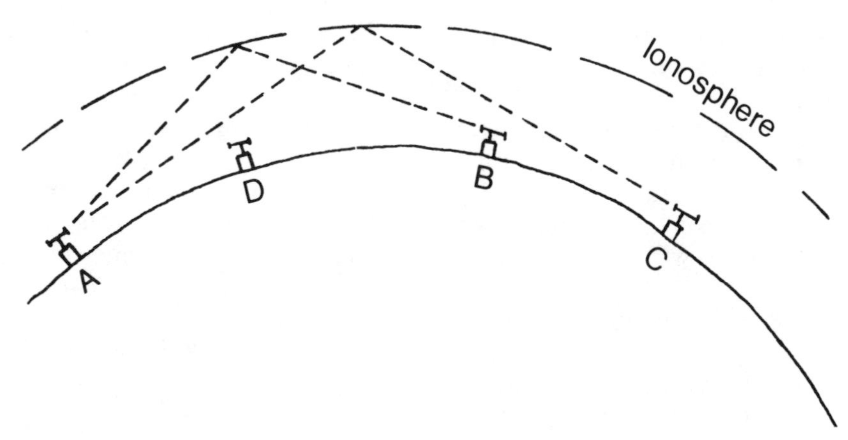

Fig. 9-8. HF propagation.

via a user who is located between the two networks and is able to access both directly (possibly only with the assistance of high power and directional antenna).

The communications on the interlinks use the packet mode. This is because the links will probably have lower signal to noise ratios than the vhf/uhf local network paths and the probability of interferences is greater.

Conventional VHF links also suffer from a routing problem. How does a message get from network G (in Fig. 9-7) to network A?

Does the originator specify the routing, the final network store and forward call, or nothing at all? Which algorithm is to be used? Fixed path? Random? Professional networks have spent thousands of dollars on this problem. Radio amateurs also cannot guarantee that all the links will be operational with a reliability of 0.999. What happens when a network node goes down for a while—are messages going to be backed up, or are they going to get lost?

HF propagation also suffers from its own characteristics. The ionosphere that reflects back the HF radio signals is a dynamic medium. Its properties change from minute to minute, are different during the day and the night, and are affected by solar activity which may enhance or detract from the reflecting properties of the ionosphere at any particular frequency. Thus the situation shown in Fig. 9-8 is typical of the conditions under which radio amateurs operate. Stations A and B are in direct contact with each other. Station C can also hear A but cannot hear B. If station B is transmitting, station A will be silent. When station A is silent, station C may try to send a packet to station A and interface with the packet that station B is sending. Station D, who cannot hear any of them at this time, may transmit to someone else and as conditions change will interfere with A, B, or C. This situation is not impossible, it is just difficult to design around.

Several techniques have been developed to minimize the QRM situation. Each station in the network can transmit at random. If a collision between two packets occurs, (i.e., one interferes with the other), the receiving station will not be able to send an acknowledgement to the sending stations so the sending station will try again later. If each station waits a different random amount of time before transmitting its packet, there is a good probability that the second time that a packet is transmitted it will get through.

Another alternative is to give each station a fixed time slot for transmission. Thus station A would always transmit during the first second of any minute, station B during the second second, and so on. If the stations are referenced to WWV or any other standard frequency and time transmission a minimal amount of interference will occur, but the throughput will go down since a station may have no packets to send but that time slot will still be reserved for it.

Adding the interference problems to the routing problems, and HF appear to be quite a problem. One solution is to use a random transmission sequence based on the probability of a successful contact. This means that messages are only originated if there is a good probability that there will be propagation to the destination or target station at that time of day.

A system in which everybody cannot hear everybody else, in which propagation is uncertain, is a difficult system to operate. The converse is also true. A system in which everybody can hear everybody else, in which propagation is 100% predictable, is ideal. This situation occurs if a communications satellite can be utilized as the relaying medium.

Figure 9-9A shows the same four stations now using a communications satellite to relay messages. They can all hear each other, and since the orbit of the satellite is known, they can compute the time when propagation will be possible between any of them. If the AMSAT Phase III or Phase IV satellites are used, each covering large areas of the world, a global network takes on the shape of a local network as sketched in Fig. 9-9B.

The satellite itself does not contain any store and forward equipment. The gateway stations on the ground each act as a local user to the satellite. They can all monitor the frequencies so can can pick up any traffic targeted at themselves. Since the satellite operates in a duplex mode, they can all monitor the downlink when uplinking and can detect errors due to noise, or due to collisions and take appropriate steps. Since the orbit is known they can determine mutual visibility and store messages until the target comes into a mutual visibility window.

Each gateway station may act as the central store and forward station or as one of the regulars in its own network, and as long as the gateway is operational any station on the network has access to the network as a whole, making the sky the limit in radio amateur digital communications.

USING THE NETWORKS

The RTTY networks can be used identically to existing RTTY channels. It does not matter if they are Baudot or ASCII, CQ random or point-to-point (auto start). Communications take place in a conventional manner. The use of NCL only becomes necessary if a message is to be stored in or retrieved from a computer. The location of the computer also does not matter. The user of the packet network will usually have a dual processor system, as shown in Fig. 9-10. The Terminal Interface Program (TIP) may or may not be part of the main computer. The TIP can operate in two modes, monitor or terminal. In the monitor mode, it can pass every packet it receives to the main computer. The destination of the packet does not matter. This mode is a good debug mode for testing the TIP, as well as providing a level of confidence in the early days. Since the packets may or may not be complete messages in themselves, the output of the TIP may or may not make sense. In the terminal mode, only messages addressed to the user will be output by the TIP.

If the TIP is a stand-alone board with an RS-232 interface, it can be connected to a terminal device and used as a dumb packet terminal. A dumb packet terminal is a terminal that can send and receive packets. It contains the basic low level software to

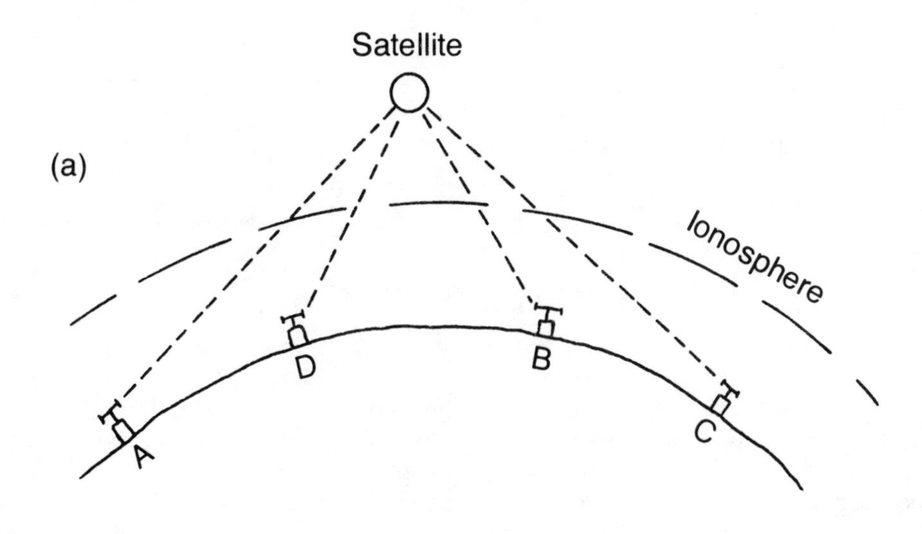

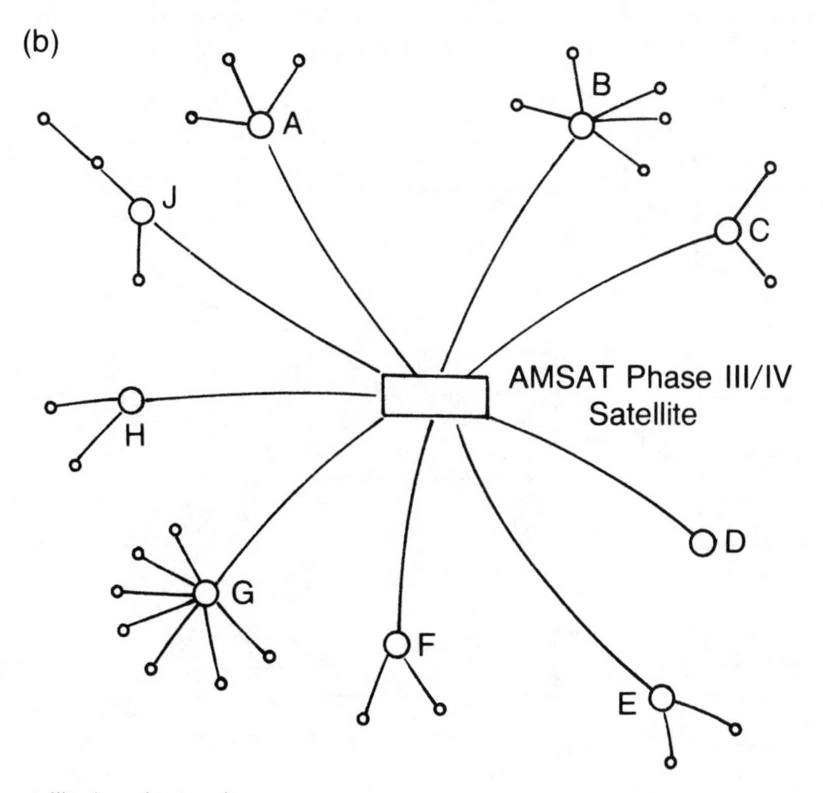

Fig. 9-9. The satellite-based network.

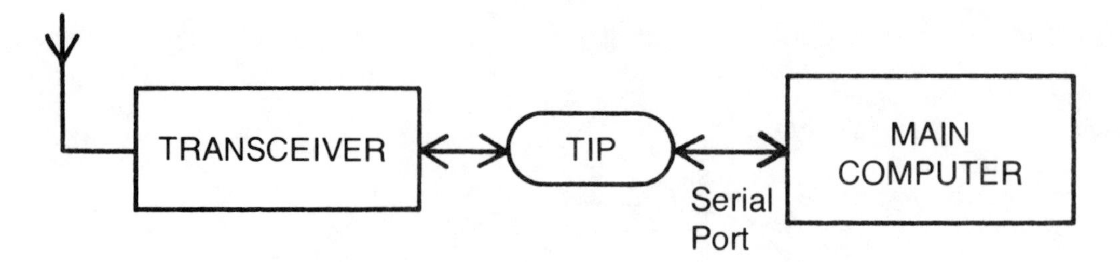

Fig. 9-10. Packet network interface.

format a packet for transmission, and acknowledges reception of, and unformats, a received packet. Such a stand alone board, microprocessor-based, is a low cost introduction to packet techniques. There is, of course, no reason why the TIP function could not be performed in software by the host machine, apart from the obvious one that it may tend to prohibit the use of the computer for other purposes. An outline of a stand-alone TIP is shown in Fig. 9-11. The breakdown shown for the TIP comprises a standard microcomputer. The control program is in PROM, the data storage area is in RAM. The more RAM that is available, the greater the length of or number of packets that can be stored in the TIP.

Most users graduate from the monitor mode to the terminal mode pretty quickly. After the novelty of receiving packet transmissions has worn off, the unit is switched to the terminal mode. Of course, the user

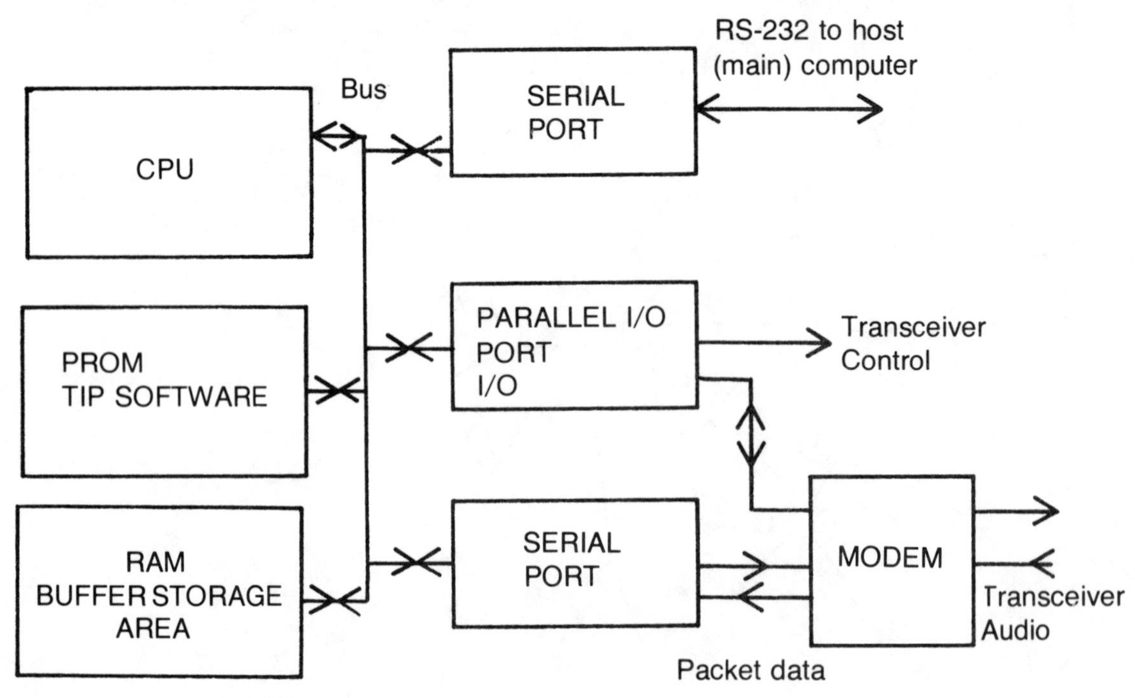

Fig. 9-11. A stand-alone TIP.

may temporarily revert to the monitor mode at any time to check that the TIP is still operating after a long period in which no messages have been received had occurred. The user (via the terminal or the host computer) can communicate with the TIP by using NCL. In this way the user does not have to learn any new control functions. The user does not really care about the mechanics of getting messages across. All he is interested in is the message—i.e., the high level protocols. The low level protocols of exactly how and when a TIP goes on the air can be left to the minority of technical hackers amidst our ranks.

The same software should be used in all the TIP's. The real world will, however, not be the same as the ideal world. The standard PROM's supplied with each TIP could be programmed with the station callsign as ***. These general call characters will respond to all callsign addresses which is the monitor mode situation. Thus, at power up the TIP outputs a sign-on message to the serial plot such as AMICOM TIP REV 3.6 QRA which identifies the network program protocol and the revision level. The TIP would then be in the monitor mode and the user would change to the terminal mode by entering :QRA: followed by the station callsign (including general call characters (i.e., :QRA: G3ZCZ). Note that a callsign such as G3ZCZ/4X would be recognized as having the 4X prefix not the, or as well as the, G3 prefix.

One advantage of a separate TIP is that it tends to maintain the integrity of the network. Consider a network in which one new user a day joins in. Given the number of radio amateurs and the number of computers in existance that is not an unreasonable assumption. If each user has to bring up software and hardware at the same time in an area in which he has not worked in before, the probability of errors occurring, bombing, or tying up the network, is high. The network could thus suffer from a lot of down time due to those newcomers not quite being able to access. If a standard board is available, the software can be provided to drive the integrated circuits on the TIP which will minimize the number of bad signals on the network frequency. If NCL is used to control the operation of the TIP, the TIP can be driven by any computer having a serial port, programmed in any language.

Once messages begin to flow into and out of the TIP, some high level control is desired. This high level control forms the user interface to the network, not at RF, as in the case of the simple RTTY network. Again, here a hierarchy is possible. The TIP can drive a simple terminal in the monitor mode or can interface a microcomputer with floppy disc capabilities for storing (and forwarding) messages.

In a simple RTTY network only one contact is taking place at a time, irrespective of how many stations are taking part in the round table. In a packet network, only one packet is being transmitted at a time, but successive packets as received at one station need not be part of the same message. There is, thus, no reason why when two stations are in contact, packets originated by other stations could not start appearing in the network and some of those packets could be addressed to one or both of the stations already in contact.

Consider for a moment the working of the TIP in its receive mode. The program must monitor the frequency and read the leaders of all the packets on the channel. Messages intended for the TIP's own station will be output at the serial port. What happens if the TIP has received a packet from one station, but that the packet was not a complete message. While the TIP is waiting for the next packet that will contain more of the message, a packet arrives addressed to the TIP but originated from station B. Does the TIP reject it because the new packet will garble the message currently being received?

If the TIP rejects it by not acknowledging it, station B may keep trying, thus slowing down the communications speed of the channel because its packets will be ignored by our TIP, yet are using up time that could be utilized by station A to complete his message or by other stations to pass different traffic. If the TIP does not reject it, a computer (either the TIP or the host) has to have some way of recognizing the different origin of the packet and storing it in the relevant bit bucket. On the other hand, the TIP could send a busy packet to station B and anybody else which means that they can either try again later, or wait until they are called back. The busy packet could contain a flag bit or byte or NCL word that instructs the calling station as to which of the choices to follow. The *call again later* technique requires minimal software in the receiving program, but can increase the amount of traffic, as the calling station keeps trying for a contact and keeps getting *try again later* responses. The *wait for me to call you* response requires some additional software in the receiving program to store a list of stations to call and notify that the TIP is ready for a new message. There the tradeoff is TIP software complexity against network traffic load.

Most disk-based BASIC (and other languages) have the capability to have more than one disc file open at a time. It is, thus, logical that a high level protocol can be used when the TIP is used together with a host computer to allocate space on a disc for more than one incoming message at a time.

The user, interested only in the whole message, does not care how many packets it took to receive the message. The network manager on the other hand, may have different interests and a distinction should be drawn between the requirements of the network user and the network manager. Once a whole message has been received, the computer can signal its own accordingly. The multimessage arrival problem is also significant in the situation in a store and forward mode, where the packets are assembled into complete messages for storage purposes.

Another situation that has to be looked into is the CQ call or rather the response to the CQ call situation. One of the major advantages of packet communications is unattended operation. A terminal can thus be programmed to respond to a CQ call so that a newcomer joining the network will have a response to his initial call. What happens if 20 stations or so try and respond to the CQ call? If multiple simultaneous received messages are allowable, a station can end up having several simultaneous QSO's. An extreme example of this condition is the effect on the network due to the appearance of a rare DX station. Consider what could happen to the network if a rare DX station signs in or originates a message.

Avid DX chasers spend a lot of time and money on working new countries. It is thus very likely that these avid DX chasers could leave their TIP's in the monitor mode and have custom DX capture software in their host computer. Thus, the appearance of a DX station can be detected if it originates even one packet into the network. The result could be a pileup. Each of the DX chasers will originate packets aimed at the DX station. Collisions will occur, due to the QRM, calls will be tried over and over again and the network will be tied up for a long time even if the DX station has gone QRT (possibly due to front end overloading of his TIP software?) as the DX chasers keep trying for a message acknowledgment. Thus, packet communications software deemed workable with few stations (initial net situation) may be less than optimum in a wide area established operational network.

Packet communications offer a revolutionary new means of passing communications to amateur radio. For optimum results, it is very advisable that the low level communication protocols and the high level software be well thought out, flexible and easily adaptable to changing circumstances.

Appendices

Appendix A

Appendix A contains a sample of the use of the programs in this book as applied to a contest.

Figure A-1 contains a copy of the station log of operation during the 1981 WPX (phone) contest using the general contest program listed in Fig. 4-1. Figure A-2 contains a selective printout of the G (prefix) stations worked. Figure A-3 contains a selective printout of the G3 prefixes. These printouts illustrate the use of the LOG-PRINT program. The checklist generated by the CKLSTGEN and CKLSTRD programs for all the contacts made in the contest is shown in Fig. A-4.

The as-run log for the 1982 WPX (phone) contest is shown in Fig. A-5. It was made using the multiband contest program listed in Fig. 4-11. Lines 84 and 131 contain fudge flags signalling bad entries to be deleted later. Note that although the entry number increment includes the fudged call and the flag entry, the number allocated to following contacts will be corrected (that is the purpose of the variable N4 in the program). In fact, the corresponding status display following the last entry was

LAST QSO = VE3JTQ TIME = 2158
Band/Freq = 15
Log file is WPX 82
THERE ARE 373 ENTRIES IN THE LOG
-?
(where -? is the prompt.)

The cleaned-up log file is given in Fig. A-6. This is the same data as was given in Fig. A-5, only it has been processed by the LOGFUDGE program.

The multiband version of the checklist is presented in Fig. A-7. The calls are printed in alphanumerical order for each band starting at 10 Meters and proceeding sequentially to 80 Meters. Blank lines are printed for the 40 and 80 Meter bands because no operation occurred on them and as such no log entries were made. Note that no duplicate contacts were made on the same band. The checklist of Fig. A-4 does

STATION LOG G3ZCZ/W3 PAGE 1 FILE WPX81

QSO #	DATE	TIME	BAND	STATION	S	R	MODE	PWR	QSL	COMMENTS
1	26 Mar 81	0147	10.	ZL4BO	59	0	SSB	1000	--	591079
2	26 Mar 81	0152	10.	AG7M	59	0	SSB	1000	--	591391
3	26 Mar 81	0155	10.	VP2MGQ	59	0	SSB	1000	--	591992
4	26 Mar 81	0158	10.	YV4BOU	59	0	SSB	1000	--	59347
5	26 Mar 81	0159	10.	ZZ5EG	59	0	SSB	1000	--	591942
6	26 Mar 81	0201	10.	N7RO	59	0	SSB	1000	--	59446
7	26 Mar 81	0201	10.	KH6XX	59	0	SSB	1000	--	594200
8	26 Mar 81	0204	10.	VK5UB	59	0	SSB	1000	--	59676
9	26 Mar 81	0206	10.	VK4WIE	59	0	SSB	1000	--	59446
10	26 Mar 81	0207	10.	WA3VUQ	59	0	SSB	1000	--	59658
11	26 Mar 81	0212	10.	KD6PY	59	0	SSB	1000	--	591292
12	26 Mar 81	0213	10.	ZL2ACP	59	0	SSB	1000	--	59627
13	26 Mar 81	0231	10.	W6OKK	59	0	SSB	1000	--	59314
14	26 Mar 81	0235	10.	CE6EZ	59	0	SSB	1000	--	591711
15	26 Mar 81	0243	10.	HP1XAT	59	0	SSB	1000	--	591261
16	26 Mar 81	0244	10.	HP2VX	59	0	SSB	1000	--	59176
17	26 Mar 81	0309	15.	5K4LRM	59	0	SSB	1000	--	591786
18	26 Mar 81	0311	15.	EL2AV	59	0	SSB	1000	--	292147
19	26 Mar 81	0313	15.	HU1SA	59	0	SSB	1000	--	59516
20	26 Mar 81	0317	15.	HC9A	59	0	SSB	1000	--	592082
21	26 Mar 81	0320	15.	KL7IRT	59	0	SSB	1000	--	591873
22	26 Mar 81	0324	15.	VE7ZZZ	59	0	SSB	1000	--	592523
23	26 Mar 81	0325	15.	VK2DDQ	59	0	SSB	1000	--	591040
24	26 Mar 81	0325	15.	KH6DQ	55	0	SSB	1000	--	55632
25	26 Mar 81	0330	15.	WB7FDQ	59	0	SSB	1000	--	591181
26	26 Mar 81	0330	15.	WB7QEL	59	0	SSB	1000	--	57666
27	26 Mar 81	0334	15.	AI6V	59	0	SSB	1000	--	581580
28	26 Mar 81	0334	15.	9Y4VU	59	0	SSB	1000	--	591816
26	26 Mar 81	0337	15.	KP4O	59	0	SSB	1000	--	591904
30	26 Mar 81	0338	15.	KA7AUH	59	0	SSB	1000	--	591339
31	26 Mar 81	0340	15.	ZL2AH	59	0	SSB	1000	--	591817
32	26 Mar 81	0341	15.	W1BIH/PJ2	59	0	SSB	1000	--	59507
33	26 Mar 81	0348	15.	VE3BMV	59	0	SSB	1000	--	591644
34	26 Mar 81	0350	15.	W3GM	59	0	SSB	1000	--	59494
35	26 Mar 81	0354	15.	KA6NGS	59	0	SSB	1000	--	56024
36	26 Mar 81	0356	15.	W7DG	59	0	SSB	1000	--	59474
37	26 Mar 81	0358	15.	4M2AMM	59	0	SSB	1000	--	591694
38	26 Mar 81	0400	15.	KG6B	59	0	SSB	1000	--	59051
39	26 Mar 81	0403	15.	ZL3BK	59	0	SSB	1000	--	59999
40	26 Mar 81	0404	15.	VP2MCL	59	0	SSB	1000	--	592847
41	26 Mar 81	0407	15.	N8UM	59	0	SSB	1000	--	59235
42	26 Mar 81	0415	20.	UK5MAF	59	0	SSB	1000	--	591689
43	26 Mar 81	0431	20.	OH8MA	59	0	SSB	1000	--	591372

Fig. A-1. Contest log for 1981 WPX contest.

44	26	Mar	81	0433	20.	SK6AW	59	0	SSB	1000	--	591292
45	26	Mar	81	0434	20.	F6DZU	59	0	SSB	1000	--	59777
46	26	Mar	81	0435	20.	YT0R	59	0	SSB	1000	--	591808
47	26	Mar	81	0437	20.	UK3FAV	59	0	SSB	1000	--	591579
48	26	Mar	81	0442	20.	HB9AOD	59	0	SSB	1000	--	591728
49	26	Mar	81	0444	20.	KN5A	59	0	SSB	1000	--	591551
50	26	Mar	81	0445	20.	F6KAW	59	0	SSB	1000	--	591769
51	26	Mar	81	0447	20.	AD8R	59	0	SSB	1000	--	59613
52	26	Mar	81	0452	20.	UK2PAP	59	0	SSB	1000	--	591533
53	26	Mar	81	0456	20.	YV3BJL	59	0	SSB	1000	--	591068
54	26	Mar	81	0456	20.	4N4Y	59	0	SSB	1000	--	592797
55	26	Mar	81	0500	20.	HG6V	59	0	SSB	1000	--	591977
56	26	Mar	81	0503	20.	GB4ANT	59	0	SSB	1000	--	592030
57	26	Mar	81	0505	20.	UK2PCR	59	0	SSB	1000	--	591951
58	26	Mar	81	0507	20.	UK3R	59	0	SSB	1000	--	591129
59	26	Mar	81	0509	20.	SP8ECV	59	0	SSB	1000	--	591516
60	26	Mar	81	0510	20.	HG1W	59	0	SSB	1000	--	592174
61	26	Mar	81	0514	20.	HA5JI	59	0	SSB	1000	--	59623
62	26	Mar	81	0516	20.	UK2GQW	59	0	SSB	1000	--	592232
63	26	Mar	81	0517	20.	DK7HW	59	0	SSB	1000	--	591152
64	26	Mar	81	0534	20.	TF3DC/OX	59	0	SSB	1000	--	592570
65	26	Mar	81	0538	20.	OE6MBG	59	0	SSB	1000	--	59142
66	26	Mar	81	0540	20.	UK2GKW	59	0	SSB	1000	--	592266
67	26	Mar	81	0544	20.	KN5H	59	0	SSB	1000	--	591162
68	26	Mar	81	0545	20.	AJ6O	59	0	SSB	1000	--	591968
69	26	Mar	81	0554	15.	KC4VA	59	0	SSB	1000	--	591363
70	26	Mar	81	0557	15.	ZS5SP	59	0	SSB	1000	--	59201
71	26	Mar	81	0600	15.	ZL2AAH	59	0	SSB	1000	--	57078
72	26	Mar	81	0603	15.	ZL1AFA	59	0	SSB	1000	--	59020
73	26	Mar	81	0604	15.	WA2PHA	59	0	SSB	1000	--	59136
74	26	Mar	81	0607	15.	ZL1AMM	59	0	SSB	1000	--	59137
75	26	Mar	81	0608	15.	KB8IO	59	0	SSB	1000	--	59105
76	26	Mar	81	0611	15.	WB8JBM	59	0	SSB	1000	--	59604
77	26	Mar	81	0614	15.	VK4VU	59	0	SSB	1000	--	591456
78	26	Mar	81	0624	20.	SM0AQD/OH0	59	0	SSB	1000	--	591923
79	26	Mar	81	0625	20.	OH5XT	59	0	SSB	1000	--	591624
80	26	Mar	81	0626	20.	VP5RFS	59	0	SSB	1000	--	593633
81	26	Mar	81	0628	20.	IV3PRK	59	0	SSB	1000	--	591571
82	26	Mar	81	0629	20.	XE1AE	59	0	SSB	1000	--	59503
83	26	Mar	81	0635	20.	PA0IJM	59	0	SSB	1000	--	59971
84	26	Mar	81	0637	20.	CS0UA	59	0	SSB	1000	--	591043
85	26	Mar	81	0639	20.	DL8PC	59	0	SSB	1000	--	591112
86	26	Mar	81	0640	20.	OH2LU	59	0	SSB	1000	--	59136
87	26	Mar	81	0643	20.	I0JX	59	0	SSB	1000	--	59179
88	26	Mar	81	0647	20.	DL0UE	59	0	SSB	1000	--	591271

89 26 Mar 81 0648 20. N4MM 59 0 SSB 1000 -- 591459
90 26 Mar 81 0653 20. I4MMV 59 0 SSB 1000 -- 591272
91 26 Mar 81 0653 20. VK2DPE 33 0 SSB 1000 -- 58013
92 26 Mar 81 0703 20. Y53TA 55 0 SSB 1000 -- 59377
93 26 Mar 81 0703 20. Y45RN 33 0 SSB 1000 -- 59362
94 26 Mar 81 0705 20. WB5UYV 59 0 SSB 1000 -- 59168
95 26 Mar 81 0716 15. TI0HE 59 0 SSB 1000 -- 59 (TI2FAG)
96 26 Mar 81 0725 20. HB9BLQ 59 0 SSB 1000 -- 591666
97 26 Mar 81 0727 20. G3VBL 59 0 SSB 1000 -- 591679
98 26 Mar 81 0730 20. GM4FDM 59 0 SSB 1000 -- 59483
99 26 Mar 81 0731 20. OZ5DD 59 0 SSB 1000 -- 59750
100 26 Mar 81 0736 20. HA4XH 59 0 SSB 1000 -- 59907
101 26 Mar 81 0740 20. VK4KA 59 0 SSB 1000 -- 57338
102 26 Mar 81 0743 20. Y24WH 59 0 SSB 1000 -- 59050
103 26 Mar 81 0757 20. K7RI 59 0 SSB 1000 -- 593023
104 26 Mar 81 0802 20. OK2KWU 59 0 SSB 1000 -- 59537
105 26 Mar 81 0804 20. G6UW 59 0 SSB 1000 -- 592405
106 26 Mar 81 0807 20. UK2RDX 59 0 SSB 1000 -- 592081
107 26 Mar 81 0810 20. HG5A 59 0 SSB 1000 -- 591848
108 26 Mar 81 0813 20. KB7LF 59 0 SSB 1000 -- 59144
109 26 Mar 81 0823 20. VE7UBC 59 0 SSB 1000 -- 592618
110 26 Mar 81 0829 20. PA0INE/LX 59 0 SSB 1000 -- 59056
111 26 Mar 81 0833 20. VP10A 59 0 SSB 1000 -- 592846
112 26 Mar 81 0835 20. K3HPG 59 0 SSB 1000 -- 59600
113 26 Mar 81 0839 20. CN8CO 59 0 SSB 1000 -- 59527
114 26 Mar 81 0842 20. OE1DH 59 0 SSB 1000 -- 59714
115 26 Mar 81 0843 20. HA7KPL 59 0 SSB 1000 -- 59868
116 26 Mar 81 0853 20. EA1PJ 59 0 SSB 1000 -- 59765
117 26 Mar 81 0900 20. N4KE 59 0 SSB 1000 -- 591119
118 26 Mar 81 0903 20. VK3QP 59 0 SSB 1000 -- 59065
119 26 Mar 81 0903 20. WB3JRU 59 0 SSB 1000 -- 59475
120 26 Mar 81 0910 20. 3L2MF 59 0 SSB 1000 -- 5915322 (WB3LPS)
121 26 Mar 81 0914 20. WA2JAS 59 0 SSB 1000 -- 59203
122 26 Mar 81 0919 20. HI3XRM 59 0 SSB 1000 -- 59803
123 26 Mar 81 0922 20. N6RM 59 0 SSB 1000 -- 59143
124 26 Mar 81 0923 20. VK4WIA 59 0 SSB 1000 -- 59045
125 26 Mar 81 0927 20. JR1RCR 59 0 SSB 1000 -- 59847
126 26 Mar 81 0939 20. K8NA 59 0 SSB 1000 -- 591022
127 26 Mar 81 0941 20. VE6OU 59 0 SSB 1000 -- 592997
128 26 Mar 81 0944 20. PY3CB 59 0 SSB 1000 -- 59285
129 26 Mar 81 0946 20. W9ZTD 59 0 SSB 1000 -- 59229
130 26 Mar 81 0950 20. I5FCK 59 0 SSB 1000 -- 591099
131 26 Mar 81 0952 20. HA0KDA 59 0 SSB 1000 -- 59418
132 26 Mar 81 0959 20. JG1GGU 59 0 SSB 1000 -- 591243
133 26 Mar 81 1001 20. Y31ZA 59 0 SSB 1000 -- 59796
134 26 Mar 81 1004 20. N5JJ 59 0 SSB 1000 -- 59469

135 26 Mar 81 1021 15. UK9FER 59 0 SSB 1000 -- 591503
136 26 Mar 81 1026 15. EA7BDK 59 0 SSB 1000 -- 59228
137 26 Mar 81 1027 15. Y44ZI/P 59 0 SSB 1000 -- 591820
138 26 Mar 81 1029 15. OZ1FRR 59 0 SSB 1000 -- 59263
139 26 Mar 81 1031 15. G3KWY 59 0 SSB 1000 -- 59158
140 26 Mar 81 1034 15. EI1AA 59 0 SSB 1000 -- 59192
141 26 Mar 81 1034 15. VK3DN 59 0 SSB 1000 -- 58024
142 26 Mar 81 1041 15. GI4KSH 59 0 SSB 1000 -- 59138
143 26 Mar 81 1041 15. YU2AAU 59 0 SSB 1000 -- 59064
144 26 Mar 81 1041 15. EA1JO 59 0 SSB 1000 -- 59053
145 26 Mar 81 1043 15. DL7MAL 59 0 SSB 1000 -- 59092
146 26 Mar 81 1046 15. LA4O 59 0 SSB 1000 -- 59774
147 26 Mar 81 1046 15. YU2OG 59 0 SSB 1000 -- 59865
148 26 Mar 81 1051 15. OE3NPW 59 0 SSB 1000 -- 59556
149 26 Mar 81 1055 15. HA3KNA 59 0 SSB 1000 -- 59717
150 26 Mar 81 1055 15. HA5NG 59 0 SSB 1000 -- 59131
151 26 Mar 81 1056 15. ZL1AAB 59 0 SSB 1000 -- 59005
152 26 Mar 81 1056 15. G3NT 59 0 SSB 1000 -- 59215
153 26 Mar 81 1057 15. YU3RM 59 0 SSB 1000 -- 59008
154 26 Mar 81 1058 15. DL7GAJ 59 0 SSB 1000 -- 59020
155 26 Mar 81 1100 15. G3JXE 59 0 SSB 1000 -- 59057
156 26 Mar 81 1101 15. YU3TWA 59 0 SSB 1000 -- 59001
157 26 Mar 81 1101 15. OK1MSN 59 0 SSB 1000 -- 59797
158 26 Mar 81 1105 15. SP9KMM 54 0 SSB 1000 -- 59718
159 26 Mar 81 1108 15. DK0LO 59 0 SSB 1000 -- 59361
160 26 Mar 81 1108 15. Y55ZG 59 0 SSB 1000 -- 59658
161 26 Mar 81 1112 15. N2WT 59 0 SSB 1000 -- 591240
162 26 Mar 81 1113 15. KF2U 59 0 SSB 1000 -- 59893
163 26 Mar 81 1115 15. W4QAW 59 0 SSB 1000 -- 591428
164 26 Mar 81 1115 15. UA6LBC 59 0 SSB 1000 -- 59660
165 26 Mar 81 1118 15. N2BA/3 59 0 SSB 1000 -- 59942
166 26 Mar 81 1120 15. YU3DDX 59 0 SSB 1000 -- 59563
167 26 Mar 81 1122 15. KA1R 59 0 SSB 1000 -- 591292
168 26 Mar 81 1128 15. K1AR 59 0 SSB 1000 -- 591356
169 26 Mar 81 1129 15. UK3APA 59 0 SSB 1000 -- 592975
170 26 Mar 81 1131 15. DK7ZT 59 0 SSB 1000 -- 59082
171 26 Mar 81 1132 15. OK1AGN 59 0 SSB 1000 -- 595746
172 26 Mar 81 1136 15. G4JJE 59 0 SSB 1000 -- 59758
173 26 Mar 81 1140 15. I4RYC 59 0 SSB 1000 -- 59207
174 26 Mar 81 1145 15. I2PHN 59 0 SSB 1000 -- 591904
175 26 Mar 81 1147 10. N4RV 59 0 SSB 1000 -- 59374
176 26 Mar 81 1149 10. WA1UZH 59 0 SSB 1000 -- 591228
177 26 Mar 81 1152 10. YO1K 59 0 SSB 1000 -- 591432
178 26 Mar 81 1156 10. I1OJE 59 0 SSB 1000 -- 591570
179 26 Mar 81 1157 10. YZ3F 59 0 SSB 1000 -- 59471
180 26 Mar 81 1159 10. YU2BOP 59 0 SSB 1000 -- 591119

181	26	Mar	81	1159	10.	I2SVA	59	0	SSB	1000	--	59630	
182	26	Mar	81	1200	10.	OK3CFA	59	0	SSB	1000	--	59470	
183	26	Mar	81	1200	10.	DL0WW	59	0	SSB	1000	--	591461	
184	26	Mar	81	1200	10.	4X0I	59	0	SSB	1000	--	591215	(4X4IL)
185	26	Mar	81	1204	10.	SV0AP	59	0	SSB	1000	--	591408	
186	26	Mar	81	1206	10.	HA4KYH	59	0	SSB	1000	--	59888	
187	26	Mar	81	1208	10.	GM3RAO	59	0	SSB	1000	--	59657	
188	26	Mar	81	1209	10.	YO6AFP	59	0	SSB	1000	--	59279	
189	26	Mar	81	1211	10.	OK1ALW	59	0	SSB	1000	--	59941	
190	26	Mar	81	1212	10.	G3WTM	59	0	SSB	1000	--	58262	
191	26	Mar	81	1213	10.	HA5KDB	59	0	SSB	1000	--	591219	
192	26	Mar	81	1216	10.	C31IU	59	0	SSB	1000	--	59837	(W8JAQ)
193	26	Mar	81	1219	10.	CS0OF	59	0	SSB	1000	--	591324	
194	26	Mar	81	1221	10.	DF0BV	59	0	SSB	1000	--	59617	
195	26	Mar	81	1223	10.	HA1KZZ	59	0	SSB	1000	--	59939	
196	26	Mar	81	1223	10.	6E1MV	59	0	SSB	1000	--	591116	
197	26	Mar	81	1228	10.	OK6OK	59	0	SSB	1000	--	592304	
198	26	Mar	81	1229	10.	K2NJ	59	0	SSB	1000	--	59047	
199	26	Mar	81	1229	10.	DF3TJ	59	0	SSB	1000	--	591425	
200	26	Mar	81	1233	10.	IN3DYG	59	0	SSB	1000	--	593644	
201	26	Mar	81	1234	10.	YZ3BO	59	0	SSB	1000	--	591639	
202	26	Mar	81	1236	10.	YU2IQ	59	0	SSB	1000	--	591294	
203	26	Mar	81	1237	10.	I5JUX	59	0	SSB	1000	--	59656	
204	26	Mar	81	1240	10.	KC1F	55	0	SSB	1000	--	591771	
205	26	Mar	81	1240	10.	I2AT	59	0	SSB	1000	--	59394	
206	26	Mar	81	1242	10.	DK5AD	59	0	SSB	1000	--	591095	
207	26	Mar	81	1246	10.	DH8AAB/A	59	0	SSB	1000	--	59562	
208	26	Mar	81	1249	10.	YU3EW	59	0	SSB	1000	--	591754	
209	26	Mar	81	1251	10.	4Z4EU	59	0	SSB	1000	--	59417	
210	26	Mar	81	1254	10.	HG0MM	59	0	SSB	1000	--	59435	
211	26	Mar	81	1257	10.	DF1DR/P	59	0	SSB	1000	--	59340	
212	26	Mar	81	1259	10.	ON7WW	59	0	SSB	1000	--	59026	
213	26	Mar	81	1301	10.	YU1AWW	59	0	SSB	1000	--	59627	
214	26	Mar	81	1303	10.	DL0SA	59	0	SSB	1000	--	59492	
215	26	Mar	81	1303	10.	HA9RVK	59	0	SSB	1000	--	59391	
216	26	Mar	81	1303	10.	PA0RRS	216	0	SSB	1000	--	59233	
217	26	Mar	81	1307	10.	DL2SAD	59	0	SSB	1000	--	59142	
218	26	Mar	81	1312	10.	F6BJA	59	0	SSB	1000	--	59047	
219	26	Mar	81	1316	15.	OZ3SK	59	0	SSB	1000	--	59961	
220	26	Mar	81	1318	15.	HB9BUN	59	0	SSB	1000	--	591263	
221	26	Mar	81	1319	15.	UK6LAZ	59	0	SSB	1000	--	592668	
222	26	Mar	81	1322	15.	OE6KDG	59	0	SSB	1000	--	59166	
223	26	Mar	81	1325	15.	G3YWI	59	0	SSB	1000	--	59368	
224	26	Mar	81	1326	15.	K5TM	59	0	SSB	1000	--	59720	
225	26	Mar	81	1328	15.	LA3WAA	59	0	SSB	1000	--	59760	
226	26	Mar	81	1330	15.	G8JC	59	0	SSB	1000	--	591175	

227 26 Mar 81 1330 15. EG1FD 59 0 SSB 1000 -- 591114
228 26 Mar 81 1335 15. AI9J 59 0 SSB 1000 -- 591428
229 26 Mar 81 1338 15. G3RRS 59 0 SSB 1000 -- 591967
230 26 Mar 81 1340 15. UB5LCV 59 0 SSB 1000 -- 591557
231 26 Mar 81 1341 15. GB3WRR 59 0 SSB 1000 -- 591553
232 26 Mar 81 1341 15. OH1AA 59 0 SSB 1000 -- 591622
233 26 Mar 81 1344 15. YU2CZA 59 0 SSB 1000 -- 591149
234 26 Mar 81 1346 15. Y63ZM 59 0 SSB 1000 -- 591553
235 26 Mar 81 1347 15. F6GCP 59 0 SSB 1000 -- 59729
236 26 Mar 81 1350 15. Y24UK 59 0 SSB 1000 -- 594037
237 26 Mar 81 1352 15. HA6KNB 59 0 SSB 1000 -- 591561
238 26 Mar 81 1355 15. SK7GC 59 0 SSB 1000 -- 57735
239 26 Mar 81 1402 10. LG5LG 59 0 SSB 1000 -- 592856
240 26 Mar 81 1417 10. F3TV 59 0 SSB 1000 -- 59037
241 26 Mar 81 1421 10. SP3KEY 59 0 SSB 1000 -- 591363
242 26 Mar 81 1423 10. DL1ZU 59 0 SSB 1000 -- 59540
243 26 Mar 81 1424 10. DJ2EH 59 0 SSB 1000 -- 59076
244 26 Mar 81 1424 10. DK3SN 59 0 SSB 1000 -- 59684
245 26 Mar 81 1428 10. ED1ABT 59 0 SSB 1000 -- 591949
246 26 Mar 81 1430 10. PA2TMS 59 0 SSB 1000 -- 592267
247 26 Mar 81 1432 10. HA5KFN 59 0 SSB 1000 -- 59842
248 26 Mar 81 1434 10. OK1DA 59 0 SSB 1000 -- 59263
249 26 Mar 81 1438 10. EG3WZ 59 0 SSB 1000 -- 591034
250 26 Mar 81 1442 10. G3NAA/M 55 0 SSB 1000 -- 56005
251 26 Mar 81 1442 10. SP9HWN 59 0 SSB 1000 -- 59750
252 26 Mar 81 1445 10. G3BLS 59 0 SSB 1000 -- 59120
253 26 Mar 81 1448 10. Y35YE 59 0 SSB 1000 -- 59628
254 26 Mar 81 1448 10. YU2CTD 59 0 SSB 1000 -- 59179
255 26 Mar 81 1450 10. PP2ZDD 59 0 SSB 1000 -- 59709
256 26 Mar 81 1452 10. ZS6BNZ 59 0 SSB 1000 -- 57336
257 26 Mar 81 1454 10. KC4UQ 59 0 SSB 1000 -- 59586
258 26 Mar 81 1456 10. HP1XWG 59 0 SSB 1000 -- 591042
259 26 Mar 81 1501 10. HGPU/P 59 0 SSB 1000 -- 59221
260 26 Mar 81 1504 10. DL8QS 59 0 SSB 1000 -- 59444
261 26 Mar 81 1504 10. G3JGW 59 0 SSB 1000 -- 59092
262 26 Mar 81 1506 10. G3VJP 59 0 SSB 1000 -- 59125
263 26 Mar 81 1506 10. HA5KDB 59 0 SSB 1000 -- 591271
264 26 Mar 81 1507 10. WD6HBR 59 0 SSB 1000 -- 591196
265 26 Mar 81 1507 10. HA7KPL 59 0 SSB 1000 -- 591038
266 26 Mar 81 1509 10. W7KXH 59 0 SSB 1000 -- 59029
267 26 Mar 81 1509 10. G3TWG 55 0 SSB 1000 -- 58013
268 26 Mar 81 1512 10. G3MDH 55 0 SSB 1000 -- 57021
269 26 Mar 81 1514 10. W6UBJ 59 0 SSB 1000 -- 59301
270 26 Mar 81 1514 10. G4HRV 59 0 SSB 1000 -- 59092
271 26 Mar 81 1515 10. G3AGF 55 0 SSB 1000 -- 59002
272 26 Mar 81 1517 10. DF4RD 55 0 SSB 1000 -- 59134

273	26	Mar	81	1519	10.	PA0LEY	55	0	SSB	1000	--	57001
274	26	Mar	81	1521	10.	6W8AR	59	0	SSB	1000	--	59044
275	26	Mar	81	1522	10.	OK3KAP	59	0	SSB	1000	--	59607
276	26	Mar	81	1524	10.	G4AGZ	59	0	SSB	1000	--	59007
277	26	Mar	81	1526	10.	G3UYW	59	0	SSB	1000	--	59021
278	26	Mar	81	1526	10.	ZS6PS	57	0	SSB	1000	--	57009
279	26	Mar	81	1528	10.	G3YPZ/M	43	0	SSB	1000	--	59005
280	26	Mar	81	1529	10.	G4JLU	59	0	SSB	1000	--	59035
281	26	Mar	81	1530	10.	G3BFC	59	0	SSB	1000	--	59001
282	26	Mar	81	1534	10.	DL0SA	59	0	SSB	1000	--	59540
283	26	Mar	81	1537	10.	UB5UKO	55	0	SSB	1000	--	59229
284	26	Mar	81	1537	10.	EA1JO	59	0	SSB	1000	--	59079
285	26	Mar	81	1540	10.	DJBUV	59	0	SSB	1000	--	59461
286	26	Mar	81	1540	10.	G4FRV	59	0	SSB	1000	--	59050
287	26	Mar	81	1543	10.	Y78XL	59	0	SSB	1000	--	59787
288	26	Mar	81	1543	10.	ON4OL	59	0	SSB	1000	--	59172
289	26	Mar	81	1543	10.	Y31ZE	59	0	SSB	1000	--	59239
290	26	Mar	81	1547	10.	Y47XF	59	0	SSB	1000	--	59279
291	26	Mar	81	1547	10.	G3CSE	59	0	SSB	1000	--	59508
292	26	Mar	81	1549	10.	ON6UC	59	0	SSB	1000	--	59094
293	26	Mar	81	1550	10.	G3COJ	59	0	SSB	1000	--	59002
294	26	Mar	81	1550	10.	ON5NQ	59	0	SSB	1000	--	59178
295	26	Mar	81	1555	10.	G3PHW	55	0	SSB	1000	--	59124
296	26	Mar	81	1556	10.	G3VSQ	59	0	SSB	1000	--	59104
297	26	Mar	81	1556	10.	G3OZF	59	0	SSB	1000	--	59081
298	26	Mar	81	1556	10.	F8RU	59	0	SSB	1000	--	59018
299	26	Mar	81	1557	10.	G3MGW	59	0	SSB	1000	--	59021
300	26	Mar	81	1559	10.	G3VYG	59	0	SSB	1000	--	59005
301	26	Mar	81	1559	10.	G3LGY	59	0	SSB	1000	--	59011
302	26	Mar	81	1600	10.	G3BGY	59	0	SSB	1000	--	59005

STATION LOG G3ZCZ PAGE 1 FILE WPX81

QSO #	DATE	TIME	BAND	STATION	S	R	MODE	PWR	QSL	COMMENTS
56	26 Mar 81	0503	20.	GB4ANT	59	0	SSB	1000	--	592030
97	26 Mar 81	0727	20.	G3VBL	59	0	SSB	1000	--	591679
98	26 Mar 81	0730	20.	GM4FDM	59	0	SSB	1000	--	59483
105	26 Mar 81	0804	20.	G6UW	59	0	SSB	1000	--	592405
139	26 Mar 81	1031	15.	G3KWY	59	0	SSB	1000	--	59158
142	26 Mar 81	1041	15.	GI4KSH	59	0	SSB	1000	--	59138
152	26 Mar 81	1056	15.	G3NT	59	0	SSB	1000	--	59215
155	26 Mar 81	1100	15.	G3JXE	59	0	SSB	1000	--	59057
172	26 Mar 81	1136	15.	G4JJE	59	0	SSB	1000	--	59758
187	26 Mar 81	1208	10.	GM3RAO	59	0	SSB	1000	--	59657
190	26 Mar 81	1212	10.	G3WTM	59	0	SSB	1000	--	58262
223	26 Mar 81	1325	15.	G3YWI	59	0	SSB	1000	--	59368
226	26 Mar 81	1330	15.	G8JC	59	0	SSB	1000	--	591175
229	26 Mar 81	1338	15.	G3RRS	59	0	SSB	1000	--	591967
231	26 Mar 81	1341	15.	GB3WRR	59	0	SSB	1000	--	591553
250	26 Mar 81	1442	10.	G3NAA/M	55	0	SSB	1000	--	56005
252	26 Mar 81	1445	10.	G3BLS	59	0	SSB	1000	--	59120
261	26 Mar 81	1504	10.	G3JGW	59	0	SSB	1000	--	59092
262	26 Mar 81	1506	10.	G3VJP	59	0	SSB	1000	--	59125
267	26 Mar 81	1509	10.	G3TWG	55	0	SSB	1000	--	58013
268	26 Mar 81	1512	10.	G3MDH	55	0	SSB	1000	--	57021
270	26 Mar 81	1514	10.	G4HRV	59	0	SSB	1000	--	59092
271	26 Mar 81	1515	10.	G3AGF	55	0	SSB	1000	--	59002
276	26 Mar 81	1524	10.	G4AGZ	59	0	SSB	1000	--	59007
277	26 Mar 81	1526	10.	G3UYW	59	0	SSB	1000	--	59021
279	26 Mar 81	1528	10.	G3YPZ/M	43	0	SSB	1000	--	59005
280	26 Mar 81	1529	10.	G4JLU	59	0	SSB	1000	--	59035
281	26 Mar 81	1530	10.	G3BFC	59	0	SSB	1000	--	59001
286	26 Mar 81	1540	10.	G4FRV	59	0	SSB	1000	--	59050
291	26 Mar 81	1547	10.	G3CSE	59	0	SSB	1000	--	59508
293	26 Mar 81	1550	10.	G3COJ	59	0	SSB	1000	--	59002
295	26 Mar 81	1555	10.	G3PHW	55	0	SSB	1000	--	59124
296	26 Mar 81	1556	10.	G3VSQ	59	0	SSB	1000	--	59104
297	26 Mar 81	1556	10.	G3OZF	59	0	SSB	1000	--	59081
299	26 Mar 81	1557	10.	G3MGW	59	0	SSB	1000	--	59021
300	26 Mar 81	1559	10.	G3VYG	59	0	SSB	1000	--	59005
301	26 Mar 81	1559	10.	G3LGY	59	0	SSB	1000	--	59011
302	26 Mar 81	1600	10.	G4BGY	59	0	SSB	1000	--	59005

Fig. A-2. Log of all contacts with stations whose callsigns begin with G.

STATION LOG G3ZCZ/4X PAGE 1 FILE WPX81

QSO #	DATE	TIME	BAND	STATION	S	R	MODE	PWR	QSL	COMMENTS
97	26 Mar 81	0727	20.	G3VBL	59	0	SSB	1000	--	591679
139	26 Mar 81	1031	15.	G3KWY	59	0	SSB	1000	--	59158
152	26 Mar 81	1056	15.	G3NT	59	0	SSB	1000	--	59215
155	26 Mar 81	1100	15.	G3JXE	59	0	SSB	1000	--	59057
190	26 Mar 81	1212	10.	G3WTM	59	0	SSB	1000	--	58262
223	26 Mar 81	1325	15.	G3YWI	59	0	SSB	1000	--	59368
229	26 Mar 81	1338	15.	G3RRS	59	0	SSB	1000	--	591967
250	26 Mar 81	1442	10.	G3NAA/M	55	0	SSB	1000	--	56005
252	26 Mar 81	1445	10.	G3BLS	59	0	SSB	1000	--	59120
261	26 Mar 81	1504	10.	G3JGW	59	0	SSB	1000	--	59092
262	26 Mar 81	1506	10.	G3VJP	59	0	SSB	1000	--	59125
267	26 Mar 81	1509	10.	G3TWG	55	0	SSB	1000	--	58013
268	26 Mar 81	1512	10.	G3MDH	55	0	SSB	1000	--	57021
271	26 Mar 81	1515	10.	G3AGF	55	0	SSB	1000	--	59002
277	26 Mar 81	1526	10.	G3UYW	59	0	SSB	1000	--	59021
279	26 Mar 81	1528	10.	G3YPZ/M	43	0	SSB	1000	--	59005
281	26 Mar 81	1530	10.	G3BFC	59	0	SSB	1000	--	59001
291	26 Mar 81	1547	10.	G3CSE	59	0	SSB	1000	--	59508
293	26 Mar 81	1550	10.	G3COJ	59	0	SSB	1000	--	59002
295	26 Mar 81	1555	10.	G3PHW	55	0	SSB	1000	--	59124
296	26 Mar 81	1556	10.	G3VSQ	59	0	SSB	1000	--	59104
297	26 Mar 81	1556	10.	G3OZF	59	0	SSB	1000	--	59081
299	26 Mar 81	1557	10.	G3MGW	59	0	SSB	1000	--	59021
300	26 Mar 81	1559	10.	G3VYG	59	0	SSB	1000	--	59005
301	26 Mar 81	1559	10.	G3LGY	59	0	SSB	1000	--	59011

Fig. A-3. Log of all contacts with stations whose callsigns begin with G3.

Check list		CK81			
3L2MF	4M2AMM	4N4Y	4X0I	4Z4EU	5K4LRM
6E1MV	6W8AR	9Y4VU	AD8R	AG7M	AI6V
AI9J	AJ6O	C31IU	CE6E7	CN8CO	CS0OF
CS0UA	DF0BV	DF1DR/P	DF3TJ	DF4RD	DH8AAR/A
DJ2EH	DJ8UV	DK0LO	DK3SN	DK5AD	DK7HW
DK7ZT	DL0SA	******	DL0UE	DL0WW	DL17U
DL2SAD	DL7GAJ	DL7MAL	DL8PC	DL8QS	EA1JO
******	EA1PJ	EA7BDK	ED1ABT	EG1FD	EG3WZ
EI1AA	EL2AV	F3TV	F6BJA	F6DZU	F6GCP
F6KAW	F8RU	G3AGF	G3BFC	G3BLS	G3COJ
G3CSE	G3JGW	G3JXE	G3KWY	G3LGY	G3MDH
G3MGW	G3NAA/M	G3NT	G3OZF	G3PHW	G3RRS
G3TWG	G3UYW	G3VBL	G3VJP	G3VSQ	G3VYG
G3WTM	G3YPZ/M	G3YWI	G4AGZ	G4BGY	G4FRV
G4HRV	G4JJE	G4JLU	G6UW	G8JC	GB3WRR
GB4ANT	GI4KSH	GM3RAO	GM4FDM	HA0KDA	HA1KZZ
HA3KNA	HA4KYH	HA4XH	HA5JI	HA5KDB	******
HA5KFN	HA5NG	HA6KNB	HA7KPL	******	HA9BVK
HB9ADO	HB9BLQ	HB9BUN	HC9A	HG0MM	HG1W
HG5A	HG6V	HGPU/P	HI3XRM	HP1XAT	HP1XWG
HP2VX	HU1SA	I0JX	I1OJE	I2AT	I2PHN
I2SVA	I4MMV	I4RYC	I5FCK	I5JUX	IN3DYG
IV3PRK	JG1GGU	JR1RCR	K1AR	K2NJ	K3HPG
K5TM	K7RI	K8NA	KA1R	KA6NGS	KA7AUH
KB7LF	KB8IO	KC1F	KC4UQ	KC4VA	KD6PY
KF2U	KG6B	KH6DQ	KH6XX	KL7IRT	KN5A
KN5H	KP4O	LA3WAA	LA4O	LG5LG	N2BA/3
N2WT	N4KE	N4MM	N4RV	N5JJ	N6RM
N7RO	N8UM	OE1DH	OE3NPW	OE6KDG	OE6MBG
OH1AA	OH2LU	OH5XT	OH8MA	OK1AGN	OK1ALW
OK1DA	OK1MSN	OK2KWU	OK3CFA	OK3KAP	OK6OK
ON4OL	ON5NQ	ON6UC	ON7WW	OZ1FRR	OZ3SK
OZ5DD	PA0IJM	PA0INE/LX	PA0LEY	PA0RRS	PA2TMS
PP2ZDD	PY3CB	SK6AW	SK7GC	SM0AQD/OH0SP3KEY	
SP8ECV	SP9HWN	SP9KMM	SV0AP	TF3DC/OX	TI0HE
UA6LBC	UB5LCV	UB5UKO	UK2GKW	UK2GQW	UK2PAP
UK2PCR	UK2RDX	UK3APA	UK3FAV	UK3R	UK5MAF
UK6LAZ	UK9FER	VE3BMV	VE6OU	VE7UBC	VE7ZZZ
VK2DDQ	VK2DPE	VK3DN	VK3QP	VK4KA	VK4VU
VK4WIA	VK4WIE	VK5UB	VP10A	VP2MCL	VP2MGQ
VP5RFS	W1BIH/PJ2	W3GM	W4QAW	W6OKK	W6UBJ
W7DG	W7KXH	W9ZTD	WA1UZH	WA2JAS	WA2PHA
WA3VUQ	WB3JRU	WB5UYV	WB7FDQ	WB7QEL	WB8JBM
WD6HBR	XE1AE	Y24UK	Y24WH	Y31ZA	Y31ZE

Fig. A-4. Contest checklist.

Y35YE	Y44ZI/P	Y45RN	Y47XF	Y53TA	Y55ZG
Y63ZM	Y78XL	YO1K	YO6AFP	YT0R	YU1AWW
YU2AAU	YU2ROP	YU2CTD	YU2CZA	YU2IQ	YU2OG
YU3DDX	YU3EW	YU3RM	YU3TWA	YV3RJL	YV4BOU
YZ3RO	YZ3F	ZL1AAB	ZL1AFA	ZL1AMM	ZL2AAH
ZL2ACP	ZL2AH	ZL3BK	ZL4RO	ZS5SP	ZS6BNZ
ZS6PS	ZZ5EG				

STATION LOG G3ZCZ/4X PAGE 1 FILE WPX82LG

QSO #	DATE	TIME	BAND	STATION	S	R	MODE	PWR	QSL	COMMENTS
1	27 Mar 82	1638	10.	HA4KYH	59	0	SSB	0	--	59337
2	27 Mar 82	1641	10.	N8AKF	59	0	SSB	200	--	59362
3	27 Mar 82	1644	10.	YU2CHI	59	0	SSB	200	--	59008
4	27 Mar 82	1645	10.	IO6FLD	59	0	SSB	200	--	59940
5	27 Mar 82	1647	10.	YU3CAB	59	0	SSB	200	--	59803
6	27 Mar 82	1648	10.	KF2O	59	0	SSB	200	--	59397
7	27 Mar 82	1650	10.	HA5KKC/7	59	0	SSB	200	--	591284
8	27 Mar 82	1645	10.	4U1ITU	59	0	SSB	200	--	591072
9	27 Mar 82	1656	10.	IV3PRK	59	0	SSB	200	--	591096
10	27 Mar 82	1659	10.	Y35LM	59	0	SSB	200	--	59576
11	27 Mar 82	1701	10.	N4NO	59	0	SSB	200	--	59222
12	27 Mar 82	1703	10.	YU3UPI	59	0	SSB	200	--	59236
13	27 Mar 82	1703	10.	Y34YF	59	0	SSB	200	--	59036
14	27 Mar 82	1705	10.	N4KG	55	0	SSB	200	--	59120
15	27 Mar 82	1705	10.	YU3TDP	59	0	SSB	200	--	59026
16	27 Mar 82	1705	10.	N2US	59	0	SSB	200	--	59233
17	27 Mar 82	1705	10.	I5MYL	59	0	SSB	200	--	59024
18	27 Mar 82	1707	10.	KE3G	59	0	SSB	200	--	59008
19	27 Mar 82	1707	10.	K3EST	59	0	SSB	200	--	59600
20	27 Mar 82	1707	10.	K3KHL	59	0	SSB	200	--	59027
21	27 Mar 82	1716	10.	Y32ZF	59	0	SSB	200	--	59023
22	27 Mar 82	1720	10.	Y56YF	59	0	SSB	200	--	59148
23	27 Mar 82	1720	10.	N2AA	55	0	SSB	200	--	59367
24	27 Mar 82	1724	10.	K2BA/4	59	0	SSB	200	--	59912
25	27 Mar 82	1725	10.	OK1KST	59	0	SSB	200	--	591114
26	27 Mar 82	1726	10.	I4ZSQ	59	0	SSB	200	--	59604
27	27 Mar 82	1728	10.	K2SS	59	0	SSB	200	--	59966
28	27 Mar 82	1736	10.	OK1ARI	59	0	SSB	200	--	59698
29	27 Mar 82	1736	10.	I2SVA	59	0	SSB	200	--	59490
30	27 Mar 82	1744	20.	HA4KYN	59	0	SSB	200	--	59790
31	27 Mar 82	1813	15.	CT7AL	59	0	SSB	200	--	59160
32	27 Mar 82	1816	15.	YU3DBC	59	0	SSB	200	--	591152
33	27 Mar 82	1818	15.	HA5KKC/7	59	0	SSB	200	--	591409
34	27 Mar 82	1819	15.	KC3N	59	0	SSB	200	--	59496
35	27 Mar 82	1823	15.	IT9WPO	59	0	SSB	200	--	59803
36	27 Mar 82	1823	15.	YU3DAW	59	0	SSB	200	--	59319
37	27 Mar 82	1825	15.	YU4YA	59	0	SSB	200	--	59287
38	27 Mar 82	1835	15.	WD0EWD	59	0	SSB	200	--	59491
39	27 Mar 82	1837	15.	AD8R	59	0	SSB	200	--	59658
40	27 Mar 82	1837	15.	KG1D	59	0	SSB	200	--	59739
41	27 Mar 82	1845	15.	K2SS	59	0	SSB	200	--	591037
42	27 Mar 82	1854	15.	4X4VE/5NB	59	0	SSB	200	--	59660

Fig. A-5. Contest log as run.

43	27	Mar	82	1859	15.	Y24UK	59	0	SSB	200	--	591460	
44	27	Mar	82	1859	15.	VE6DU	59	0	SSB	200	--	591097	
45	27	Mar	82	1859	15.	OK3CRH	59	0	SSB	200	--	59260	
46	27	Mar	82	1903	15.	OK1AJN	59	0	SSB	200	--	59292	
47	27	Mar	82	1903	15.	NP4A	59	0	SSB	200	--	51220	
48	27	Mar	82	1906	15.	UK3ADZ	59	0	SSB	200	--	591348	
49	27	Mar	82	1907	15.	YU3VM	59	0	SSB	200	--	59833	
50	27	Mar	82	1909	15.	EA1ABT	59	0	SSB	200	--	591139	
51	27	Mar	82	1911	15.	Y33ZB	59	0	SSB	200	--	59925	
52	27	Mar	82	1913	15.	AH2M	59	0	SSB	200	--	59150	
53	27	Mar	82	1914	15.	UK9FER	59	0	SSB	200	--	591004	
54	27	Mar	82	1917	15.	AC3A	59	0	SSB	200	--	591239	
55	27	Mar	82	1917	15.	VE3PCA	59	0	SSB	200	--	591379	
56	27	Mar	82	1920	15.	UA3VAS	59	0	SSB	200	--	59049	
57	27	Mar	82	1920	15.	K6IR	59	0	SSB	200	--	59256	
58	27	Mar	82	1927	15.	UW3HV	59	0	SSB	200	--	591018	
59	27	Mar	82	1927	15.	DF4NP	59	0	SSB	200	--	59168	
60	27	Mar	82	1930	15.	YU2CT	59	0	SSB	200	--	59258	
61	27	Mar	82	1937	15.	EI7DJ	59	0	SSB	200	--	59466	
62	27	Mar	82	1945	10.	PJ2VR	59	0	SSB	200	--	59646	
63	27	Mar	82	1956	10.	VP2EC	59	0	SSB	200	--	592649	
64	27	Mar	82	2005	15.	DL6FAW	59	0	SSB	200	--	59252	
65	27	Mar	82	2008	15.	DL8PC	59	0	SSB	200	--	591196	
66	27	Mar	82	2013	15.	I1JHS	59	0	SSB	200	--	59312	
67	27	Mar	82	2013	15.	VE5UF	59	0	SSB	200	--	59773	
68	27	Mar	82	2013	15.	VE7BTV	59	0	SSB	200	--	591157	
69	27	Mar	82	2016	15.	AI9J	59	0	SSB	200	--	591166	
70	27	Mar	82	2018	15.	DL4SAR	59	0	SSB	200	--	59405	
71	27	Mar	82	2020	15.	KQ2M	59	0	SSB	200	--	59852	
72	27	Mar	82	2024	15.	VE3BMV	59	0	SSB	200	--	591163	
73	27	Mar	82	2026	15.	AI7B	59	0	SSB	200	--	591369	
74	27	Mar	82	2028	15.	IZ1ARI	59	0	SSB	200	--	591304	Q=I1VEH
75	27	Mar	82	2030	15.	I0KWX	59	0	SSB	200	--	591200	
76	27	Mar	82	2043	15.	HB9BMY/4X	59	0	SSB	200	--	59005	
77	27	Mar	82	2045	15.	AB0I	59	0	SSB	200	--	591248	
78	27	Mar	82	2052	15.	F6BFN	59	0	SSB	200	--	59813	
79	27	Mar	82	2057	15.	KQ8M	59	0	SSB	200	--	59600	
80	27	Mar	82	2057	15.	KA1YQ	59	0	SSB	200	--	59220	
81	27	Mar	82	2115	15.	VP2MGQ	59	0	SSB	200	--	591287	
82	27	Mar	82	2118	15.	KJ9W	59	0	SSB	200	--	591841	
83	27	Mar	82	2120	20.	UK3QAW	59	0	SSB	200	--	59988	
84	27	Mar	82	2120	20.	/*	59	0	SSB	200	--		
85	27	Mar	82	2120	20.	UK3QAE	59	0	SSB	200	--	59988	
86	27	Mar	82	2125	20.	UK6LAA	59	0	SSB	200	--	591326	
87	27	Mar	82	2125	20.	UK0QAA	59	0	SSB	200	--	592219	
88	27	Mar	82	2131	20.	YU7KMN	59	0	SSB	200	--	59679	

```
 89 27 Mar 82 2137  20.   UK5QBE    59  0  SSB  200  --  59840
 90 27 Mar 82 2140  20.   KL7RA     59  0  SSB  200  --  59579
 91 27 Mar 82 2143  20.   UK5UDX    59  0  SSB  200  --  591396
 92 27 Mar 82 2143  20.   FC6FPH    59  0  SSB  200  --  59087 (G8KA)
 93 27 Mar 82 2148  20.   HB9AGC    59  0  SSB  200  --  59511
 94 27 Mar 82 2150  15.   YV3BJL    59  0  SSB  200  --  591144
 95 27 Mar 82 2155  15.   KB8LH     59  0  SSB  200  --  59256
 96 27 Mar 82 2157  15.   K3ZJ      59  0  SSB  200  --  59999
 97 27 Mar 82 2158  15.   WB4KRH    59  0  SSB  200  --  59350
 98 27 Mar 82 2158  15.   P42J      59  0  SSB  200  --  591513 Curaco
 99 27 Mar 82 2206  15.   KC1F      59  0  SSB  200  --  591314
100 27 Mar 82 2208  15.   N4WW      59  0  SSB  200  --  591904
101 27 Mar 82 2213  15.   N4XD      59  0  SSB  200  --  59243
102 27 Mar 82 2215  15.   AK1A      59  0  SSB  200  --  591145
103 27 Mar 82 2217  15.   WB9TIY    59  0  SSB  200  --  59184
104 27 Mar 82 2220  20.   Y22OM/A   59  0  SSB  200  --  591111
105 27 Mar 82 2221  20.   LZ2DB     59  0  SSB  200  --  59193
106 27 Mar 82 2222  20.   UK5MAF    59  0  SSB  200  --  591844
107 27 Mar 82 2222  20.   UK3DBG    59  0  SSB  200  --  59708
108 27 Mar 82 2224  20.   UK2BBB    59  0  SSB  200  --  591873
109 27 Mar 82 2226  20.   R6L       59  0  SSB  200  --  592876
110 27 Mar 82 2228  20.   UK2PCR    59  0  SSB  200  --  591939
111 27 Mar 82 2232  20.   UK3QAA    59  0  SSB  200  --  59117
112 27 Mar 82 2234  20.   UK6LTG    59  0  SSB  200  --  59135
113 27 Mar 82 2234  20.   F6KAW     59  0  SSB  200  --  591491
114 27 Mar 82 2236  20.   UK6XAA    59  0  SSB  200  --  591199
115 27 Mar 82 2240  20.   UK5QAV    59  0  SSB  200  --  59812
116 27 Mar 82 2243  15.   W1IHN     59  0  SSB  200  --  59190
117 27 Mar 82 2245  15.   KK5I      59  0  SSB  200  --  591112
118 27 Mar 82 2245  15.   K2IJL     59  0  SSB  200  --  59130
119 27 Mar 82 2250  15.   KI2P      59  0  SSB  200  --  59940
120 27 Mar 82 2254  15.   WA3GSC    59  0  SSB  200  --  59121
121 27 Mar 82 2256  15.   KA4SUN    59  0  SSB  200  --  59019
122 27 Mar 82 2259  15.   KV5Q      59  0  SSB  200  --  59924
123 27 Mar 82 2302  15.   KB2MG     59  0  SSB  200  --  59931
124 27 Mar 82 2302  15.   HK5BCZ    55  0  SSB  200  --  59182
125 28 Mar 82 0416  15.   VK2ERT    59  0  SSB  200  --  59836
126 28 Mar 82 0425  20.   AC3A      59  0  SSB  200  --  591630
127 28 Mar 82 0428  20.   UK2GKW    59  0  SSB  200  --  592283
128 28 Mar 82 0430  20.   HA9KPU    59  0  SSB  200  --  59756
129 28 Mar 82 0447  10.   RA3DKE    55  0  SSB  200  --  59327
130 28 Mar 82 0453  10.   UA3TDK    59  0  SSB  200  --  59147
131 28 Mar 82 0456  10.   /*        59  0  SSB  200  --
132 28 Mar 82 0456  10.   UA3TBK    59  0  SSB  200  --  59147
133 28 Mar 82 0501  10.   UK3DAH    59  0  SSB  200  --  59868
134 28 Mar 82 0516  10.   UK3ADZ    59  0  SSB  200  --  591685
```

```
135 28 Mar 82 0516  10.   UA3ACY/U9J 55   0   SSB  200  --  55579
136 28 Mar 82 0521  10.   JH0BBE     55   0   SSB  200  --  59805
137 28 Mar 82 0524  10.   UK3AAC     59   0   SSB  200  --  591203
138 28 Mar 82 0526  10.   UA0ZDD     55   0   SSB  200  --  59808
139 28 Mar 82 0528  10.   UK4FAV     59   0   SSB  200  --  591748
140 28 Mar 82 0530  10.   OK1ALQ     59   0   SSB  200  --  59068
141 28 Mar 82 0534  10.   UA3AHA     59   0   SSB  200  --  59189
142 28 Mar 82 0534  10.   F6BXQ      59   0   SSB  200  --  59014
143 28 Mar 82 0534  10.   I4VOS      59   0   SSB  200  --  59864
144 28 Mar 82 0536  10.   RA3RBU     59   0   SSB  200  --  59132
145 28 Mar 82 0536  10.   OK1DLA     59   0   SSB  200  --  59483
146 28 Mar 82 0538  10.   UA3AGL     59   0   SSB  200  --  59009
147 28 Mar 82 0538  10.   UK4WAA     59   0   SSB  200  --  59595
148 28 Mar 82 0549  15.   VK2APK     59   0   SSB  200  --  59640
149 28 Mar 82 0549  15.   FR0FLO     59   0   SSB  200  --  59919
150 28 Mar 82 0552  15.   VK3AJJ     59   0   SSB  200  --  59380
151 28 Mar 82 0556  15.   OE1WO      59   0   SSB  200  --  59199
152 28 Mar 82 0557  15.   ED3VM      59   0   SSB  200  --  59489
153 28 Mar 82 0559  15.   UK5AAA     44   0   SSB  200  --  59366
154 28 Mar 82 0559  15.   UA6LGP     59   0   SSB  200  --  59237
155 28 Mar 82 0602  15.   UW3ZU      59   0   SSB  200  --  59048
156 28 Mar 82 0603  15.   YU7NZR     59   0   SSB  200  --  59230
157 28 Mar 82 0611  20.   UK5GKW     59   0   SSB  200  --  591180
158 28 Mar 82 0613  20.   N4WW       59   0   SSB  200  --  592398
159 28 Mar 82 0616  20.   OH1IG      59   0   SSB  200  --  591169
160 28 Mar 82 0619  20.   K2BA/4     59   0   SSB  200  --  591454
161 28 Mar 82 0619  20.   HB9BUN     59   0   SSB  200  --  591008
162 28 Mar 82 0624  20.   Y22JJ      59   0   SSB  200  --  59606
163 28 Mar 82 0627  20.   I0JBL      59   0   SSB  200  --  59170
164 28 Mar 82 0630  20.   YO9HP      59   0   SSB  200  --  59180
165 28 Mar 82 0633  20.   DK0MM      59   0   SSB  200  --  591201
166 28 Mar 82 0638  15.   G3XBY      59   0   SSB  200  --  59229
167 28 Mar 82 0640  15.   HA5KKG     59   0   SSB  200  --  59531
168 28 Mar 82 0643  10.   Y22OM/A    59   0   SSB  200  --  591156
169 28 Mar 82 0652  10.   Y37XJ      59   0   SSB  200  --  59189
170 28 Mar 82 0700  20.   OK2RZ      59   0   SSB  200  --  59215
171 28 Mar 82 0703  20.   LZ1BM      59   0   SSB  200  --  59374
172 28 Mar 82 0704  20.   LZ2KIM     59   0   SSB  200  --  59882
173 28 Mar 82 0707  20.   LZ2KRR     59   0   SSB  200  --  591462
174 28 Mar 82 0709  20.   K2NG       59   0   SSB  200  --  55104
175 28 Mar 82 0714  15.   F8WE       59   0   SSB  200  --  59692
176 28 Mar 82 0716  15.   LZ13C      59   0   SSB  200  --  59650
177 28 Mar 82 0718  15.   UB5MBZ     59   0   SSB  200  --  59914
178 28 Mar 82 0719  15.   R6L        59   0   SSB  200  --  593794
179 28 Mar 82 0721  15.   YU3EF      59   0   SSB  200  --  59567
180 28 Mar 82 0722  15.   OH6AC      59   0   SSB  200  --  591357
```

181	28	Mar	82	0724	10.	OK3JW	59	0	SSB	200	--	59788
182	28	Mar	82	0724	10.	OK3CFA	59	0	SSB	200	--	59557
183	28	Mar	82	0729	10.	OK3CSC	59	0	SSB	200	--	59624
184	28	Mar	82	0732	10.	HA9KPU	59	0	SSB	200	--	59837
185	28	Mar	82	0732	10.	I0SKK	59	0	SSB	200	--	59197
186	28	Mar	82	0745	10.	HG5A	59	0	SSB	200	--	592255
187	28	Mar	82	0747	10.	HA4ZB	59	0	SSB	200	--	59058
188	28	Mar	82	0753	10.	JI1QQI	55	0	SSB	200	--	591150
189	28	Mar	82	0753	10.	UY5XE	59	0	SSB	200	--	59238
190	28	Mar	82	0759	10.	Y47XF	59	0	SSB	200	--	59457
191	28	Mar	82	0800	10.	UA3QAE	59	0	SSB	200	--	591241
192	28	Mar	82	0802	10.	YO5KAD	59	0	SSB	200	--	59350
193	28	Mar	82	0803	10.	YU1NZW	59	0	SSB	200	--	59278
194	28	Mar	82	0803	10.	UK2AAF	55	0	SSB	200	--	59276
195	28	Mar	82	0805	10.	YU3EO	59	0	SSB	200	--	59229
196	28	Mar	82	0807	10.	UB5DBV	59	0	SSB	200	--	59085
197	28	Mar	82	0810	10.	YU2RVL	59	0	SSB	200	--	59028
198	28	Mar	82	0813	10.	UK5MAF	59	0	SSB	200	--	592325
199	28	Mar	82	0817	10.	YU7AV	59	0	SSB	200	--	59270
200	28	Mar	82	0821	20.	UC2OBP	59	0	SSB	200	--	59509
201	28	Mar	82	0826	20.	EA6GP	59	0	SSB	200	--	59842
202	28	Mar	82	0830	20.	HA0DU	59	0	SSB	200	--	59571
203	28	Mar	82	0833	15.	UK5OAA	59	0	SSB	200	--	59721
204	28	Mar	82	0837	15.	IS0QDV	59	0	SSB	200	--	591560
205	28	Mar	82	0841	15.	HA4XX	55	0	SSB	200	--	57508
206	28	Mar	82	0850	10.	DK0MM	59	0	SSB	200	--	591366
207	28	Mar	82	0855	10.	YU3DBC	59	0	SSB	200	--	591856
208	28	Mar	82	0857	10.	YU7KWX	59	0	SSB	200	--	591827
209	28	Mar	82	0858	10.	IT9WPO	59	0	SSB	200	--	591157
210	28	Mar	82	0901	10.	UA9MAF	59	0	SSB	200	--	59280
211	28	Mar	82	0905	20.	4N6HN	59	0	SSB	200	--	59242
212	28	Mar	82	0907	20.	UK6YAB	59	0	SSB	200	--	59176
213	28	Mar	82	0909	20.	HG5A	59	0	SSB	200	--	592338
214	28	Mar	82	0910	20.	UK3XAB	59	0	SSB	200	--	59876
215	28	Mar	82	0912	20.	OH2AW	55	0	SSB	200	--	59592
216	28	Mar	82	0914	20.	UG6GDS	59	0	SSB	200	--	59278
217	28	Mar	82	0918	15.	UK2GBL	59	0	SSB	200	--	592342
218	28	Mar	82	0919	15.	Y23DL	59	0	SSB	200	--	591173
219	28	Mar	82	0920	15.	UK2RDX	59	0	SSB	200	--	592360
220	28	Mar	82	0920	15.	UQ2GFN	59	0	SSB	200	--	591548
221	28	Mar	82	0922	15.	OK2BQL	59	0	SSB	200	--	59197
222	28	Mar	82	0922	15.	UK3SAB	59	0	SSB	200	--	591221
223	28	Mar	82	0925	10.	UK9AAN	59	0	SSB	200	--	59020
224	28	Mar	82	0928	10.	RB5CCO	59	0	SSB	200	--	59546
225	28	Mar	82	0928	10.	EA2AEA	55	0	SSB	200	--	59191
226	28	Mar	82	0929	10.	I2QMU	59	0	SSB	200	--	59145

227	28	Mar	82	0933	10.	UC2OBZ	59	0	SSB	200	--	59094
228	28	Mar	82	0935	10.	UK5WBG	59	0	SSB	200	--	59287
229	28	Mar	82	0939	10.	JN1XEV	55	0	SSB	200	--	59153
230	28	Mar	82	0947	10.	DK5AD	59	0	SSB	200	--	591060
231	28	Mar	82	0948	20.	HA4XX	59	0	SSB	200	--	59566
232	28	Mar	82	0950	20.	OK2BTI	59	0	SSB	200	--	59417
233	28	Mar	82	0951	20.	SV1MO	59	0	SSB	200	--	59562
234	28	Mar	82	0953	20.	DL0IV	59	0	SSB	200	--	59348
235	28	Mar	82	0955	20.	UG6LQ	59	0	SSB	200	--	59724
236	28	Mar	82	0958	15.	UA1FV	59	0	SSB	200	--	59184
237	28	Mar	82	1003	15.	VE1BRB	59	0	SSB	200	--	591119
238	28	Mar	82	1006	15.	VE1DXA	59	0	SSB	200	--	593280
239	28	Mar	82	1006	15.	YO6KEA	59	0	SSB	200	--	59982
240	28	Mar	82	1007	15.	4X6CJ	59	0	SSB	200	--	59043
241	28	Mar	82	1014	10.	YU7AU	59	0	SSB	200	--	591299
242	28	Mar	82	1027	10.	UA3TN	56	0	SSB	200	--	57463
243	28	Mar	82	1032	10.	ZS4SP	55	0	SSB	200	--	59040
244	28	Mar	82	1034	10.	DL0UE	59	0	SSB	200	--	591626
245	28	Mar	82	1035	10.	UA9CRR	59	0	SSB	200	--	59790
246	28	Mar	82	1036	10.	I2JSB	59	0	SSB	200	--	59333
247	28	Mar	82	1038	10.	UA0WAM	59	0	SSB	200	--	591260
248	28	Mar	82	1040	10.	4N0SM	59	0	SSB	200	--	592331
249	28	Mar	82	1041	10.	YU2JL	59	0	SSB	200	--	59068
250	28	Mar	82	1044	10.	HA4KYN	59	0	SSB	200	--	591010
251	28	Mar	82	1047	20.	I5MXX	59	0	SSB	200	--	59195
252	28	Mar	82	1050	20.	UK5FAD	59	0	SSB	200	--	59530
253	28	Mar	82	1055	20.	EA1ABT	59	0	SSB	200	--	591600
254	28	Mar	82	1101	15.	LA7JD	59	0	SSB	200	--	591166
255	28	Mar	82	1102	15.	EA3WZ	59	0	SSB	200	--	592078
256	28	Mar	82	1105	15.	YU7MY	59	0	SSB	200	--	59093
257	28	Mar	82	1109	15.	UK2PAP	59	0	SSB	200	--	592001
258	28	Mar	82	1112	10.	UK0QAA	59	0	SSB	200	--	593502
259	28	Mar	82	1116	10.	YU7QDT	48	0	SSB	200	--	59407
260	28	Mar	82	1119	10.	Y27FN	59	0	SSB	200	--	59715
261	28	Mar	82	1121	10.	YT3L	59	0	SSB	200	--	59142
262	28	Mar	82	1124	10.	OK1KUR/P	59	0	SSB	200	--	59613
263	28	Mar	82	1143	10.	EG3SF	59	0	SSB	200	--	59449
264	28	Mar	82	1145	10.	NU4Y	59	0	SSB	200	--	59909
265	28	Mar	82	1150	10.	HG1KZC	59	0	SSB	200	--	59200
266	28	Mar	82	1201	10.	PA2TMS	59	0	SSB	200	--	592433
267	28	Mar	82	1202	10.	YU3TLA	59	0	SSB	200	--	59656
268	28	Mar	82	1208	10.	HG6V	59	0	SSB	200	--	593204
269	28	Mar	82	1211	10.	DK7JQ	59	0	SSB	200	--	59039
270	28	Mar	82	1215	10.	DL5NBC	55	0	SSB	200	--	57032
271	28	Mar	82	1221	20.	YZ6G	59	0	SSB	200	--	59742
272	28	Mar	82	1222	20.	4N0SM	59	0	SSB	200	--	592413

273 28 Mar 82 1223 20. IS0WON 59 0 SSB 200 -- 59815
274 28 Mar 82 1223 20. YU4EXA 59 0 SSB 200 -- 59068
275 28 Mar 82 1226 20. IV3WMP 59 0 SSB 200 -- 59095
276 28 Mar 82 1232 15. 4U1ITU 59 0 SSB 200 -- 591911
277 28 Mar 82 1241 10. DL6RAI 59 0 SSB 200 -- 59494
278 28 Mar 82 1243 10. I2YKV 59 0 SSB 200 -- 59147
279 28 Mar 82 1244 10. HG3KHO 59 0 SSB 200 -- 59060
280 28 Mar 82 1248 10. UA3DUA 59 0 SSB 200 -- 59099
281 28 Mar 82 1248 10. I2ARN 59 0 SSB 200 -- 59130
282 28 Mar 82 1335 10. LU8FEU 59 0 SSB 200 -- 591233
283 28 Mar 82 1340 10. UK5UDX 59 0 SSB 200 -- 591992
284 28 Mar 82 1344 20. UK3IAA 59 0 SSB 200 -- 59738
285 28 Mar 82 1345 20. YU7AD 59 0 SSB 200 -- 59275
286 28 Mar 82 1347 20. OE1DH 59 0 SSB 200 -- 58775
287 28 Mar 82 1350 15. LZ2CC 59 0 SSB 200 -- 591315
288 28 Mar 82 1352 15. UP2NK 59 0 SSB 200 -- 591949
289 28 Mar 82 1400 10. DF2AO/A 55 0 SSB 200 -- 55790
290 28 Mar 82 1403 10. YU1BAU 59 0 SSB 200 -- 59094
291 28 Mar 82 1415 10. YU3EY 59 0 SSB 200 -- 592340
292 28 Mar 82 1418 10. YU1FW 59 0 SSB 200 -- 59616
293 28 Mar 82 1422 20. UY5YB 59 0 SSB 200 -- 59341
294 28 Mar 82 1426 20. DJ1BZ 59 0 SSB 200 -- 59464
295 28 Mar 82 1429 15. OH3OQ 59 0 SSB 200 -- 59238
296 28 Mar 82 1431 15. I6CXD 59 0 SSB 200 -- 59005
297 28 Mar 82 1431 15. LZ2KIM 59 0 SSB 200 -- 591101
298 28 Mar 82 1433 15. YO6AFP 59 0 SSB 200 -- 59176
299 28 Mar 82 1438 15. HA8ZB 59 0 SSB 200 -- 591105
300 28 Mar 82 1440 15. UB5VAZ 59 0 SSB 200 -- 591519
301 28 Mar 82 1449 10. UB5MPD 59 0 SSB 200 -- 59403
302 28 Mar 82 1500 20. 4X4UH 59 0 SSB 200 -- 591303
303 28 Mar 82 1503 20. ED3VM 59 0 SSB 200 -- 59896
304 28 Mar 82 1505 20. 9H4B 59 0 SSB 200 -- 59603
305 28 Mar 82 1513 20. G3VZT 59 0 SSB 200 -- 59768
306 28 Mar 82 1530 10. UB5EJA 59 0 SSB 200 -- 59228
307 28 Mar 82 1532 10. LU5FGG 59 0 SSB 200 -- 591733
308 28 Mar 82 1543 20. LZ1RN 59 0 SSB 200 -- 59155
309 28 Mar 82 1545 20. UB5KAN 59 0 SSB 200 -- 59415
310 28 Mar 82 1557 20. UK5OAA 59 0 SSB 200 -- 59907
311 28 Mar 82 1618 15. OK1MSN 59 0 SSB 200 -- 591262
312 28 Mar 82 1621 15. YU1DW 59 0 SSB 200 -- 591185
313 28 Mar 82 1621 15. N2WT 59 0 SSB 200 -- 59486
314 28 Mar 82 1628 15. UB5AAL 59 0 SSB 200 -- 59100
315 28 Mar 82 1630 15. DF5IY 55 0 SSB 200 -- 59215
316 28 Mar 82 1635 15. YU7KWX 59 0 SSB 200 -- 592248
317 28 Mar 82 1638 20. JA9YBA 59 0 SSB 200 -- 592802
318 28 Mar 82 1647 20. UB5JK 59 0 SSB 200 -- 59209

No.	Day	Month	Year	Time	Band	Call	RST sent		Mode			Received
319	28	Mar	82	1648	20.	YU7ECD	59	0	SSB	200	--	59530
320	28	Mar	82	1650	20.	UK0AMM	59	0	SSB	200	--	593127
321	28	Mar	82	1650	20.	OK1MSN	59	0	SSB	200	--	591283
322	28	Mar	82	1656	15.	ZS5IV	55	0	SSB	200	--	58493
323	28	Mar	82	1701	15.	DJ8UV	59	0	SSB	200	--	59204
324	28	Mar	82	1704	15.	UK5FAA	59	0	SSB	200	--	59667
325	28	Mar	82	1707	15.	UK5UDX	59	0	SSB	200	--	592167
326	28	Mar	82	1716	10.	OE3KTA	59	0	SSB	200	--	59179
327	28	Mar	82	1716	10.	W8UA	55	0	SSB	200	--	551017
328	28	Mar	82	1720	10.	I5JUX	59	0	SSB	200	--	59205
329	28	Mar	82	1723	10.	N4WW	59	0	SSB	200	--	593033
330	28	Mar	82	1730	10.	ED3VM	59	0	SSB	200	--	59806
331	28	Mar	82	1736	20.	HA5KKB	59	0	SSB	200	--	59346
332	28	Mar	82	1746	15.	JR1WHW	55	0	SSB	200	--	591123
333	28	Mar	82	1748	15.	F6GKH	59	0	SSB	200	--	59503
334	28	Mar	82	1749	15.	F5RU	59	0	SSB	200	--	59171
335	28	Mar	82	1751	15.	HG5A	59	0	SSB	200	--	592956
336	28	Mar	82	1800	20.	IT9WPO	59	0	SSB	200	--	591526
337	28	Mar	82	1817	15.	LU7MAY	55	0	SSB	200	--	591666
338	28	Mar	82	1825	15.	UP2PBW	59	0	SSB	200	--	59675
339	28	Mar	82	1827	15.	KS2G	55	0	SSB	200	--	55084
340	28	Mar	82	1853	15.	ZY5EG	55	0	SSB	200	--	593376
341	28	Mar	82	1857	15.	K4KZZ	59	0	SSB	200	--	591200
342	28	Mar	82	1900	20.	HB9BLQ	57	0	SSB	200	--	571286
343	28	Mar	82	1905	20.	UA9XWU	59	0	SSB	200	--	59616
344	28	Mar	82	1905	20.	UL7QF	59	0	SSB	200	--	59588
345	28	Mar	82	1912	15.	W9LT	59	0	SSB	200	--	591971
346	28	Mar	82	1916	15.	VE7WJ	59	0	SSB	200	--	593325
347	28	Mar	82	1919	15.	UK2GAB	59	0	SSB	200	--	591940
348	28	Mar	82	1926	20.	JG1ILF	59	0	SSB	200	--	591700
349	28	Mar	82	1939	15.	6Y5HN	59	0	SSB	200	--	551641
350	28	Mar	82	1941	15.	XK5XK	59	0	SSB	200	--	594165
351	28	Mar	82	1949	15.	WB1GZE	59	0	SSB	200	--	55144
352	28	Mar	82	1954	15.	OH2AA	59	0	SSB	200	--	593176
353	28	Mar	82	2003	20.	OK1FV	59	0	SSB	200	--	59455
354	28	Mar	82	2008	20.	UK5IBB	59	0	SSB	200	--	593956
355	28	Mar	82	2020	15.	YU2BST	59	0	SSB	200	--	59256
356	28	Mar	82	2032	15.	4Z4JS	59	0	SSB	200	--	59002
357	28	Mar	82	2038	20.	YU7KWX	59	0	SSB	200	--	592516
358	28	Mar	82	2041	20.	UK7NAQ	59	0	SSB	200	--	59248
359	28	Mar	82	2049	15.	KM5R	59	0	SSB	200	--	592245
360	28	Mar	82	2051	15.	Y22JJ	59	0	SSB	200	--	591045
361	28	Mar	82	2054	15.	YU3EY	59	0	SSB	200	--	593169
362	28	Mar	82	2056	15.	VE7UBC	59	0	SSB	200	--	592026
363	28	Mar	82	2100	15.	VE1AI	59	0	SSB	200	--	59812

```
364 28 Mar 82 2104  20.   I0NKN    59  0  SSB  200  --  591467
365 28 Mar 82 2116  15.   OZ5EV    59  0  SSB  200  --  59971
366 28 Mar 82 2124  15.   IP1TTM   59  0  SSB  200  --  591554
367 28 Mar 82 2135  15.   DJ3HJ    59  0  SSB  200  --  592220
368 28 Mar 82 2137  15.   OZ5EDR   59  0  SSB  200  --  59568
369 28 Mar 82 2145  15.   W3XU     59  0  SSB  200  --  59165
370 28 Mar 82 2148  15.   W1NG     59  0  SSB  200  --  59286
371 28 Mar 82 2152  15.   GW4HSH   59  0  SSB  200  --  59163
372 28 Mar 82 2154  15.   KI3L     59  0  SSB  200  --  59462
373 28 Mar 82 2158  15.   VE3JTQ   59  0  SSB  200  --  59027
```

STATION LOG G3ZCZ/4X PAGE 1 FILE WPX82FX

QSO #	DATE	TIME	BAND	STATION	S	R	MODE	PWR	QSL	COMMENTS
1	27 Mar 82	1638	10.	HA4KYH	59	0	SSB	0	--	59337
2	27 Mar 82	1641	10.	N8AKF	59	0	SSB	200	--	59362
3	27 Mar 82	1644	10.	YU2CHI	59	0	SSB	200	--	59008
4	27 Mar 82	1645	10.	IO6FLD	59	0	SSB	200	--	59940
5	27 Mar 82	1647	10.	YU3CAB	59	0	SSB	200	--	59803
6	27 Mar 82	1648	10.	KF2O	59	0	SSB	200	--	59397
7	27 Mar 82	1650	10.	HA5KKC/7	59	0	SSB	200	--	591284
8	27 Mar 82	1645	10.	4U1ITU	59	0	SSB	200	--	591072
9	27 Mar 82	1656	10.	IV3PRK	59	0	SSB	200	--	591096
10	27 Mar 82	1659	10.	Y35LM	59	0	SSB	200	--	59576
11	27 Mar 82	1701	10.	N4NO	59	0	SSB	200	--	59222
12	27 Mar 82	1703	10.	YU3UPI	59	0	SSB	200	--	59236
13	27 Mar 82	1703	10.	Y34YF	59	0	SSB	200	--	59036
14	27 Mar 82	1705	10.	N4KG	55	0	SSB	200	--	59120
15	27 Mar 82	1705	10.	YU3TDP	59	0	SSB	200	--	59026
16	27 Mar 82	1705	10.	N2US	59	0	SSB	200	--	59233
17	27 Mar 82	1705	10.	I5MYL	59	0	SSB	200	--	59024
18	27 Mar 82	1707	10.	KE3G	59	0	SSB	200	--	59008
19	27 Mar 82	1707	10.	K3EST	59	0	SSB	200	--	59600
20	27 Mar 82	1707	10.	K3KHL	59	0	SSB	200	--	59027
21	27 Mar 82	1716	10.	Y32ZF	59	0	SSB	200	--	59023
22	27 Mar 82	1720	10.	Y56YF	59	0	SSB	200	--	59148
23	27 Mar 82	1720	10.	N2AA	55	0	SSB	200	--	59367
24	27 Mar 82	1724	10.	K2BA/4	59	0	SSB	200	--	59912
25	27 Mar 82	1725	10.	OK1KST	59	0	SSB	200	--	591114
26	27 Mar 82	1726	10.	I4ZSQ	59	0	SSB	200	--	59604
27	27 Mar 82	1728	10.	K2SS	59	0	SSB	200	--	59966
28	27 Mar 82	1736	10.	OK1ARI	59	0	SSB	200	--	59698
29	27 Mar 82	1736	10.	I2SVA	59	0	SSB	200	--	59490
30	27 Mar 82	1744	20.	HA4KYN	59	0	SSB	200	--	59790
31	27 Mar 82	1813	15.	CT7AL	59	0	SSB	200	--	59160
32	27 Mar 82	1816	15.	YU3DBC	59	0	SSB	200	--	591152
33	27 Mar 82	1818	15.	HA5KKC/7	59	0	SSB	200	--	591409
34	27 Mar 82	1819	15.	KC3N	59	0	SSB	200	--	59496
35	27 Mar 82	1823	15.	IT9WPO	59	0	SSB	200	--	59803
36	27 Mar 82	1823	15.	YU3DAW	59	0	SSB	200	--	59319
37	27 Mar 82	1825	15.	YU4YA	59	0	SSB	200	--	59287
38	27 Mar 82	1835	15.	WD0EWD	59	0	SSB	200	--	59491
39	27 Mar 82	1837	15.	AD8R	59	0	SSB	200	--	59658
40	27 Mar 82	1837	15.	KG1D	59	0	SSB	200	--	59739
41	27 Mar 82	1845	15.	K2SS	59	0	SSB	200	--	591037
42	27 Mar 82	1854	15.	4X4VE/5N8	59	0	SSB	200	--	59660
43	27 Mar 82	1859	15.	Y24UK	59	0	SSB	200	--	591460

Fig. A-6. Cleaned up (fudged) contest log.

44	27 Mar 82	1859	15.	VE6OU	59	0	SSB	200	--	591097
45	27 Mar 82	1859	15.	OK3CRH	59	0	SSB	200	--	59260
46	27 Mar 82	1903	15.	OK1AJN	59	0	SSB	200	--	59292
47	27 Mar 82	1903	15.	NP4A	59	0	SSB	200	--	51220
48	27 Mar 82	1906	15.	UK3ADZ	59	0	SSB	200	--	591348
49	27 Mar 82	1907	15.	YU3VM	59	0	SSB	200	--	59833
50	27 Mar 82	1909	15.	EA1ABT	59	0	SSB	200	--	591139
51	27 Mar 82	1911	15.	Y33ZB	59	0	SSB	200	--	59825
52	27 Mar 82	1913	15.	AH2M	59	0	SSB	200	--	59150
53	27 Mar 82	1914	15.	UK9FER	59	0	SSB	200	--	591004
54	27 Mar 82	1917	15.	AC3A	59	0	SSB	200	--	591239
55	27 Mar 82	1917	15.	VE3PCA	59	0	SSB	200	--	591379
56	27 Mar 82	1920	15.	UA3VAS	59	0	SSB	200	--	59049
57	27 Mar 82	1920	15.	K6IR	59	0	SSB	200	--	59256
58	27 Mar 82	1927	15.	UW3HV	59	0	SSB	200	--	591018
59	27 Mar 82	1927	15.	DF4NP	59	0	SSB	200	--	59168
60	27 Mar 82	1930	15.	YU2CT	59	0	SSB	200	--	59258
61	27 Mar 82	1937	15.	EI7DJ	59	0	SSB	200	--	59466
62	27 Mar 82	1945	10.	PJ2VR	59	0	SSB	200	--	59646
63	27 Mar 82	1956	10.	VP2EC	59	0	SSB	200	--	592649
64	27 Mar 82	2005	15.	DL6FAW	59	0	SSB	200	--	59252
65	27 Mar 82	2008	15.	DL8PC	59	0	SSB	200	--	591196
66	27 Mar 82	2013	15.	I1JHS	59	0	SSB	200	--	59312
67	27 Mar 82	2013	15.	VE5UF	59	0	SSB	200	--	59773
68	27 Mar 82	2013	15.	VE7BTV	59	0	SSB	200	--	591157
69	27 Mar 82	2016	15.	AI9J	59	0	SSB	200	--	591166
70	27 Mar 82	2018	15.	DL4SAR	59	0	SSB	200	--	59405
71	27 Mar 82	2020	15.	KQ2M	59	0	SSB	200	--	59852
72	27 Mar 82	2024	15.	VE3BMV	59	0	SSB	200	--	591163
73	27 Mar 82	2026	15.	AI7B	59	0	SSB	200	--	591369
74	27 Mar 82	2028	15.	IZ1ARI	59	0	SSB	200	--	591304 Q=I1VEH
75	27 Mar 82	2030	15.	I0KWX	59	0	SSB	200	--	591200
76	27 Mar 82	2043	15.	HB9BMY/4X	59	0	SSB	200	--	59005
77	27 Mar 82	2045	15.	AB0I	59	0	SSB	200	--	591248
78	27 Mar 82	2052	15.	F6BFN	59	0	SSB	200	--	59813
79	27 Mar 82	2057	15.	KQ8M	59	0	SSB	200	--	59600
80	27 Mar 82	2057	15.	KA1YQ	59	0	SSB	200	--	59220
81	27 Mar 82	2115	15.	VP2MGQ	59	0	SSB	200	--	591287
82	27 Mar 82	2118	15.	KJ9W	59	0	SSB	200	--	591841
83	27 Mar 82	2120	20.	UK3QAE	59	0	SSB	200	--	59988
84	27 Mar 82	2125	20.	UK6LAA	59	0	SSB	200	--	591326
85	27 Mar 82	2125	20.	UK0QAA	59	0	SSB	200	--	592219
86	27 Mar 82	2131	20.	YU7KMN	59	0	SSB	200	--	59679
87	27 Mar 82	2137	20.	UK5QBE	59	0	SSB	200	--	59840
88	27 Mar 82	2140	20.	KL7RA	59	0	SSB	200	--	59579
89	27 Mar 82	2143	20.	UK5UDX	59	0	SSB	200	--	591396

90	27	Mar	82	2143	20.	FC6FPH	59	0	SSB	200	--	59087 (G8KA)
91	27	Mar	82	2148	20.	HB9AGC	59	0	SSB	200	--	59511
92	27	Mar	82	2150	15.	YV3BJL	59	0	SSB	200	--	591144
93	27	Mar	82	2155	15.	KB8LH	59	0	SSB	200	--	59256
94	27	Mar	82	2157	15.	K3ZJ	59	0	SSB	200	--	59999
95	27	Mar	82	2158	15.	WB4KRH	59	0	SSB	200	--	59350
96	27	Mar	82	2158	15.	P42J	59	0	SSB	200	--	591513 Curaco
97	27	Mar	82	2206	15.	KC1F	59	0	SSB	200	--	591314
98	27	Mar	82	2208	15.	N4WW	59	0	SSB	200	--	591904
99	27	Mar	82	2213	15.	N4XD	59	0	SSB	200	--	59243
100	27	Mar	82	2215	15.	AK1A	59	0	SSB	200	--	591145
101	27	Mar	82	2217	15.	WB9TIY	59	0	SSB	200	--	59184
102	27	Mar	82	2220	20.	Y22OM/A	59	0	SSB	200	--	591111
103	27	Mar	82	2221	20.	LZ2DB	59	0	SSB	200	--	59193
104	27	Mar	82	2222	20.	UK5MAF	59	0	SSB	200	--	591844
105	27	Mar	82	2222	20.	UK3DBG	59	0	SSB	200	--	59708
106	27	Mar	82	2224	20.	UK2BBB	59	0	SSB	200	--	591873
107	27	Mar	82	2226	20.	R6L	59	0	SSB	200	--	592876
108	27	Mar	82	2228	20.	UK2PCR	59	0	SSB	200	--	591939
109	27	Mar	82	2232	20.	UK3QAA	59	0	SSB	200	--	59117
110	27	Mar	82	2234	20.	UK6LTG	59	0	SSB	200	--	59135
111	27	Mar	82	2234	20.	F6KAW	59	0	SSB	200	--	591491
112	27	Mar	82	2236	20.	UK6XAA	59	0	SSB	200	--	591199
113	27	Mar	82	2240	20.	UK5QAV	59	0	SSB	200	--	59812
114	27	Mar	82	2243	15.	W1IHN	59	0	SSB	200	--	59190
115	27	Mar	82	2245	15.	KK5I	59	0	SSB	200	--	591112
116	27	Mar	82	2245	15.	K2IJL	59	0	SSB	200	--	59130
117	27	Mar	82	2250	15.	KI2P	59	0	SSB	200	--	59940
118	27	Mar	82	2254	15.	WA3GSC	59	0	SSB	200	--	59121
119	27	Mar	82	2256	15.	KA4SUN	59	0	SSB	200	--	59019
120	27	Mar	82	2259	15.	KV5Q	59	0	SSB	200	--	59924
121	27	Mar	82	2302	15.	KB2MG	59	0	SSB	200	--	59931
122	27	Mar	82	2302	15.	HK5BCZ	55	0	SSB	200	--	59182
123	28	Mar	82	0416	15.	VK2ERT	59	0	SSB	200	--	59836
124	28	Mar	82	0425	20.	AC3A	59	0	SSB	200	--	591630
125	28	Mar	82	0428	20.	UK2GKW	59	0	SSB	200	--	592283
126	28	Mar	82	0430	20.	HA9KPU	59	0	SSB	200	--	59756
127	28	Mar	82	0447	10.	RA3DKE	55	0	SSB	200	--	59327
128	28	Mar	82	0456	10.	UA3TBK	59	0	SSB	200	--	59147
129	28	Mar	82	0501	10.	UK3DAH	59	0	SSB	200	--	59868
130	28	Mar	82	0516	10.	UK3ADZ	59	0	SSB	200	--	591685
131	28	Mar	82	0516	10.	UA3ACY/U9J	55	0	SSB	200	--	55579
132	28	Mar	82	0521	10.	JH0BBE	55	0	SSB	200	--	59805
133	28	Mar	82	0524	10.	UK3AAC	59	0	SSB	200	--	591203
134	28	Mar	82	0526	10.	UA0ZDD	55	0	SSB	200	--	59808

180	28 Mar 82	0732	10.	HA9KPU	59	0	SSB	200	--	59837
181	28 Mar 82	0732	10.	I0SKK	59	0	SSB	200	--	59197
182	28 Mar 82	0745	10.	HG5A	59	0	SSB	200	--	592255
183	28 Mar 82	0747	10.	HA4ZB	59	0	SSB	200	--	59058
184	28 Mar 82	0753	10.	JI1QQI	55	0	SSB	200	--	591150
185	28 Mar 82	0753	10.	UY5XE	59	0	SSB	200	--	59238
186	28 Mar 82	0759	10.	Y47XF	59	0	SSB	200	--	59457
187	28 Mar 82	0800	10.	UA3QAE	59	0	SSB	200	--	591241
188	28 Mar 82	0802	10.	YO5KAD	59	0	SSB	200	--	59350
189	28 Mar 82	0803	10.	YU1NZW	59	0	SSB	200	--	59278
190	28 Mar 82	0803	10.	UK2AAF	55	0	SSB	200	--	59276
191	28 Mar 82	0805	10.	YU3EO	59	0	SSB	200	--	59229
192	28 Mar 82	0807	10.	UB5DBV	59	0	SSB	200	--	59085
193	28 Mar 82	0810	10.	YU2RVL	59	0	SSB	200	--	59028
194	28 Mar 82	0813	10.	UK5MAF	59	0	SSB	200	--	592325
195	28 Mar 82	0817	10.	YU7AV	59	0	SSB	200	--	59270
196	28 Mar 82	0821	20.	UC2OBP	59	0	SSB	200	--	59509
197	28 Mar 82	0826	20.	EA6GP	59	0	SSB	200	--	59842
198	28 Mar 82	0830	20.	HA0DU	59	0	SSB	200	--	59571
199	28 Mar 82	0833	15.	UK5OAA	59	0	SSB	200	--	59721
200	28 Mar 82	0837	15.	IS0QDV	59	0	SSB	200	--	591560
201	28 Mar 82	0841	15.	HA4XX	55	0	SSB	200	--	57508
202	28 Mar 82	0850	10.	DK0MM	59	0	SSB	200	--	591366
203	28 Mar 82	0855	10.	YU3DBC	59	0	SSB	200	--	591856
204	28 Mar 82	0857	10.	YU7KWX	59	0	SSB	200	--	591827
205	28 Mar 82	0858	10.	IT9WPO	59	0	SSB	200	--	591157
206	28 Mar 82	0901	10.	UA9MAF	59	0	SSB	200	--	59280
207	28 Mar 82	0905	20.	4N6HN	59	0	SSB	200	--	59242
208	28 Mar 82	0907	20.	UK6YAB	59	0	SSB	200	--	59176
209	28 Mar 82	0909	20.	HG5A	59	0	SSB	200	--	592338
210	28 Mar 82	0910	20.	UK3XAB	59	0	SSB	200	--	59876
211	28 Mar 82	0912	20.	OH2AW	55	0	SSB	200	--	59592
212	28 Mar 82	0914	20.	UG6GDS	59	0	SSB	200	--	59278
213	28 Mar 82	0918	15.	UK2GBL	59	0	SSB	200	--	592342
214	28 Mar 82	0919	15.	Y23DL	59	0	SSB	200	--	591173
215	28 Mar 82	0920	15.	UK2RDX	59	0	SSB	200	--	592360
216	28 Mar 82	0920	15.	UQ2GFN	59	0	SSB	200	--	591548
217	28 Mar 82	0922	15.	OK2BQL	59	0	SSB	200	--	59197
218	28 Mar 82	0922	15.	UK3SAB	59	0	SSB	200	--	591221
219	28 Mar 82	0925	10.	UK9AAN	59	0	SSB	200	--	59020
220	28 Mar 82	0928	10.	RB5CCO	59	0	SSB	200	--	59546
221	28 Mar 82	0928	10.	EA2AEA	55	0	SSB	200	--	59191
222	28 Mar 82	0929	10.	I2QMU	59	0	SSB	200	--	59145
223	28 Mar 82	0933	10.	UC2OBZ	59	0	SSB	200	--	59094
224	28 Mar 82	0935	10.	UK5WBG	59	0	SSB	200	--	59287

135	28	Mar	82	0528	10.	UK4FAV	59	0	SSB	200	--	591748
136	28	Mar	82	0530	10.	OK1ALQ	59	0	SSB	200	--	59068
137	28	Mar	82	0534	10.	UA3AHA	59	0	SSB	200	--	59189
138	28	Mar	82	0534	10.	F6BXQ	59	0	SSB	200	--	59014
139	28	Mar	82	0534	10.	I4VOS	59	0	SSB	200	--	59864
140	28	Mar	82	0536	10.	RA3RBU	59	0	SSB	200	--	59132
141	28	Mar	82	0536	10.	OK1DLA	59	0	SSB	200	--	59483
142	28	Mar	82	0538	10.	UA3AGL	59	0	SSB	200	--	59009
143	28	Mar	82	0538	10.	UK4WAA	59	0	SSB	200	--	59595
144	28	Mar	82	0549	15.	VK2APK	59	0	SSB	200	--	59640
145	28	Mar	82	0549	15.	FR0FLO	59	0	SSB	200	--	59919
146	28	Mar	82	0552	15.	VK3AJJ	59	0	SSB	200	--	59380
147	28	Mar	82	0556	15.	OE1WO	59	0	SSB	200	--	59199
148	28	Mar	82	0557	15.	ED3VM	59	0	SSB	200	--	59489
149	28	Mar	82	0559	15.	UK5AAA	44	0	SSB	200	--	59366
150	28	Mar	82	0559	15.	UA6LGP	59	0	SSB	200	--	59237
151	28	Mar	82	0602	15.	UW3ZU	59	0	SSB	200	--	59048
152	28	Mar	82	0603	15.	YU7NZR	59	0	SSB	200	--	59230
153	28	Mar	82	0611	20.	UK5GKW	59	0	SSB	200	--	591180
154	28	Mar	82	0613	20.	N4WW	59	0	SSB	200	--	592398
155	28	Mar	82	0616	20.	OH1IG	59	0	SSB	200	--	591169
156	28	Mar	82	0619	20.	K2BA/4	59	0	SSB	200	--	591454
157	28	Mar	82	0619	20.	HB9BUN	59	0	SSB	200	--	591008
158	28	Mar	82	0624	20.	Y22JJ	59	0	SSB	200	--	59606
159	28	Mar	82	0627	20.	I0JBL	59	0	SSB	200	--	59170
160	28	Mar	82	0630	20.	YO9HP	59	0	SSB	200	--	59180
161	28	Mar	82	0633	20.	DK0MM	59	0	SSB	200	--	591201
162	28	Mar	82	0638	15.	G3XBY	59	0	SSB	200	--	59229
163	28	Mar	82	0640	15.	HA5KKG	59	0	SSB	200	--	59531
164	28	Mar	82	0643	10.	Y22OM/A	59	0	SSB	200	--	591156
165	28	Mar	82	0652	10.	Y37XJ	59	0	SSB	200	--	59189
166	28	Mar	82	0700	20.	OK2RZ	59	0	SSB	200	--	59215
167	28	Mar	82	0703	20.	LZ1BM	59	0	SSB	200	--	59374
168	28	Mar	82	0704	20.	LZ2KIM	59	0	SSB	200	--	59882
169	28	Mar	82	0707	20.	LZ2KRR	59	0	SSB	200	--	591462
170	28	Mar	82	0709	20.	K2NG	59	0	SSB	200	--	55104
171	28	Mar	82	0714	15.	F8WE	59	0	SSB	200	--	59692
172	28	Mar	82	0716	15.	LZ13C	59	0	SSB	200	--	59650
173	28	Mar	82	0718	15.	UB5MBZ	59	0	SSB	200	--	59914
174	28	Mar	82	0719	15.	R6L	59	0	SSB	200	--	593794
175	28	Mar	82	0721	15.	YU3EF	59	0	SSB	200	--	59567
176	28	Mar	82	0722	15.	OH6AC	59	0	SSB	200	--	591357
177	28	Mar	82	0724	10.	OK3JW	59	0	SSB	200	--	59788
178	28	Mar	82	0724	10.	OK3CFA	59	0	SSB	200	--	59557
179	28	Mar	82	0729	10.	OK3CSC	59	0	SSB	200	--	59624

225	28	Mar	82	0939	10.	JN1XEV	55	0	SSB	200	--	59153
226	28	Mar	82	0947	10.	DK5AD	59	0	SSB	200	--	591060
227	28	Mar	82	0948	20.	HA4XX	59	0	SSB	200	--	59566
228	28	Mar	82	0950	20.	OK2BTI	59	0	SSB	200	--	59417
229	28	Mar	82	0951	20.	SV1MO	59	0	SSB	200	--	59562
230	28	Mar	82	0953	20.	DL0IV	59	0	SSB	200	--	59348
231	28	Mar	82	0955	20.	UG6LQ	59	0	SSB	200	--	59724
232	28	Mar	82	0958	15.	UA1FV	59	0	SSB	200	--	59184
233	28	Mar	82	1003	15.	VE1BRB	59	0	SSB	200	--	591119
234	28	Mar	82	1006	15.	VE1DXA	59	0	SSB	200	--	593280
235	28	Mar	82	1006	15.	YO6KEA	59	0	SSB	200	--	59982
236	28	Mar	82	1007	15.	4X6CJ	59	0	SSB	200	--	59043
237	28	Mar	82	1014	10.	YU7AU	59	0	SSB	200	--	591299
238	28	Mar	82	1027	10.	UA3TN	56	0	SSB	200	--	57463
239	28	Mar	82	1032	10.	ZS4SP	55	0	SSB	200	--	59040
240	28	Mar	82	1034	10.	DL0UE	59	0	SSB	200	--	591626
241	28	Mar	82	1035	10.	UA9CRR	59	0	SSB	200	--	59790
242	28	Mar	82	1036	10.	I2JSB	59	0	SSB	200	--	59333
243	28	Mar	82	1038	10.	UA0WAM	59	0	SSB	200	--	591260
244	28	Mar	82	1040	10.	4N0SM	59	0	SSB	200	--	592331
245	28	Mar	82	1041	10.	YU2JL	59	0	SSB	200	--	59068
246	28	Mar	82	1044	10.	HA4KYN	59	0	SSB	200	--	591010
247	28	Mar	82	1047	20.	I5MXX	59	0	SSB	200	--	59195
248	28	Mar	82	1050	20.	UK5FAD	59	0	SSB	200	--	59530
249	28	Mar	82	1055	20.	EA1ABT	59	0	SSB	200	--	591600
250	28	Mar	82	1101	15.	LA7JO	59	0	SSB	200	--	591166
251	28	Mar	82	1102	15.	EA3WZ	59	0	SSB	200	--	592078
252	28	Mar	82	1105	15.	YU7MY	59	0	SSB	200	--	59093
253	28	Mar	82	1109	15.	UK2PAP	59	0	SSB	200	--	592001
254	28	Mar	82	1112	10.	UK0QAA	59	0	SSB	200	--	593502
255	28	Mar	82	1116	10.	YU7QDT	48	0	SSB	200	--	59407
256	28	Mar	82	1119	10.	Y27FN	59	0	SSB	200	--	59715
257	28	Mar	82	1121	10.	YT3L	59	0	SSB	200	--	59142
258	28	Mar	82	1124	10.	OK1KUR/P	59	0	SSB	200	--	59613
259	28	Mar	82	1143	10.	EG3SF	59	0	SSB	200	--	59449
260	28	Mar	82	1145	10.	NU4Y	59	0	SSB	200	--	59909
261	28	Mar	82	1150	10.	HG1KZC	59	0	SSB	200	--	59200
262	28	Mar	82	1201	10.	PA2TMS	59	0	SSB	200	--	592433
263	28	Mar	82	1202	10.	YU3TLA	59	0	SSB	200	--	59656
264	28	Mar	82	1208	10.	HG6V	59	0	SSB	200	--	593204
265	28	Mar	82	1211	10.	DK7JQ	59	0	SSB	200	--	59039
266	28	Mar	82	1215	10.	DL5NBC	55	0	SSB	200	--	57032
267	28	Mar	82	1221	20.	YZ6G	59	0	SSB	200	--	59742
268	28	Mar	82	1222	20.	4N0SM	59	0	SSB	200	--	592413
269	28	Mar	82	1223	20.	IS0WON	59	0	SSB	200	--	59815

270	28	Mar	82	1223	20.	YU4EXA	59	0	SSB	200	--	59068
271	28	Mar	82	1226	20.	IV3WMP	59	0	SSB	200	--	59095
272	28	Mar	82	1232	15.	4U1ITU	59	0	SSB	200	--	591911
273	28	Mar	82	1241	10.	DL6RAI	59	0	SSB	200	--	59494
274	28	Mar	82	1243	10.	I2YKV	59	0	SSB	200	--	59147
275	28	Mar	82	1244	10.	HG3KHO	59	0	SSB	200	--	59060
276	28	Mar	82	1248	10.	UA3DUA	59	0	SSB	200	--	59099
277	28	Mar	82	1248	10.	I2ARN	59	0	SSB	200	--	59130
278	28	Mar	82	1335	10.	LU8FEU	59	0	SSB	200	--	591233
279	28	Mar	82	1340	10.	UK5UDX	59	0	SSB	200	--	591992
280	28	Mar	82	1344	20.	UK3IAA	59	0	SSB	200	--	59738
281	28	Mar	82	1345	20.	YU7AD	59	0	SSB	200	--	59275
282	28	Mar	82	1347	20.	OE1DH	59	0	SSB	200	--	58775
283	28	Mar	82	1350	15.	LZ2CC	59	0	SSB	200	--	591315
284	28	Mar	82	1352	15.	UP2NK	59	0	SSB	200	--	591949
285	28	Mar	82	1400	10.	DF2AO/A	55	0	SSB	200	--	55790
286	28	Mar	82	1403	10.	YU1BAU	59	0	SSB	200	--	59094
287	28	Mar	82	1415	10.	YU3EY	59	0	SSB	200	--	592340
288	28	Mar	82	1418	10.	YU1FW	59	0	SSB	200	--	59616
289	28	Mar	82	1422	20.	UY5YB	59	0	SSB	200	--	59341
290	28	Mar	82	1426	20.	DJ1BZ	59	0	SSB	200	--	59464
291	28	Mar	82	1429	15.	OH3OQ	59	0	SSB	200	--	59238
292	28	Mar	82	1431	15.	I6CXD	59	0	SSB	200	--	59005
293	28	Mar	82	1431	15.	LZ2KIM	59	0	SSB	200	--	591101
294	28	Mar	82	1433	15.	YO6AFP	59	0	SSB	200	--	59176
295	28	Mar	82	1438	15.	HA8ZB	59	0	SSB	200	--	591105
296	28	Mar	82	1440	15.	UB5VAZ	59	0	SSB	200	--	591519
297	28	Mar	82	1449	10.	UB5MPD	59	0	SSB	200	--	59403
298	28	Mar	82	1500	20.	4X4UH	59	0	SSB	200	--	591303
299	28	Mar	82	1503	20.	ED3VM	59	0	SSB	200	--	59896
300	28	Mar	82	1505	20.	9H4B	59	0	SSB	200	--	59603
301	28	Mar	82	1513	20.	G3VZT	59	0	SSB	200	--	59768
302	28	Mar	82	1530	10.	UB5EJA	59	0	SSB	200	--	59228
303	28	Mar	82	1532	10.	LU5FGG	59	0	SSB	200	--	591733
304	28	Mar	82	1543	20.	LZ1RN	59	0	SSB	200	--	59155
305	28	Mar	82	1545	20.	UB5KAN	59	0	SSB	200	--	59415
306	28	Mar	82	1557	20.	UK5OAA	59	0	SSB	200	--	59907
307	28	Mar	82	1618	15.	OK1MSN	59	0	SSB	200	--	591262
308	28	Mar	82	1621	15.	YU1DW	59	0	SSB	200	--	591185
309	28	Mar	82	1621	15.	N2WT	59	0	SSB	200	--	59486
310	28	Mar	82	1628	15.	UB5AAL	59	0	SSB	200	--	59100
311	28	Mar	82	1630	15.	DF5IY	55	0	SSB	200	--	59215
312	28	Mar	82	1635	15.	YU7KWX	59	0	SSB	200	--	592248
313	28	Mar	82	1638	20.	JA9YBA	59	0	SSB	200	--	592802
314	28	Mar	82	1647	20.	UB5JK	59	0	SSB	200	--	59209

315	28	Mar	82	1648	20.	YU7ECD	59	0	SSB	200	--	59530
316	28	Mar	82	1650	20.	UK0AMM	59	0	SSB	200	--	593127
317	28	Mar	82	1650	20.	OK1MSN	59	0	SSB	200	--	591283
318	28	Mar	82	1656	15.	ZS5IV	55	0	SSB	200	--	58493
319	28	Mar	82	1701	15.	DJ8UV	59	0	SSB	200	--	59204
320	28	Mar	82	1704	15.	UK5FAA	59	0	SSB	200	--	59667
321	28	Mar	82	1707	15.	UK5UDX	59	0	SSB	200	--	592167
322	28	Mar	82	1716	10.	OE3KTA	59	0	SSB	200	--	59179
323	28	Mar	82	1716	10.	W8UA	55	0	SSB	200	--	551017
324	28	Mar	82	1720	10.	I5JUX	59	0	SSB	200	--	59205
325	28	Mar	82	1723	10.	N4WW	59	0	SSB	200	--	593033
326	28	Mar	82	1730	10.	ED3VM	59	0	SSB	200	--	59806
327	28	Mar	82	1736	20.	HA5KKB	59	0	SSB	200	--	59346
328	28	Mar	82	1746	15.	JR1WHW	55	0	SSB	200	--	591123
329	28	Mar	82	1748	15.	F6GKH	59	0	SSB	200	--	59503
330	28	Mar	82	1749	15.	F5RU	59	0	SSB	200	--	59171
331	28	Mar	82	1751	15.	HG5A	59	0	SSB	200	--	592956
332	28	Mar	82	1800	20.	IT9WPO	59	0	SSB	200	--	591526
333	28	Mar	82	1817	15.	LU7MAY	55	0	SSB	200	--	591666
334	28	Mar	82	1825	15.	UP2PBW	59	0	SSB	200	--	59675
335	28	Mar	82	1827	15.	KS2G	55	0	SSB	200	--	55084
336	28	Mar	82	1853	15.	ZY5EG	55	0	SSB	200	--	593376
337	28	Mar	82	1857	15.	K4KZZ	59	0	SSB	200	--	591200
338	28	Mar	82	1900	20.	HB9BLQ	57	0	SSB	200	--	571286
339	28	Mar	82	1905	20.	UA9XWU	59	0	SSB	200	--	59616
340	28	Mar	82	1905	20.	UL7QF	59	0	SSB	200	--	59588
341	28	Mar	82	1912	15.	W9LT	59	0	SSB	200	--	591971
342	28	Mar	82	1916	15.	VE7WJ	59	0	SSB	200	--	593325
343	28	Mar	82	1919	15.	UK2GAB	59	0	SSB	200	--	591940
344	28	Mar	82	1926	20.	JG1ILF	59	0	SSB	200	--	591700
345	28	Mar	82	1939	15.	6Y5HN	59	0	SSB	200	--	551641
346	28	Mar	82	1941	15.	XK5XK	59	0	SSB	200	--	594165
347	28	Mar	82	1949	15.	WB1GZE	59	0	SSB	200	--	55144
348	28	Mar	82	1954	15.	OH2AA	59	0	SSB	200	--	593176
349	28	Mar	82	2003	20.	OK1FV	59	0	SSB	200	--	59455
350	28	Mar	82	2008	20.	UK5IBB	59	0	SSB	200	--	593956
351	28	Mar	82	2020	15.	YU2BST	59	0	SSB	200	--	59256
352	28	Mar	82	2032	15.	4Z4JS	59	0	SSB	200	--	59002
353	28	Mar	82	2038	20.	YU7KWX	59	0	SSB	200	--	592516
354	28	Mar	82	2041	20.	UK7NAQ	59	0	SSB	200	--	59248
355	28	Mar	82	2049	15.	KM5R	59	0	SSB	200	--	592245
356	28	Mar	82	2051	15.	Y22JJ	59	0	SSB	200	--	591045
357	28	Mar	82	2054	15.	YU3EY	59	0	SSB	200	--	593169
358	28	Mar	82	2056	15.	VE7UBC	59	0	SSB	200	--	592026
359	28	Mar	82	2100	15.	VE1AI	59	0	SSB	200	--	59812

360	28	Mar	82	2104	20.	I0NKN	59	0	SSB	200	--	591467
361	28	Mar	82	2116	15.	OZ5EV	59	0	SSB	200	--	59971
362	28	Mar	82	2124	15.	IP1TTM	59	0	SSB	200	--	591554
363	28	Mar	82	2135	15.	DJ3HJ	59	0	SSB	200	--	592220
364	28	Mar	82	2137	15.	OZ5EDR	59	0	SSB	200	--	59568
365	28	Mar	82	2145	15.	W3XU	59	0	SSB	200	--	59165
366	28	Mar	82	2148	15.	W1NG	59	0	SSB	200	--	59286
367	28	Mar	82	2152	15.	GW4HSH	59	0	SSB	200	--	59163
368	28	Mar	82	2154	15.	KI3L	59	0	SSB	200	--	59462
369	28	Mar	82	2158	15.	VE3JTQ	59	0	SSB	200	--	59027

BAND= 10

4N0SM	4U1ITU	DF2AO/A	DK0MM	DK5AD	DK7JQ
DL0UE	DL5NBC	DL6RAI	EA2AEA	ED3VM	EG3SF
F6BXQ	HA4KYH	HA4KYN	HA4ZB	HA5KKC/7	HA9KPU
HG1KZC	HG3KHO	HG5A	HG6V	I0SKK	I2ARN
I2JSB	I2QMU	I2SVA	I2YKV	I4VOS	I4ZSQ
I5JUX	I5MYL	IO6FLD	IT9WPO	IV3PRK	JH0BBE
JI1QQI	JN1XEV	K2BA/4	K2SS	K3EST	K3KHL
KE3G	KF2O	LU5FGG	LU8FEU	N2AA	N2US
N4KG	N4NO	N4WW	N8AKF	NU4Y	OE3KTA
OK1ALQ	OK1ARI	OK1DLA	OK1KST	OK1KUR/P	OK3CFA
OK3CSC	OK3JW	PA2TMS	PJ2VR	RA3DKE	RA3RBU
RB5CCO	UA0WAM	UA0ZDD	UA3ACY/U9JUA3AGL		UA3AHA
UA3DUA	UA3QAE	UA3TBK	UA3TN	UA9CRR	UA9MAF
UB5DBV	UB5EJA	UB5MPD	UC2OBZ	UK0QAA	UK2AAF
UK3AAC	UK3ADZ	UK3DAH	UK4FAV	UK4WAA	UK5MAF
UK5UDX	UK5WBG	UK9AAN	UY5XE	VP2EC	W8UA
Y22OM/A	Y27FN	Y32ZF	Y34YF	Y35LM	Y37XJ
Y47XF	Y56YF	YO5KAD	YT3L	YU1BAU	YU1FW
YU1NZW	YU2CHI	YU2JL	YU2RVL	YU3CAB	YU3DBC
YU3EO	YU3EY	YU3TDP	YU3TLA	YU3UPI	YU7AU
YU7AV	YU7KWX	YU7QDT	ZS4SP		

BAND= 15

4U1ITU	4X4VE/5NB	4X6CJ	4Z4JS	6Y5HN	AB0I
AC3A	AD8R	AH2M	AI7B	AI9J	AK1A
CT7AL	DF4NP	DF5IY	DJ3HJ	DJ8UV	DL4SAR
DL6FAW	DL8PC	EA1ABT	EA3WZ	ED3VM	EI7DJ
F5RU	F6BFN	F6GKH	F8WE	FR0FLO	G3XBY
GW4HSH	HA4XX	HA5KKC/7	HA5KKG	HA8ZB	HB9BMY/4X
HG5A	HK5BCZ	I0KWX	I1JHS	I6CXD	IP1TTM
IS0QDV	IT9WPO	IZ1ARI	JR1WHW	K2IJL	K2SS
K3ZJ	K4KZZ	K6IR	KA1YQ	KA4SUN	KB2MG
KB8LH	KC1F	KC3N	KG1D	KI2P	KI3L
KJ9W	KK5I	KM5R	KQ2M	KQ8M	KS2G
KV5Q	LA7JO	LU7MAY	LZ13C	LZ2CC	LZ2KIM
N2WT	N4WW	N4XD	NP4A	OE1WO	OH2AA
OH3OQ	OH6AC	OK1AJN	OK1MSN	OK2BQL	OK3CRH
OZ5EDR	OZ5EV	P42J	R6L	UA1FV	UA3VAS
UA6LGP	UB5AAL	UB5MBZ	UB5VAZ	UK2GAB	UK2GBL
UK2PAP	UK2RDX	UK3ADZ	UK3SAB	UK5AAA	UK5FAA
UK5OAA	UK5UDX	UK9FER	UP2NK	UP2PBW	UQ2GFN
UW3HV	UW3ZU	VE1AI	VE1BRB	VE1DXA	VE3BMV
VE3JTQ	VE3PCA	VE5UF	VE6OU	VE7BTV	VE7UBC

Fig. A-7. Contest checklist by band.

VE7WJ VK2APK VK2ERT VK3AJJ VP2MGQ W1IHN
W1NG W3XU W9LT WA3GSC WB1GZE WB4KRH
WB9TIY WD0EWD XK5XK Y22JJ Y23DL Y24UK
Y33ZB YO6AFP YO6KEA YU1DW YU2BST YU2CT
YU3DAW YU3DBC YU3EF YU3EY YU3VM YU4YA
YU7KWX YU7MY YU7NZR YV3BJL ZS5IV ZY5EG

BAND= 20

4N0SM 4N6HN 4X4UH 9H4B AC3A DJ1BZ
DK0MM DL0IV EA1ABT EA6GP ED3VM F6KAW
FC6FPH G3VZT HA0DU HA4KYN HA4XX HA5KKB
HA9KPU HB9AGC HB9BLQ HB9BUN HG5A I0JBL
I0NKN I5MXX IS0WON IT9WPO IV3WMP JA9YBA
JG1ILF K2BA/4 K2NG KL7RA LZ1BM LZ1RN
LZ2DB LZ2KIM LZ2KRR N4WW OE1DH OH1IG
OH2AW OK1FV OK1MSN OK2BTI OK2RZ R6L
SV1MO UA9XWU UB5JK UB5KAN UC2OBP UG6GDS
UG6LQ UK0AMM UK0QAA UK2BBB UK2GKW UK2PCR
UK3DBG UK3IAA UK3QAA UK3QAE UK3XAB UK5FAD
UK5GKW UK5IBB UK5MAF UK5OAA UK5QAV UK5QBE
UK5UDX UK6LAA UK6LTG UK6XAA UK6YAB UK7NAQ
UL7QF UY5YB Y22JJ Y22OM/A YO9HP YU4EXA
YU7AD YU7ECD YU7KMN YU7KWX YZ6G

BAND= 40

BAND= 80

TOTAL

4N0SM ****** 4N6HN 4U1ITU ****** 4X4UH
4X4VE/5N8 4X6CJ 4Z4JS 6Y5HN 9H4B AB0I
AC3A ****** AD8R AH2M AI7B AI9J
AK1A CT7AL DF2AD/A DF4NP DF5IY DJ1BZ
DJ3HJ DJ8UV DK0MM ****** DK5AD DK7JQ
DL0IV DL0UE DL4SAR DL5NBC DL6FAW DL6RAI
DL8PC EA1ABT ****** EA2AEA EA3WZ EA6GP
ED3VM ****** ****** EG3SF EI7DJ F5RU
F6BFN F6BXQ F6GKH F6KAW F8WE FC6FPH
FR0FLO G3VZT G3XBY GW4HSH HA0DU HA4KYH
HA4KYN ****** HA4XX ****** HA4ZB HA5KKB
HA5KKC/7 ****** HA5KKG HA8ZB HA9KPU ******
HB9AGC HB9BLQ HB9BMY/4X HB9BUN HG1KZC HG3KHO
HG5A ****** ****** HG6V HK5BCZ I0JBL
I0KWX I0NKN I0SKK I1JHS I2ARN I2JSB
I2QMU I2SVA I2YKV I4VDS I4ZSQ I5JUX
I5MXX I5MYL I6CXD IO6FLD IP1TTM IS0QDV

IS0WDN IT9WPO ****** ****** IV3PRK IV3WMP
IZ1ARI JA9YBA JG1ILF JH0BBE JI1QQI JN1XEV
JR1WHW K2BA/4 ****** K2IJL K2NG K2SS
****** K3EST K3KHL K3ZJ K4KZZ K6IR
KA1YQ KA4SUN KB2MG KB8LH KC1F KC3N
KE3G KF2O KG1D KI2P KI3L KJ9W
KK5I KL7RA KM5R KQ2M KQ8M KS2G
KV5Q LA7JO LU5FGG LU7MAY LU8FEU LZ13C
LZ1BM LZ1RN LZ2CC LZ2DB LZ2KIM ******
LZ2KRR N2AA N2US N2WT N4KG N4NO
N4WW ****** ****** N4XD N8AKF NP4A
NU4Y OE1DH OE1WO OE3KTA OH1IG OH2AA
OH2AW OH3OQ OH6AC OK1AJN OK1ALQ OK1ARI
OK1DLA OK1FV OK1KST OK1KUR/P OK1MSN ******
OK2BQL OK2BTI OK2RZ OK3CFA OK3CRH OK3CSC
OK3JW OZ5EDR OZ5EV P42J PA2TMS PJ2VR
R6L ****** RA3DKE RA3RBU RB5CCO SV1MO
UA0WAM UA0ZDD UA1FV UA3ACY/U9 JUA3AGL UA3AHA
UA3DUA UA3QAE UA3TBK UA3TN UA3VAS UA6LGP
UA9CRR UA9MAF UA9XWU UB5AAL UB5DBV UB5EJA
UB5JK UB5KAN UB5MBZ UB5MPD UB5VAZ UC2OBP
UC2OBZ UG6GDS UG6LQ UK0AMM UK0QAA ******
UK2AAF UK2BBB UK2GAB UK2GBL UK2GKW UK2PAP
UK2PCR UK2RDX UK3AAC UK3ADZ ****** UK3DAH
UK3DBG UK3IAA UK3QAA UK3QAE UK3SAB UK3XAB
UK4FAV UK4WAA UK5AAA UK5FAA UK5FAD UK5GKW
UK5IBB UK5MAF ****** UK5OAA ****** UK5QAV
UK5QBE UK5UDX ****** ****** UK5WBG UK6LAA
UK6LTG UK6XAA UK6YAB UK7NAQ UK9AAN UK9FER
UL7QF UP2NK UP2PBW UQ2GFN UW3HV UW3ZU
UY5XE UY5YB VE1AI VE1BRB VE1DXA VE3BMV
VE3JTQ VE3PCA VE5UF VE6OU VE7BTV VE7UBC
VE7WJ VK2APK VK2ERT VK3AJJ VP2EC VP2MGQ
W1IHN W1NG W3XU W8UA W9LT WA3GSC
WB1GZE WB4KRH WB9TIY WD0EWD XK5XK Y22JJ
****** Y22OM/A ****** Y23DL Y24UK Y27FN
Y32ZF Y33ZB Y34YF Y35LM Y37XJ Y47XF
Y56YF YO5KAD YO6AFP YO6KEA YO9HP YT3L
YU1BAU YU1DW YU1FW YU1NZW YU2BST YU2CHI
YU2CT YU2JL YU2RVL YU3CAB YU3DAW YU3DBC
****** YU3EF YU3EO YU3EY ****** YU3TDP
YU3TLA YU3UPI YU3VM YU4EXA YU4YA YU7AD
YU7AU YU7AV YU7ECD YU7KMN YU7KWX ******
****** YU7MY YU7NZR YU7QDT YV3BJL YZ6G
ZS4SP ZS5IV ZY5EG

```
WPX82         82/03/30      PAGE  1

    PX   CALL          DATE      TIME  BAND MODE   QSL STATUS
-------------------------------------------------------------------

 1  4U   4U1ITU     82/03/27  1645   10  SSB   WORKED
 2  4X   4Z4JS      82/03/28  2032   15  SSB   WORKED
 3  5N   4X4VE/5N8  82/03/27  1854   15  SSB   WORKED
 4  6Y   6Y5HN      82/03/28  1939   15  SSB   WORKED
 5  9H   9H4B       82/03/28  1505   20  SSB   WORKED
 6  CT   CT7AL      82/03/27  1813   15  SSB   WORKED
 7  DJ   DF4NP      82/03/27  1927   15  SSB   WORKED
 8  DM   Y35LM      82/03/27  1659   10  SSB   WORKED
 9  EA   EA1ABT     82/03/27  1909   27  SSB   WORKED
10  EA6  EA6GP      82/03/28  0826   20  SSB   WORKED
11  EI   EI7DJ      82/03/27  1937   15  SSB   WORKED
12  F    F6BFN      82/03/27  2052   15  SSB   WORKED
13  FC   FC6FPH     82/03/27  2143   20  SSB   WORKED
14  FR   FR0FLO     82/03/28  0549   15  SSB   WORKED
15  G    G3XBY      82/03/28  0638   15  SSB   WORKED
16  GW   GW4HSH     82/03/28  2152   15  SSB   WORKED
17  HA   HA4KYH     82/03/27  1638   10  SSB   WORKED
18  HB   HB9AGC     82/03/27  2148   20  SSB   WORKED
19  HK   HK5BCZ     82/03/27  2302   15  SSB   WORKED
20  I    I06FLD     82/03/27  1645   10  SSB   WORKED
21  IS   IS0QDV     82/03/28  0837   15  SSB   WORKED
22  JA   JH0BBE     82/03/28  0521   10  SSB   WORKED
23  KL7  KL7RA      82/03/27  2140   20  SSB   WORKED
24  KP4  NP4A       82/03/27  1903   15  SSB   WORKED
25  LA   LA7JO      82/03/28  1101   15  SSB   WORKED
26  LU   LU8FEU     82/03/28  1335   10  SSB   WORKED
27  LZ   LZ2DB      82/03/27  2221   20  SSB   WORKED
28  OE   OE1WO      82/03/28  0556   15  SSB   WORKED
29  OH   OH1IG      82/03/28  0616   20  SSB   WORKED
30  OK   OK1KST     82/03/27  1725   10  SSB   WORKED
31  OZ   OZ5EV      82/03/28  2116   15  SSB   WORKED
32  PA   PA2TMS     82/03/28  1201   10  SSB   WORKED
33  PJ   P42J       82/03/27  2158   15  SSB   WORKED
34  PY   ZY5EG      82/03/28  0015   15  SSB   WORKED
35  SV   SV1MO      82/03/28  0951   20  SSB   WORKED
36  UA0  UA0ZDD     82/03/28  0526   10  SSB   WORKED
37  UA3  UA3VAS     82/03/28  1920   15  SSB   WORKED
38  UB5  UB5MBZ     82/03/28  0718   15  SSB   WORKED
39  UC2  UC2OBP     82/03/28  0821   20  SSB   WORKED
40  UG6  UG6GDS     82/03/28  0914   20  SSB   WORKED
```

Fig. A-8. Stations worked in order by prefix.

	PX	CALL	DATE	TIME	BAND	MODE	QSL STATUS
41	UL7	UL7QF	82/03/28	1905	20	SSB	WORKED
42	UP2	UP2NK	82/03/28	1352	15	SSB	WORKED
43	UQ2	UQ2GFN	82/03/28	0920	15	SSB	WORKED
44	VE	VE6OU	82/03/27	1859	15	SSB	WORKED
45	VK	VK2ERT	82/03/28	0416	15	SSB	WORKED
46	VP2E	VP2EC	82/03/27	1956	10	SSB	WORKED
47	VP2M	VP2MGQ	82/03/27	2115	15	SSB	WORKED
48	W	N8AKF	82/03/27	1641	10	SSB	WORKED
49	XE	XK5XK	82/03/28	1941	15	SSB	WORKED
50	YO	YO9HP	82/03/28	0630	20	SSB	WORKED

WPX82 82/03/30 PAGE 2

	PX	CALL	DATE	TIME	BAND	MODE	QSL STATUS
51	YU	YU2CHI	82/03/27	1644	10	SSB	WORKED
52	YV	YV3BJL	82/03/27	2150	15	SSB	WORKED
53	ZS	ZS4SP	82/03/28	1032	10	SSB	WORKED

WPX82 82/03/30 PAGE 3

PX	CALL	DATE	TIME	BAND	MODE	QSL STATUS

SUMMARY

TOTAL = 53

QSL'S RECEIVED = 0

QSL'S SENT DIRECT = 0

QSL'S SENT VIA BUREAU = 0

WORKED BUT NO CARD SENT = 53

show duplicates but does not distinguish between contacts on different bands. The total part of the list shown in Fig. A-7 can be used to detect if any stations were contacted on all bands. It can be seen that, in fact, several were.

A quick first look at the DXCC status of the WPX82 contest log is shown in Fig. A-8. Here the DXCCGEN and DXCCREAD programs have been used to generate the list. In the future the programs can be modified to perform the WPX and DXCC totaling by themselves.

The best computer program is no substitute for high power, a good antenna, and good operating practices.

Appendix B

Appendix B contains the data, antenna direction, and distance information for various prefixes as seen from different cities. The tables have been calculated using the programs described in chapter 6.

Figure B-1 shows latitude and longitude data for various prefixes. Figure B-2 lists data for G3ZCZ/W3 in Washington DC. Figure B-3 gives data for G3ZCZ/W0 in Denver. Figure B-4 includes data for G3ZCZ/W6 in San Francisco. Figure B-5 lists data for G3ZCZ in London. Figure B-6 includes data for G3ZCZ/4X in Jerusalem. Figure B-7 gives data for G3ZCZ/VK3 in Melbourne. Figure B-8 lists data for G3ZCZ/ZS in Capetown.

Many prefixes (such as W0 or UA0) cover large geographical areas, and as such the directions and distances are only approximate. For such large prefix areas, if the exact location chosen is close to the selected prefix the data may be very approximate (such as the pointing and distance data for the G prefix as seen from London, England). For actual data the exact latitude and longitude information will have to be entered into the program as was done in Chapter 6.

New prefixes are constantly appearing as countries change their status or celebrate special occasions. These new prefixes can be added at will by entering the data into the QTHPX table of Fig. B-1 and running the relevant QTH calculation program.

Data in the tables are in the same format as in Fig. 6-4, i.e., prefix (call area), distance in miles, distance in kilometers, and lastly the bearing or direction with respect to north.

PX	LATITUDE	LONGITUDE (E)	(W)
1A0KM	41.90 N	12.50	347.50
3A	45.80 N	7.50	352.50
3B6	10.50 S	56.00	304.00
3B8	20.00 S	58.00	302.00
3B9	19.00 S	65.00	295.00
3C	1.50 N	10.00	350.00
3D2	18.00 S	178.00	182.00
3D6	26.50 S	28.50	331.50
3V8	35.00 N	10.00	350.00
3X	10.00 N	348.00	12.00
3Y	55.00 S	4.00	356.00
4N	44.00 N	20.00	340.00
4S7	13.00 N	81.50	278.50
4U1ITU	46.18 N	6.15	353.85
4U1UN	40.70 N	286.20	73.80
4W	16.00 N	44.00	316.00
4X	32.00 N	35.00	325.00
4Z	32.00 N	35.00	325.00
5A	30.00 N	15.00	345.00
5B4	35.00 N	33.00	327.00
5H	6.00 S	321.00	39.00
5L	7.00 N	350.00	10.00
5N	7.00 N	5.00	355.00
5R	20.00 S	47.00	313.00
5T	20.00 N	347.00	13.00
5U	15.00 N	5.00	355.00
5V	6.00 N	1.50	358.50
5W	14.00 S	189.00	171.00
5X	.00 N	33.50	326.50
5Z	.00 N	38.00	322.00
6O	5.00 N	45.00	315.00
6W	15.00 N	348.00	12.00
6Y	18.00 N	283.00	77.00
7J	26.00 N	130.00	230.00
7O	13.00 N	45.00	315.00
7P	30.00 S	28.00	332.00
7Q	14.00 N	35.00	325.00
7X	35.00 N	5.00	355.00
7Z	24.00 N	45.00	315.00
8P	13.00 N	301.00	59.00
8Q	5.00 N	74.00	286.00
8R	5.00 N	301.50	58.50
8Z	29.00 N	46.00	314.00

PX	LATITUDE	LONGITUDE (E)	(W)
9G	7.00 N	358.50	1.50
9H	35.90 N	14.20	345.80
9J	13.00 S	28.00	332.00
9K	29.00 N	52.50	307.50
9L	8.00 N	347.00	13.00
9M2	4.00 N	103.00	257.00
9M6	6.00 N	118.00	242.00
9M8	2.50 N	113.00	247.00
9N	27.50 N	85.00	275.00
9Q	5.00 S	24.00	336.00
9U5	3.00 S	30.00	330.00
9V1	1.30 N	104.00	256.00
9X	2.00 S	30.00	330.00
9Y4	11.00 N	299.00	61.00
A2	25.00 S	25.00	335.00
A3	21.20 S	175.00	185.00
A4	23.00 N	58.00	302.00
A5	27.00 N	90.00	270.00
A6	25.00 N	55.00	305.00
A7	26.00 N	51.50	308.50
A9	26.00 N	50.75	309.25
AP	30.00 N	70.00	290.00
BV	32.00 N	114.00	246.00
BY	25.00 N	121.50	238.50
C2	.55 S	168.55	191.45
C3	42.70 N	1.50	358.50
C5	13.00 N	345.00	15.00
C6	25.00 N	283.00	77.00
C9	15.00 S	40.00	320.00
CE	35.00 S	288.00	72.00
CE0A	26.00 S	251.00	109.00
CE0X	26.00 S	279.50	80.50
CE0Z	33.00 S	279.50	80.50
CE9	65.00 S	300.00	60.00
CE9	63.00 S	300.00	60.00
CN	35.00 N	354.50	5.50
CO	22.50 N	280.00	80.00
CP	17.50 S	295.00	65.00
CR3	12.00 N	345.00	15.00
CR9	22.20 N	113.60	246.40
CT1	40.00 N	352.00	8.00
CT2	39.00 N	332.50	27.50
CT3	32.75 N	348.00	12.00

Fig. B-1. Geographic coordinates for various prefixes.

PX	LATITUDE	LONGITUDE (E)	(W)
CX	34.00 S	305.00	55.00
CZ6	53.00 N	248.00	112.00
D2	10.00 S	15.00	345.00
D4	15.00 N	335.00	25.00
D6	11.50 S	43.50	316.50
DA	50.00 N	9.00	351.00
DB	50.00 N	9.00	351.00
DF	50.00 N	9.00	351.00
DJ	50.00 N	9.00	351.00
DL	50.00 N	9.00	351.00
DM	52.00 N	13.50	346.50
DT	52.00 N	13.00	347.00
DU	14.00 N	121.00	239.00
EA	40.00 N	356.00	4.00
EA6	39.00 N	3.00	357.00
EA8	28.00 N	343.00	17.00
EA9	25.00 N	346.00	14.00
EI	53.00 N	353.50	6.50
EL	7.00 N	350.00	10.00
EP	35.50 N	51.00	309.00
ET	15.00 N	40.00	320.00
F	46.00 N	2.00	358.00
FB8W	46.50 S	52.00	308.00
FB8X	49.50 S	69.50	290.50
FB8Y	68.00 S	140.00	220.00
FB8Z	38.00 S	77.50	282.50
FC	42.00 N	9.00	351.00
FG7	16.00 N	298.50	61.50
FH8	12.50 S	45.00	315.00
FK8	21.00 S	165.00	195.00
FM7	14.50 N	298.70	61.30
FO8	17.50 S	210.50	149.50
FO8 Clprtn	10.00 S	252.00	108.00
FP8	47.00 N	303.70	56.30
FR7	21.00 S	55.50	304.50
FS7	17.60 N	297.00	63.00
FW8	14.00 S	182.00	178.00
FY7	5.00 N	308.00	52.00
G	51.00 N	358.50	1.50
GD	54.20 N	355.50	4.50
GI	54.50 N	354.00	6.00
GJ	49.15 N	357.80	2.20
GM	56.00 N	357.00	3.00
GU	49.50 N	357.40	2.60

PX	LATITUDE	LONGITUDE (E)	(W)
GW	52.50 N	356.50	3.50
H4	6.00 S	157.00	203.00
HA	47.00 N	20.00	340.00
HB	47.00 N	9.00	351.00
HB0	47.15 N	9.60	350.40
HC	1.00 S	282.00	78.00
HC8	.00 N	270.00	90.00
HH	18.00 N	287.50	72.50
HI	18.00 N	290.00	70.00
HK	5.00 N	285.00	75.00
HK0	12.50 N	278.00	82.00
HL	37.50 N	127.00	233.00
HP	9.00 N	280.00	80.00
HR	15.00 N	273.00	87.00
HR0	17.40 N	276.00	84.00
HS	15.00 N	100.00	260.00
HT	12.50 N	274.00	86.00
HV	41.90 N	12.50	347.50
HZ	25.00 N	47.00	313.00
I	42.00 N	14.00	346.00
IC	40.50 N	14.20	345.80
IH	36.60 N	12.00	348.00
IS	40.00 N	9.00	351.00
IT	37.50 N	14.00	346.00
J28	11.50 N	43.00	317.00
J3	11.50 N	298.70	61.30
JA	36.00 N	138.00	222.00
JD1	28.00 N	135.00	225.00
JE	40.00 N	142.00	218.00
JF	35.00 N	138.00	222.00
JG	35.00 N	136.00	224.00
JH	32.00 N	133.00	227.00
JI	32.00 N	134.00	226.00
JJ	38.00 N	140.00	220.00
JL	36.00 N	138.00	222.00
JR	35.00 N	138.00	222.00
JR6	26.50 N	128.00	232.00
JT1	43.00 N	107.00	253.00
JW	79.00 N	18.00	342.00
JX	71.00 N	351.00	9.00
JY	32.00 N	36.00	324.00
K0	40.00 N	255.00	105.00
K1	40.00 N	287.00	73.00
K2	41.00 N	286.00	74.00

PX	LATITUDE		LONGITUDE (E)	(W)
K3	39.00	N	284.00	76.00
K4	35.00	N	275.00	85.00
K5	32.00	N	267.00	93.00
K6	35.00	N	240.00	120.00
K7	40.00	N	245.00	115.00
K8	43.00	N	277.00	83.00
K9	43.00	N	273.00	87.00
KA	36.00	N	138.00	222.00
KB6	3.00	S	189.00	171.00
KC4	17.80	N	284.70	75.30
KC4A	79.00	S	165.00	195.00
KC4A	90.00	S	.00	195.00
KC6	5.00	N	150.00	210.00
KG4	20.00	N	284.88	75.12
KG6	13.33	N	144.75	215.25
KH3	17.00	N	169.50	190.50
KH6	21.00	N	203.00	157.00
KH6 Kure	28.15	N	178.00	182.00
KL7	60.00	N	210.00	150.00
KM6	26.00	N	185.00	175.00
KP4	18.75	N	293.50	66.50
KP6	3.00	N	199.00	161.00
KS6	15.00	S	190.00	170.00
KV4	18.75	N	295.40	64.60
KW6	19.30	N	166.58	193.42
KX6	9.00	N	167.00	193.00
LA	60.00	N	10.00	350.00
LU	35.00	S	300.00	60.00
LX	49.60	N	6.00	354.00
LZ	43.00	N	26.00	334.00
M1	43.90	N	12.40	347.60
N0	40.00	N	255.00	105.00
N1	40.00	N	287.00	73.00
N2	41.00	N	286.00	74.00
N3	39.00	N	284.00	76.00
N4	35.00	N	275.00	85.00
N5	32.00	N	267.00	93.00
N6	35.00	N	240.00	120.00
N7	40.00	N	245.00	115.00
N8	43.00	N	277.00	83.00
N9	43.00	N	273.00	87.00
OA	12.00	S	283.00	77.00
OD	34.00	N	35.50	324.50
OE	48.00	N	14.00	346.00

PX	LATITUDE		LONGITUDE (E)	(W)
OH	60.00	N	24.00	336.00
OH0	60.00	N	20.00	340.00
OJ0	60.00	N	20.00	340.00
OK	49.50	N	16.00	344.00
ON	51.00	N	4.00	356.00
OR	51.00	N	4.00	356.00
OX	64.00	N	320.00	40.00
OY	62.00	N	353.00	7.00
OZ	56.00	N	9.00	351.00
P29	5.00	S	145.00	215.00
P5	40.00	N	126.00	234.00
PA	52.00	N	6.00	354.00
PJ2	12.00	N	291.40	68.60
PJ3	12.10	N	290.00	70.00
PJ4	12.00	N	292.00	68.00
PJ5	17.50	N	297.00	63.00
PJ6	17.50	N	296.50	63.50
PJ7	18.00	N	297.00	63.00
PP	15.00	S	315.00	45.00
PT	15.00	S	315.00	45.00
PY	15.00	S	315.00	45.00
PY0 St Pl	.50	N	331.00	29.00
PZ	5.00	N	304.00	56.00
RA0	60.00	N	120.00	240.00
RA1	60.00	N	36.00	324.00
RA2	54.50	N	21.00	339.00
RA3	56.00	N	38.00	322.00
RA4	53.00	N	48.00	312.00
RA6	45.00	N	45.00	315.00
RA9	55.00	N	73.00	287.00
RB5	50.00	N	36.00	324.00
RC2	54.00	N	28.00	332.00
RD6	41.00	N	48.00	312.00
RF6	44.00	N	44.00	316.00
RG6	44.00	N	55.00	305.00
RH8	40.00	N	60.00	300.00
RI8	40.00	N	65.00	295.00
RJ8	39.00	N	72.00	288.00
RL7	50.00	N	70.00	290.00
RM7	43.00	N	75.00	285.00
RN1	65.00	N	35.00	325.00
RO5	47.00	N	29.00	331.00
RP2	55.00	N	24.00	336.00
RQ2	57.00	N	24.00	336.00

PX	LATITUDE	LONGITUDE (E)	(W)
RR2	59.00 N	26.00	334.00
S2	23.00 N	90.00	270.00
S7	5.00 S	56.00	304.00
S8	33.00 S	30.00	330.00
S9	1.00 N	8.00	352.00
SK	58.00 N	17.00	343.00
SL	58.00 N	17.00	343.00
SM	58.00 N	17.00	343.00
SP	52.00 N	20.00	340.00
ST	15.50 N	32.50	327.50
SU	30.00 N	30.00	330.00
SV	38.00 N	22.00	338.00
T3	2.00 N	174.00	186.00
T5	5.00 N	45.00	315.00
TA	40.00 N	30.00	330.00
TF	64.00 N	338.00	22.00
TG	15.00 N	270.00	90.00
TI	10.00 N	277.00	83.00
TI9	5.00 N	273.50	86.50
TJ	5.00 N	10.00	350.00
TL	7.50 N	20.00	340.00
TN	.00 N	17.00	343.00
TR	.00 N	10.00	350.00
TT	15.00 N	18.00	342.00
TU	6.00 N	355.00	5.00
TY	8.00 N	2.50	357.50
TZ	17.00 N	357.00	3.00
UA0	60.00 N	120.00	240.00
UA1	60.00 N	36.00	324.00
UA1 F J L	80.00 N	55.00	305.00
UA2	54.50 N	21.00	339.00
UA3	56.00 N	38.00	322.00
UA4	53.00 N	48.00	312.00
UA6	45.00 N	45.00	315.00
UA9	55.00 N	73.00	287.00
UB5	50.00 N	36.00	324.00
UC2	54.00 N	28.00	332.00
UD6	41.00 N	48.00	312.00
UF6	44.00 N	44.00	316.00
UG6	44.00 N	55.00	305.00
UH8	40.00 N	60.00	300.00
UI8	40.00 N	65.00	295.00
UJ8	39.00 N	72.00	288.00
UK0	60.00 N	120.00	240.00

PX	LATITUDE	LONGITUDE (E)	(W)
UK1	60.00 N	36.00	324.00
UK2	54.50 N	21.00	339.00
UK3	56.00 N	38.00	322.00
UK4	53.00 N	48.00	312.00
UK5	50.00 N	36.00	324.00
UK6	45.00 N	45.00	315.00
UK7	50.00 N	70.00	290.00
UK8	40.00 N	65.00	295.00
UK9	55.00 N	73.00	287.00
UL7	50.00 N	70.00	290.00
UM8	43.00 N	75.00	285.00
UN1	65.00 N	35.00	325.00
UO5	47.00 N	29.00	331.00
UP2	55.00 N	24.00	336.00
UQ2	57.00 N	24.00	336.00
UR2	59.00 N	26.00	334.00
UT5	50.00 N	36.00	324.00
UW1	60.00 N	36.00	324.00
UW2	54.50 N	21.00	339.00
UW3	56.00 N	38.00	322.00
UW4	53.00 N	48.00	312.00
UW6	45.00 N	45.00	315.00
V3A	17.00 N	272.00	88.00
VE1	60.00 N	240.00	120.00
VE2	47.00 N	288.00	72.00
VE3	45.00 N	283.00	77.00
VE4	50.00 N	262.00	98.00
VE5	50.00 N	255.00	105.00
VE6	53.00 N	248.00	112.00
VE7	50.00 N	235.00	125.00
VE8	62.00 N	225.00	135.00
VK0	70.00 S	80.00	280.00
VK0 Heard	53.00 S	73.33	286.67
VK1	35.50 S	149.00	211.00
VK2	33.00 S	152.00	208.00
VK2 L H I	32.00 S	158.50	201.50
VK3	37.00 S	145.00	215.00
VK4	25.00 S	150.00	210.00
VK5	35.00 S	138.00	222.00
VK6	33.00 S	117.00	243.00
VK7	43.00 S	147.00	213.00
VK8	12.00 S	132.00	228.00
VK9N	29.00 S	108.00	252.00
VK9X	10.50 S	168.00	192.00

PX	LATITUDE	LONGITUDE (E)	(W)
VK9Y	11.50 S	97.00	263.00
VK9Z	16.50 S	150.00	210.00
VO1	47.00 N	308.00	52.00
VO2	53.00 N	300.00	60.00
VP2	15.00 N	298.00	62.00
VP5	22.00 N	288.50	71.50
VP8	51.50 S	300.00	60.00
VP9	32.30 N	295.20	64.80
VQ9 Chagos	6.00 S	72.00	288.00
VR1	2.00 S	180.00	180.00
VR3	5.00 N	200.00	160.00
VR6	25.10 S	230.00	130.00
VR7	5.00 S	155.00	205.00
VS5	5.00 N	115.00	245.00
VS6	22.00 N	115.00	245.00
VS9	12.50 N	54.00	306.00
VU	15.00 N	77.50	282.50
VU7 Andamn	10.00 N	93.00	267.00
VU7 Lacadv	12.00 N	72.50	287.50
W0	40.00 N	255.00	105.00
W1	40.00 N	287.00	73.00
W2	41.00 N	286.00	74.00
W3	39.00 N	284.00	76.00
W4	35.00 N	275.00	85.00
W5	32.00 N	267.00	93.00
W6	35.00 N	240.00	120.00
W7	40.00 N	245.00	115.00
W8	43.00 N	277.00	83.00
W9	43.00 N	273.00	87.00
XE	20.00 N	260.00	100.00
XT	12.00 N	357.00	3.00
XU	12.00 N	105.00	255.00

PX	LATITUDE	LONGITUDE (E)	(W)
XV	15.00 N	108.00	252.00
XW	17.00 N	105.00	255.00
XZ	20.00 N	95.00	265.00
Y	52.00 N	13.50	346.50
YA	35.00 N	69.00	291.00
YB	5.00 S	115.00	245.00
YI	33.00 N	44.00	316.00
YJ	18.00 S	166.00	194.00
YK	34.00 N	37.00	323.00
YN	12.00 N	273.00	87.00
YO	44.50 N	26.00	334.00
YS	13.50 N	271.00	89.00
YU	44.00 N	20.00	340.00
YV	10.00 N	295.00	65.00
Z2	18.00 S	30.00	330.00
ZA	41.00 N	20.00	340.00
ZB	36.10 N	354.70	5.30
ZC4	35.00 N	33.00	327.00
ZD7	16.00 S	354.00	6.00
ZD8	8.00 S	346.00	14.00
ZD9	38.00 S	348.00	12.00
ZF1	19.00 N	278.90	81.10
ZK1	18.00 S	195.00	165.00
ZK2	19.50 S	190.00	170.00
ZL	42.00 S	174.00	186.00
ZM2	10.00 S	190.00	170.00
ZP	25.00 S	303.00	57.00
ZS	33.00 S	23.00	337.00
ZS3	25.00 S	15.00	345.00
ZV	10.00 S	305.00	55.00
ZW	10.00 S	305.00	55.00
ZY	10.00 S	305.00	55.00
ZZ	10.00 S	305.00	55.00

1A0KM	4482	7213	55
3A	4134	6653	53
3B6	8946	14397	69
3B8	9466	15234	77
3B9	9784	15745	70
3C	5990	9640	91
3D2	7783	12525	265
3D6	8136	13093	103
3V8	4616	7429	63
3X	4444	7152	98
3Y	8040	12939	141
4N	4725	7604	50
4S7	8589	13822	26
4U1ITU	4065	6542	53
4U1UN	210	338	53
4W	7063	11366	57
4X	5878	9459	52
4Z	5878	9459	52
5A	5050	8127	65
5B4	5652	9096	51
5H	3939	6339	133
5L	4681	7533	99
5N	5484	8825	90
5R	8882	14294	85
5T	3981	6407	89
5U	5146	8281	83
5V	5339	8592	93
5W	7032	11317	261
5X	7309	11762	77
5Z	7543	12139	74
6O	7657	12322	65
6W	4234	6814	94
6Y	1444	2324	180
7J	7623	12268	334
7O	7260	11684	59
7P	8236	13254	107
7Q	6736	10840	65
7X	4374	7039	65
7Z	6698	10779	51
8P	2100	3380	143
8Q	8880	14291	38
8R	2612	4203	149
8Z	6479	10427	47
9G	5135	8264	94
9H	4780	7692	60
9J	7579	12197	91
9K	6726	10824	43
9L	4478	7206	100
9M2	9470	15240	180
9M6	9186	14783	339
9M8	9503	15293	345
9N	7702	12395	17
9Q	7024	11304	87
9U5	7258	11680	82
9V1	9656	15539	358
9X	7214	11609	81
9Y4	2164	3483	149
A2	7897	12709	103
A3	8069	12985	264
A4	7262	11687	42
A5	7804	12559	13
A6	7040	11329	44
A7	6853	11029	45
A9	6824	10982	46
AP	7222	11622	29
BV	7485	12046	350
BY	7860	12649	342
C2	7540	12134	285
C3	3959	6371	59
C5	4157	6690	97
C6	960	1545	180
C9	8306	13367	85
CE	5114	8230	176
CE0A	4936	7944	210
CE0X	4488	7223	183
CE0Z	4971	8000	183
CE9	7235	11643	173
CE9	7101	11428	172
CN	3852	6199	70
CO	1147	1846	190
CP	3972	6392	166
CR3	4199	6757	98
CR9	8157	13127	349
CT1	3584	5768	65
CT2	2625	4224	74
CT3	3588	5774	75
CX	5229	8415	161
CZ6	1915	3082	312

Fig. B-2. Direction distance table for G3ZCZ/W3 located at 38.9 — 77.

D2	6756	10872	97
D4	3543	5702	103
D6	8345	13430	79
DA	4071	6551	48
DB	4071	6551	48
DF	4071	6551	48
DJ	4071	6551	48
DL	4071	6551	48
DM	4187	6738	45
DT	4168	6708	45
DU	8599	13838	339
EA	3778	6080	64
EA6	4143	6667	62
EA8	3476	5594	83
EA9	3741	6020	85
EI	3375	5431	49
EL	4681	7533	99
EP	6317	10166	40
ET	6928	11149	61
F	3889	6259	55
FB8W	9838	15832	119
FB8X	10639	17121	125
FB8Y	9987	16072	203
FB8Z	11057	17794	95
FC	4322	6955	57
FG7	1839	2960	145
FH8	8467	13626	79
FK8	8598	13837	270
FM7	1937	3117	146
FO8	6080	9785	246
FO8 Clprtn	3922	6312	217
FP8	1182	1902	55
FR7	9374	15086	80
FS7	1695	2728	146
FW8	7406	11918	266
FY7	2816	4532	140
G	3617	5821	50
GD	3436	5530	47
GI	3372	5427	47
GJ	3628	5839	53
GM	3465	5576	44
GU	3603	5798	52
GW	3505	5641	49
H4	8383	13491	290
HA	4611	7420	47
HB	4161	6696	52
HB0	4182	6730	51
HC	2757	4437	182
HC8	2810	4522	200
HH	1469	2364	168
HI	1504	2420	162
HK	2345	3774	176
HK0	1849	2976	191
HL	6940	11169	341
HP	2074	3338	186
HR	1759	2831	203
HR0	1544	2485	198
HS	8705	14009	4
HT	1905	3066	199
HV	4482	7213	55
HZ	6729	10829	49
I	4544	7313	55
IC	4608	7416	56
IH	4651	7485	60
IS	4390	7065	59
IT	4710	7580	59
J2B	7241	11653	61
J3	2124	3418	149
JA	6799	10942	332
JD1	7372	11864	331
JE	6453	10385	331
JF	6861	11041	332
JG	6910	11120	333
JH	7171	11540	334
JI	7147	11502	333
JJ	6625	10662	332
JL	6799	10942	332
JR	6861	11041	332
JR6	7636	12289	336
JT1	6770	10895	357
JW	3653	5879	14
JX	3210	5166	25
JY	5920	9527	52
K0	1489	2396	282
K1	226	364	69
K2	215	346	47
K3	54	87	82
K4	517	832	241
K5	1017	1637	247
K6	2367	3809	277
K7	2013	3240	284
K8	422	679	314

K9	593	954	302
KA	6799	10942	332
KB6	6562	10560	270
KC4	1461	2351	176
KC4A	9209	14820	193
KC4A	8903	14328	180
KC6	8171	13150	304
KG4	1310	2108	175
KG6	7933	12767	314
KH3	6665	10726	298
KH6	4796	7718	281
KH6 Kure	5746	9247	301
KL7	3376	5433	321
KM6	5509	8866	295
KP4	1527	2457	153
KP6	5764	9276	269
KS6	7021	11299	260
KV4	1577	2538	148
KW6	6690	10766	302
KX6	7170	11539	294
LA	3844	6186	37
LU	5219	8399	166
LX	3960	6373	50
LZ	5010	8063	48
M1	4408	7094	53
N0	1489	2396	282
N1	226	364	69
N2	215	346	47
N3	54	87	82
N4	517	832	241
N5	1017	1637	247
N6	2367	3809	277
N7	2013	3240	284
N8	422	679	314
N9	593	954	302
OA	3516	5658	180
OD	5803	9339	50
OE	4336	6978	49
OH	4282	6891	34
OH0	4161	6696	35
OJ0	4161	6696	35
OK	4365	7025	47
ON	3842	6183	49
OR	3842	6183	49
OX	2292	3689	29

OY	3260	5246	37
OZ	3907	6288	42
P29	8918	14352	301
P5	6793	10932	342
PA	3896	6270	47
PJ2	1928	3103	162
PJ3	1900	3058	165
PJ4	1938	3119	161
PJ5	1701	2737	146
PJ6	1687	2715	147
PJ7	1671	2689	146
PP	4258	6852	144
PT	4258	6852	144
PY	4258	6852	144
PY0 St Pl	4023	6474	119
PZ	2685	4321	145
RA0	5534	8906	351
RA1	4623	7440	30
RA2	4377	7044	40
RA3	4857	7816	33
RA4	5286	8507	30
RA6	5611	9030	37
RA9	5709	9187	17
RB5	5070	8159	38
RC2	4633	7456	38
RD6	5919	9525	38
RF6	5629	9059	39
RG6	5972	9611	32
RH8	6345	10211	32
RI8	6478	10425	28
RJ8	6705	10790	24
RL7	5973	9612	21
RM7	6511	10478	20
RN1	4387	7060	26
RO5	4957	7977	44
RP2	4461	7179	39
RQ2	4388	7062	37
RR2	4378	7046	34
S2	8075	12995	13
S7	8682	13972	64
S8	8440	13582	109
S9	5904	9501	92
SK	4127	6642	38
SL	4127	6642	38
SM	4127	6642	38

SP	4429	7128	43
ST	6543	10530	66
SU	5753	9258	57
SV	5049	8125	54
T3	7140	11490	284
T5	7657	12322	65
TA	5301	8531	49
TF	2799	4504	34
TG	1830	2945	209
TI	2030	3267	192
TI9	2416	3888	197
TJ	5839	9397	88
TL	6264	10081	80
TN	6431	10349	87
TR	6055	9744	92
TT	5832	9385	75
TU	4991	8032	97
TY	5307	8541	90
TZ	4636	7461	86
UA0	5534	8906	351
UA1	4623	7440	30
UA1 F J L	4015	6461	9
UA2	4377	7044	40
UA3	4857	7816	33
UA4	5286	8507	30
UA6	5611	9030	37
UA9	5709	9187	17
UB5	5070	8159	38
UC2	4633	7456	38
UD6	5919	9525	38
UF6	5629	9059	39
UG6	5972	9611	32
UH8	6345	10211	32
UI8	6478	10425	28
UJ8	6705	10790	24
UK0	5534	8906	351
UK1	4623	7440	30
UK2	4377	7044	40
UK3	4857	7816	33
UK4	5286	8507	30
UK5	5070	8159	38
UK6	5611	9030	37
UK7	5973	9612	21
UK8	6478	10425	28
UK9	5709	9187	17
UL7	5973	9612	21
UM8	6511	10478	20
UN1	4387	7060	26
UO5	4957	7977	44
UP2	4461	7179	39
UQ2	4388	7062	37
UR2	4378	7046	34
UT5	5070	8159	38
UW1	4623	7440	30
UW2	4377	7044	40
UW3	4857	7816	33
UW4	5286	8507	30
UW6	5611	9030	37
V3A	1652	2659	207
VE1	2353	3787	322
VE2	614	988	23
VE3	421	678	0
VE4	1281	2062	314
VE5	1566	2520	308
VE6	1915	3082	312
VE7	2441	3928	304
VE8	2872	4622	323
VK0	10128	16299	166
VK0 Heard	10726	17261	135
VK1	9918	15961	261
VK2	9696	15604	263
VK2 L H I	9322	15002	261
VK3	10162	16354	260
VK4	9551	15370	275
VK5	10494	16888	267
VK6	11552	18591	293
VK7	10147	16330	248
VK8	9862	15871	308
VK9N	11692	18816	336
VK9X	8010	12890	278
VK9Y	10505	16906	13
VK9Z	9215	14830	285
VO1	1374	2211	58
VO2	1264	2034	34
VP2	1885	3034	147
VP5	1212	1950	163
VP8	6328	10184	170
VP9	822	1323	120
VQ9 Chagos	9450	15208	48
VR1	7001	11267	277

VR3	5624	9051	270
VR6	5590	8996	227
VR7	8437	13578	292
VS5	9305	14975	343
VS6	8155	13124	347
VS9	7685	12367	52
VU	8362	13457	29
VU7 Andamn	8995	14476	13
VU7 Lacadv	8405	13526	36
W0	1489	2396	282
W1	226	364	69
W2	215	346	47
W3	54	87	82
W4	517	832	241
W5	1017	1637	247
W6	2367	3809	277
W7	2013	3240	284
W8	422	679	314
W9	593	954	302
XE	1891	3043	233
XT	4842	7792	90
XU	8915	14347	357
XV	8696	13994	354
XW	8570	13792	358
XZ	8332	13409	9
Y	4187	6738	45
YA	6886	11082	28
YB	9974	16051	339
YI	6193	9966	46
YJ	8427	13562	272
YK	5865	9439	49
YN	1957	3149	201
YO	4947	7961	47
YS	1901	3059	206
YU	4725	7604	50
YV	2129	3426	156
Z2	7890	12697	94
ZA	4842	7792	53
ZB	3829	6162	69
ZC4	5652	9096	51
ZD7	5937	9554	114
ZD8	5165	8312	114
ZD9	6722	10818	134
ZF1	1396	2247	191
ZK1	6885	11080	255
ZK2	7208	11600	256
ZL	8805	14170	242
ZM2	6809	10958	264
ZP	4599	7401	160
ZS	8087	13014	112
ZS3	7381	11878	109
ZV	3662	5893	152
ZW	3662	5893	152
ZY	3662	5893	152
ZZ	3662	5893	152

1A0KM	5572	8967	42
3A	5206	8378	42
3B6	10103	16259	35
3B8	10740	17284	41
3B9	10884	17516	25
3C	7457	12001	72
3D2	6346	10213	248
3D6	9627	15493	86
3V8	5820	9366	48
3X	5934	9550	80
3Y	9106	14654	133
4N	5715	9197	36
4S7	8769	14112	352
4U1ITU	5140	8272	42
4U1UN	1639	2638	77
4W	8099	13034	34
4X	6858	11037	34
4Z	6858	11037	34
5A	6271	10092	49
5B4	6621	10655	33
5H	5239	8431	110
5L	6172	9933	81
5N	6946	11178	72
5R	10289	16558	59
5T	5447	8766	73
5U	6569	10571	66
5V	6815	10967	75
5W	5624	9051	244
5X	8647	13916	54
5Z	8835	14218	50
6O	8803	14167	39
6W	5715	9197	77
6Y	2244	3611	124
7J	6679	10749	312
7O	8314	13380	34
7P	9718	15639	91
7Q	7920	12746	43
7X	5617	9039	51
7Z	7627	12274	29
8P	3340	5375	110
8Q	9345	15039	1
8R	3752	6038	117
8Z	7335	11804	26
9G	6615	10646	76
9H	5937	9554	45
9J	9043	14553	71
9K	7472	12025	21
9L	5970	9608	83
9M2	8930	14371	323
9M6	8256	13286	309
9M8	8655	13928	311
9N	7747	12467	350
9Q	8465	13623	67
9U5	8649	13919	60
9V1	9060	14580	320
9X	8598	13837	59
9Y4	3336	5369	114
A2	9387	15106	86
A3	6638	10683	247
A4	7966	12820	17
A5	7726	12433	346
A6	7782	12524	20
A7	7649	12310	23
A9	7634	12285	23
AP	7608	12244	5
BV	6887	11083	327
BY	7054	11352	318
C2	6052	9739	267
C3	5124	8246	47
C5	5646	9086	80
C6	1911	3075	114
C9	9720	15642	61
CE	5572	8967	153
CE0A	4545	7314	184
CE0X	4805	7733	157
CE0Z	5260	8465	159
CE9	7627	12274	161
CE9	7515	12094	160
CN	5166	8314	57
CO	1885	3034	122
CP	4716	7589	139
CR3	5689	9155	81
CR9	7486	12047	323
CT1	4849	7803	54
CT2	3999	6436	64
CT3	4962	7985	62
CX	6007	9667	141
CZ6	976	1571	343

Fig. B-3. Direction distance table for G3ZCZ/W0 located at 39.7 255.

D2	8242	13264	78
D4	5034	8101	84
D6	9696	15604	53
DA	5060	8143	38
DB	5060	8143	38
DF	5060	8143	38
DJ	5060	8143	38
DL	5060	8143	38
DM	5104	8214	34
DT	5089	8190	35
DU	7691	12377	311
EA	5018	8075	52
EA6	5350	8610	49
EA8	4912	7905	69
EA9	5185	8344	70
EI	4406	7091	42
EL	6172	9933	81
EP	7019	11296	20
ET	8035	12931	38
F	4997	8042	44
FB8W	11187	18003	120
FB8X	11705	18837	160
FB8Y	9364	15069	209
FB8Z	12254	19720	310
FC	5440	8755	44
FG7	3078	4953	109
FH8	9813	15792	53
FK8	7131	11476	254
FM7	3156	5079	111
FO8	4880	7853	225
FO8 Clprtn	3438	5533	184
FP8	2456	3952	62
FR7	10703	17224	47
FS7	2930	4715	109
FW8	5964	9598	248
FY7	4066	6543	112
G	4659	7498	41
GD	4431	7131	40
GI	4369	7031	40
GJ	4709	7578	43
GM	4413	7102	38
GU	4681	7533	43
GW	4530	7290	41
H4	6906	11114	270
HA	5556	8941	35

HB	5201	8370	40
HB0	5215	8392	40
HC	3288	5291	142
HC8	2901	4669	157
HH	2449	3941	118
HI	2568	4133	115
HK	3039	4891	134
HK0	2344	3772	137
HL	6165	9921	321
HP	2621	4218	137
HR	2023	3256	142
HR0	1988	3199	135
HS	8326	13399	332
HT	2207	3552	143
HV	5572	8967	42
HZ	7614	12253	27
I	5620	9044	41
IC	5703	9178	42
IH	5818	9363	46
IS	5537	8911	45
IT	5848	9411	44
J28	8344	13428	37
J3	3298	5307	114
JA	5848	9411	314
JD1	6374	10258	310
JE	5499	8850	314
JF	5899	9493	313
JG	5976	9617	314
JH	6245	10050	314
JI	6207	9989	313
JJ	5672	9128	314
JL	5848	9411	314
JR	5899	9493	313
JR6	6730	10831	314
JT1	6381	10269	337
JW	3926	6318	11
JX	3779	6082	23
JY	6887	11083	33
K0	0	0	0
K1	1688	2716	79
K2	1626	2617	77
K3	1543	2483	82
K4	1142	1838	100
K5	856	1378	125
K6	884	1423	253

K7	530	853	275
K8	1160	1867	72
K9	959	1543	70
KA	5848	9411	314
KB6	5097	8203	252
KC4	2330	3750	122
KC4A	8899	14321	194
KC4A	8959	14418	180
KC6	6783	10916	284
KG4	2230	3589	119
KG6	6660	10718	293
KH3	5239	8431	281
KH6	3305	5319	263
KH6 Kure	4352	7004	285
KL7	2375	3822	321
KM6	4068	6547	279
KP4	2704	4352	111
KP6	4311	6938	249
KS6	5626	9054	242
KV4	2799	4504	109
KW6	5293	8518	284
KX6	5715	9197	276
LA	4628	7448	29
LU	5902	9498	144
LX	4978	8011	39
LZ	5953	9580	34
M1	5469	8801	41
N0	0	0	0
N1	1688	2716	79
N2	1626	2617	77
N3	1543	2483	82
N4	1142	1838	100
N5	856	1378	125
N6	884	1423	253
N7	530	853	275
N8	1160	1867	72
N9	959	1543	70
OA	3998	6434	147
OD	6752	10866	32
OE	5318	8558	37
OH	4965	7990	24
OH0	4875	7845	26
OJ0	4875	7845	26
OK	5305	8537	35
ON	4849	7803	39
OR	4849	7803	39
OX	3054	4915	35
OY	4081	6568	33
OZ	4782	7696	33
P29	7496	12063	279
P5	6059	9751	323
PA	4870	7837	38
PJ2	2928	4712	121
PJ3	2859	4601	122
PJ4	2956	4757	120
PJ5	2935	4723	109
PJ6	2910	4683	110
PJ7	2913	4688	109
PP	5394	8681	121
PT	5394	8681	121
PY	5394	8681	121
PY0 St Pl	5453	8776	99
PZ	3871	6230	115
RA0	5089	8190	338
RA1	5199	8367	19
RA2	5186	8346	29
RA3	5476	8813	20
RA4	5830	9382	16
RA6	6294	10129	21
RA9	5891	9480	1
RB5	5800	9334	24
RC2	5386	8668	26
RD6	6606	10631	20
RF6	6338	10200	22
RG6	6519	10491	14
RH8	6847	11019	12
RI8	6892	11091	8
RJ8	6994	11255	2
RL7	6230	10026	3
RM7	6721	10816	0
RN1	4886	7863	17
RO5	5809	9348	30
RP2	5235	8425	27
RQ2	5126	8249	26
RR2	5062	8146	24
S2	7995	12866	345
S7	9760	15707	31
S8	9914	15955	94
S9	7379	11875	74
SK	4904	7892	28

SL	4904	7892	28
SM	4904	7892	28
SP	5291	8515	31
ST	7742	12459	45
SU	6819	10974	38
SV	6104	9823	39
T3	5650	9093	266
T5	8803	14167	39
TA	6241	10044	33
TF	3594	5784	33
TG	1932	3109	148
TI	2457	3954	141
TI9	2660	4281	149
TJ	7291	11733	70
TL	7649	12310	60
TN	7878	12678	68
TR	7527	12113	73
TT	7173	11544	57
TU	6478	10425	79
TY	6773	10900	73
TZ	6083	9789	69
UA0	5089	8190	338
UA1	5199	8367	19
UA1 F J L	4128	6643	4
UA2	5186	8346	29
UA3	5476	8813	20
UA4	5830	9382	16
UA6	6294	10129	21
UA9	5891	9480	1
UB5	5800	9334	24
UC2	5386	8668	26
UD6	6606	10631	20
UF6	6338	10200	22
UG6	6519	10491	14
UH8	6847	11019	12
UI8	6892	11091	8
UJ8	6994	11255	2
UK0	5089	8190	338
UK1	5199	8367	19
UK2	5186	8346	29
UK3	5476	8813	20
UK4	5830	9382	16
UK5	5800	9334	24
UK6	6294	10129	21
UK7	6230	10026	3
UK8	6892	11091	8
UK9	5891	9480	1
UL7	6230	10026	3
UM8	6721	10816	0
UN1	4886	7863	17
UO5	5809	9348	30
UP2	5235	8425	27
UQ2	5126	8249	26
UR2	5062	8146	24
UT5	5800	9334	24
UW1	5199	8367	19
UW2	5186	8346	29
UW3	5476	8813	20
UW4	5830	9382	16
UW6	6294	10129	21
V3A	1870	3009	142
VE1	1545	2486	340
VE2	1718	2765	62
VE3	1467	2361	66
VE4	789	1270	23
VE5	711	1144	0
VE6	976	1571	343
VE7	1204	1938	313
VE8	1985	3194	331
VK0	10332	16627	183
VK0 Heard	11511	18525	176
VK1	8491	13665	249
VK2	8257	13288	250
VK2 L H I	7899	12712	248
VK3	8736	14059	249
VK4	8066	12981	259
VK5	9030	14532	254
VK6	10073	16210	268
VK7	8810	14178	241
VK8	8487	13658	282
VK9N	10425	16777	281
VK9X	6521	10494	260
VK9Y	10066	16199	319
VK9Z	7726	12433	266
VO1	2657	4276	61
VO2	2287	3680	51
VP2	3098	4986	111
VP5	2312	3721	112
VP8	6857	11035	154
VP9	2284	3676	90

VQ9 Chagos	10098	16251	5
VR1	5513	8872	259
VR3	4166	6704	250
VR6	4757	7655	204
VR7	6968	11214	273
VS5	8434	13573	311
VS6	7452	11993	321
VS9	8583	13813	25
VU	8651	13922	357
VU7 Andamn	8811	14180	337
VU7 Lacadv	8858	14255	3
W0	0	0	0
W1	1688	2716	79
W2	1626	2617	77
W3	1543	2483	82
W4	1142	1838	100
W5	856	1378	125
W6	884	1423	253
W7	530	853	275
W8	1160	1867	72
W9	959	1543	70
XE	1393	2242	166
XT	6310	10155	73
XU	8376	13479	325
XV	8099	13034	324
XW	8066	12981	328
XZ	8112	13055	339
Y	5104	8214	34
YA	7259	11682	5
YB	8980	14452	303
YI	7034	11320	26
YJ	6951	11186	255
YK	6794	10934	31
YN	2206	3550	145
YO	5869	9445	33
YS	2054	3306	147
YU	5715	9197	36
YV	3195	5142	119
Z2	9368	15076	74
ZA	5876	9456	38
ZB	5128	8252	56
ZC4	6621	10655	33
ZD7	7388	11890	97
ZD8	6622	10657	96
ZD9	7954	12800	120
ZF1	2016	3244	128
ZK1	5547	8927	237
ZK2	5847	9410	239
ZL	7581	12200	231
ZM2	5382	8661	246
ZP	5433	8743	137
ZS	9544	15359	98
ZS3	8856	14252	93
ZV	4690	7548	125
ZW	4690	7548	125
ZY	4690	7548	125
ZZ	4690	7548	125

1A0KM	6242	10045	32
3A	5878	9459	32
3B6	10545	16970	3
3B8	11203	18029	358
3B9	11058	17796	339
3C	8370	13470	59
3D2	5452	8774	237
3D6	10576	17020	74
3V8	6556	10551	37
3X	6877	11067	69
3Y	9705	15618	134
4N	6316	10164	26
4S7	8595	13832	331
4U1ITU	5818	9363	33
4U1UN	2579	4150	70
4W	8615	13864	16
4X	7399	11907	20
4Z	7399	11907	20
5A	7006	11275	37
5B4	7165	11531	20
5H	6118	9846	99
5L	7116	11452	70
5N	7857	12644	59
5R	11052	17786	30
5T	6368	10248	62
5U	7447	11984	54
5V	7739	12454	63
5W	4760	7660	231
5X	9408	15140	36
5Z	9542	15356	30
6O	9370	15079	18
6W	6647	10697	66
6Y	3057	4920	104
7J	5998	9653	301
7O	8832	14213	15
7P	10667	17166	81
7Q	8569	13790	27
7X	6385	10275	41
7Z	8088	13016	13
8P	4229	6806	96
8Q	9292	14954	337
8R	4594	7393	103
8Z	7759	12487	11
9G	7544	12141	64
9H	6638	10683	34
9J	9939	15995	55
9K	7808	12565	5
9L	6916	11130	71
9M2	8333	13410	304
9M6	7518	12099	294
9M8	7934	12768	295
9N	7584	12205	334
9Q	9340	15031	51
9U5	9467	15235	43
9V1	8427	13562	301
9X	9408	15140	42
9Y4	4203	6764	99
A2	10337	16635	75
A3	5746	9247	236
A4	8233	13249	359
A5	7487	12049	330
A6	8092	13022	3
A7	8009	12889	6
A9	8004	12881	7
AP	7681	12361	348
BV	6395	10291	315
BY	6434	10354	305
C2	5102	8211	256
C3	5856	9424	38
C5	6588	10602	69
C6	2794	4496	95
C9	10529	16944	39
CE	5978	9620	141
CE0A	4493	7231	167
CE0X	5179	8335	141
CE0Z	5586	8990	145
CE9	7851	12635	156
CE9	7761	12490	154
CN	5988	9636	47
CO	2721	4379	100
CP	5336	8587	124
CR3	6633	10674	70
CR9	6918	11133	309
CT1	5649	9091	45
CT2	4876	7847	55
CT3	5825	9374	52
CX	6581	10591	130
CZ6	1164	1873	22

Fig. B-4. Direction distance table for G3ZCZ/W6 located at 37.8 237.5.

D2	9177	14769	65
D4	5983	9628	74
D6	10421	16771	29
DA	5689	9155	29
DB	5689	9155	29
DF	5689	9155	29
DJ	5689	9155	29
DL	5689	9155	29
DM	5688	9154	26
DT	5677	9136	26
DU	6988	11246	298
EA	5800	9334	43
EA6	6101	9818	39
EA8	5815	9358	59
EA9	6091	9802	59
EI	5089	8190	34
EL	7116	11452	70
EP	7353	11833	6
ET	8613	13861	21
F	5701	9175	35
FB8W	11770	18941	157
FB8X	11429	18393	213
FB8Y	8797	14157	208
FB8Z	11345	18258	263
FC	6133	9870	34
FG7	3972	6392	95
FH8	10523	16935	27
FK8	6208	9991	243
FM7	4043	6506	96
FO8	4203	6764	210
FO8 Clprtn	3432	5523	161
FP8	3334	5365	57
FR7	11266	18130	6
FS7	3828	6160	94
FW8	5069	8158	237
FY7	4943	7955	98
G	5338	8590	34
GD	5093	8196	33
GI	5035	8103	33
GJ	5408	8703	35
GM	5050	8127	31
GU	5378	8655	35
GW	5204	8375	33
H4	5957	9587	259
HA	6134	9871	25
HB	5855	9422	31
HB0	5863	9435	30
HC	3898	6273	123
HC8	3330	5359	134
HH	3298	5307	100
HI	3432	5523	98
HK	3742	6022	115
HK0	3040	4892	114
HL	5606	9022	311
HP	3304	5317	116
HR	2676	4306	116
HR0	2717	4372	110
HS	7863	12654	314
HT	2844	4577	118
HV	6242	10045	32
HZ	8042	12942	11
I	6279	10105	31
IC	6372	10254	31
IH	6530	10509	35
IS	6245	10050	35
IT	6540	10525	33
J28	8901	14324	18
J3	4166	6704	99
JA	5197	8364	304
JD1	5674	9131	300
JE	4861	7823	306
JF	5238	8430	304
JG	5328	8574	305
JH	5593	9001	304
JI	5547	8927	303
JJ	5027	8090	305
JL	5197	8364	304
JR	5238	8430	304
JR6	6068	9765	302
JT1	6047	9731	326
JW	4211	6777	8
JX	4249	6838	20
JY	7417	11936	19
K0	951	1530	75
K1	2631	4234	71
K2	2565	4128	69
K3	2491	4009	73
K4	2080	3347	84
K5	1711	2754	95
K6	238	383	144

K7	431	694	67
K8	2089	3362	67
K9	1887	3037	68
KA	5197	8364	304
KB6	4186	6737	239
KC4	3156	5079	103
KC4A	8551	13761	193
KC4A	8827	14205	180
KC6	5869	9445	272
KG4	3079	4955	100
KG6	5802	9337	282
KH3	4315	6944	271
KH6	2359	3796	250
KH6 Kure	3454	5559	277
KL7	1948	3135	331
KM6	3142	5056	271
KP4	3593	5782	95
KP6	3418	5501	235
KS6	4772	7680	230
KV4	3696	5948	94
KW6	4385	7057	275
KX6	4774	7683	265
LA	5160	8304	22
LU	6425	10340	133
LX	5629	9059	31
LZ	6512	10480	23
M1	6125	9857	31
N0	951	1530	75
N1	2631	4234	71
N2	2565	4128	69
N3	2491	4009	73
N4	2080	3347	84
N5	1711	2754	95
N6	238	383	144
N7	431	694	67
N8	2089	3362	67
N9	1887	3037	68
OA	4526	7284	130
OD	7276	11709	19
OE	5931	9545	28
OH	5414	8713	16
OH0	5349	8608	18
OJ0	5349	8608	18
OK	5893	9484	26
ON	5498	8848	31
OR	5498	8848	31
OX	3689	5937	33
OY	4668	7512	27
OZ	5358	8623	25
P29	6564	10563	268
P5	5532	8903	314
PA	5500	8851	29
PJ2	3754	6041	103
PJ3	3676	5916	104
PJ4	3786	6093	103
PJ5	3832	6167	94
PJ6	3805	6123	94
PJ7	3812	6135	94
PP	6191	9963	109
PT	6191	9963	109
PY	6191	9963	109
PY0 St Pl	6386	10277	87
PZ	4728	7609	101
RA0	4808	7738	332
RA1	5568	8961	11
RA2	5700	9173	20
RA3	5853	9419	11
RA4	6135	9873	6
RA6	6661	10720	9
RA9	5958	9588	351
RB5	6229	10024	14
RC2	5853	9419	17
RD6	6957	11196	7
RF6	6720	10814	10
RG6	6781	10913	2
RH8	7057	11357	358
RI8	7038	11326	354
RJ8	7049	11344	349
RL7	6321	10172	352
RM7	6745	10855	347
RN1	5229	8415	10
RO5	6317	10166	19
RP2	5724	9212	19
RQ2	5600	9012	18
RR2	5507	8862	16
S2	7733	12445	328
S7	10165	16359	3
S8	10856	17471	86
S9	8300	13357	62
SK	5414	8713	21

SL	5414	8713	21
SM	5414	8713	21
SP	5832	9385	22
ST	8410	13534	29
SU	7424	11947	25
SV	6731	10832	27
T3	4701	7565	255
T5	9370	15079	18
TA	6784	10917	21
TF	4200	6759	30
TG	2531	4073	120
TI	3110	5005	118
TI9	3202	5153	126
TJ	8190	13180	57
TL	8481	13648	46
TN	8767	14109	54
TR	8445	13591	61
TT	7977	12837	43
TU	7417	11936	68
TY	7689	12374	61
TZ	6985	11241	58
UA0	4808	7738	332
UA1	5568	8961	11
UA1 F J L	4296	6914	0
UA2	5700	9173	20
UA3	5853	9419	11
UA4	6135	9873	6
UA6	6661	10720	9
UA9	5958	9588	351
UB5	6229	10024	14
UC2	5853	9419	17
UD6	6957	11196	7
UF6	6720	10814	10
UG6	6781	10913	2
UH8	7057	11357	358
UI8	7038	11326	354
UJ8	7049	11344	349
UK0	4808	7738	332
UK1	5568	8961	11
UK2	5700	9173	20
UK3	5853	9419	11
UK4	6135	9873	6
UK5	6229	10024	14
UK6	6661	10720	9
UK7	6321	10172	352
UK8	7038	11326	354
UK9	5958	9588	351
UL7	6321	10172	352
UM8	6745	10855	347
UN1	5229	8415	10
UO5	6317	10166	19
UP2	5724	9212	19
UQ2	5600	9012	18
UR2	5507	8862	16
UT5	6229	10024	14
UW1	5568	8961	11
UW2	5700	9173	20
UW3	5853	9419	11
UW4	6135	9873	6
UW6	6661	10720	9
V3A	2534	4078	115
VE1	1537	2473	3
VE2	2605	4192	59
VE3	2377	3825	63
VE4	1471	2367	47
VE5	1206	1941	40
VE6	1164	1873	22
VE7	852	1371	352
VE8	1755	2824	346
VK0	10060	16190	193
VK0 Heard	11138	17924	211
VK1	7586	12208	240
VK2	7347	11824	241
VK2 L H I	7001	11267	238
VK3	7829	12599	240
VK4	7126	11468	248
VK5	8101	13037	245
VK6	9123	14682	257
VK7	7945	12786	234
VK8	7564	12173	270
VK9N	9494	15279	266
VK9X	5579	8978	249
VK9Y	9396	15121	296
VK9Z	6776	10905	255
VO1	3529	5679	56
VO2	3097	4984	49
VP2	3985	6413	96
VP5	3199	5148	95
VP8	7227	11630	145
VP9	3233	5203	78

VQ9 Chagos	10055	16182	334
VR1	4575	7363	247
VR3	3268	5259	236
VR6	4372	7036	188
VR7	6021	9690	261
VS5	7714	12414	295
VS6	6869	11054	308
VS9	8951	14405	4
VU	8562	13779	337
VU7 Andamn	8414	13541	318
VU7 Lacadv	8859	14257	341
W0	951	1530	75
W1	2631	4234	71
W2	2565	4128	69
W3	2491	4009	73
W4	2080	3347	84
W5	1711	2754	95
W6	238	383	144
W7	431	694	67
W8	2089	3362	67
W9	1887	3037	68
XE	1824	2935	126
XT	7229	11634	62
XU	7822	12588	308
XV	7534	12124	308
XW	7553	12155	312
XZ	7757	12483	322
Y	5688	9154	26
YA	7351	11830	350

YB	8181	13166	287
YI	7466	12015	12
YJ	6022	9691	245
YK	7302	11751	18
YN	2820	4538	120
YO	6418	10328	22
YS	2652	4268	120
YU	6316	10164	26
YV	4031	6487	103
Z2	10283	16548	58
ZA	6498	10457	27
ZB	5942	9562	46
ZC4	7165	11531	20
ZD7	8325	13397	87
ZD8	7562	12170	85
ZD9	8740	14065	113
ZF1	2800	4506	106
ZK1	4739	7626	224
ZK2	5016	8072	227
ZL	6805	10951	222
ZM2	4504	7248	233
ZP	6066	9762	124
ZS	10474	16856	91
ZS3	9802	15774	83
ZV	5454	8777	112
ZW	5454	8777	112
ZY	5454	8777	112
ZZ	5454	8777	112

1A0KM	881	1418	134
3A	516	830	137
3B6	5415	8714	124
3B8	6043	9725	127
3B9	6234	10032	121
3C	3501	5634	167
3D2	10116	16280	4
3D6	5656	9102	155
3V8	1240	1996	153
3X	2948	4744	198
3Y	7359	11843	178
4N	1052	1693	112
4S7	5145	8280	89
4U1ITU	457	735	141
4U1UN	3465	5576	288
4W	3430	5520	119
4X	2210	3557	114
4Z	2210	3557	114
5A	1668	2684	147
5B4	1980	3186	112
5H	4596	7396	223
5L	3128	5034	194
5N	3086	4966	173
5R	5690	9157	136
5T	2290	3685	203
5U	2535	4080	172
5V	3144	5060	178
5W	9797	15766	346
5X	4052	6521	140
5Z	4182	6730	135
6O	4108	6611	125
6W	2610	4200	199
6Y	4702	7567	272
7J	6277	10102	44
7O	3639	5856	120
7P	5878	9459	156
7Q	3229	5196	130
7X	1165	1875	166
7Z	3026	4870	112
8P	4204	6765	253
8Q	5254	8455	100
8R	4628	7448	248
8Z	2807	4517	106
9G	3075	4949	182
9H	1280	2060	142
9J	4758	7657	151
9K	3080	4957	100
9L	3097	4984	199
9M2	6547	10536	77
9M6	7043	11334	64
9M8	7042	11333	70
9N	4540	7306	76
9Q	4149	6677	152
9U5	4152	6682	145
9V1	6736	10840	78
9X	4087	6577	145
9Y4	4396	7074	254
A2	5495	8843	157
A3	10322	16611	10
A4	3616	5819	100
A5	4772	7680	73
A6	3386	5449	101
A7	3188	5130	103
A9	3157	5081	104
AP	3781	6085	85
BV	5414	8713	52
BY	6068	9765	51
C2	8849	14241	15
C3	611	983	174
C5	2791	4492	203
C6	4342	6988	277
C9	5179	8335	140
CE	7392	11896	235
CE0A	8412	13537	267
CE0X	7226	11629	246
CE0Z	7596	12224	242
CE9	8657	13932	207
CE9	8553	13764	208
CN	1174	1889	196
CO	4598	7400	277
CP	6161	9915	240
CR3	2858	4599	203
CR9	5953	9580	58
CT1	885	1424	209
CT2	1584	2549	248
CT3	1433	2306	210
CX	6784	10917	223
CZ6	4219	6790	320

Fig. B-5. Direction distance table for G3ZCZ located at 51.5 .17.

HB	505	813	125
HB0	519	835	122
HC	5765	9278	260
HC8	6224	10016	270
HH	4509	7256	268
HI	4402	7084	266
HK	5310	8545	261
HK0	5208	8381	272
HL	5499	8850	40
HP	5309	8544	268
HR	5289	8512	277
HR0	5034	8101	277
HS	5820	9366	73
HT	5379	8656	275
HV	881	1418	134
HZ	3054	4915	109
I	924	1487	130
IC	1012	1629	133
IH	1182	1902	146
IS	900	1448	149
IT	1179	1897	140
J28	3653	5879	123
J3	4381	7050	254
JA	5873	9451	33
JD1	6296	10132	39
JE	5708	9186	29
JF	5935	9551	33
JG	5887	9474	35
JH	5995	9648	39
JI	6022	9691	38
JJ	5792	9321	31
JL	5873	9451	33
JR	5935	9551	33
JR6	6187	9957	45
JT1	4580	7371	50
JW	1948	3135	7
JX	1377	2216	351
JY	2250	3621	113
K0	4681	7533	307
K1	3462	5571	287
K2	3460	5568	289
K3	3629	5840	288
K4	4182	6730	290
K5	4650	7483	293
K6	5449	8769	314

D2	4339	6983	164
D4	2880	4635	218
D6	5062	8146	135
DA	399	642	102
DB	399	642	102
DF	399	642	102
DJ	399	642	102
DL	399	642	102
DM	570	917	81
DT	549	884	81
DU	6693	10771	57
EA	819	1318	196
EA6	874	1407	170
EA8	1851	2979	215
EA9	1977	3182	208
EI	300	483	293
EL	3128	5034	194
EP	2710	4361	94
ET	3341	5377	124
F	389	626	167
FB8W	7434	11964	145
FB8X	8074	12993	137
FB8Y	10683	17192	146
FB8Z	7734	12446	124
FC	777	1250	144
FG7	4146	6672	258
FH8	5169	8318	135
FK8	10173	16371	27
FM7	4218	6788	256
FO8	9559	15383	314
FO8 Clprtn	7535	12126	278
FP8	2501	4025	286
FR7	6018	9685	130
FS7	4124	6637	260
FW8	9841	15837	357
FY7	4376	7042	242
G	80	129	245
GD	270	435	316
GI	330	531	311
GJ	193	311	214
GM	337	542	339
GU	184	296	222
GW	171	275	295
H4	9022	14519	31
HA	943	1518	101

K7	5010	8063	313
K8	3729	6001	296
K9	3882	6247	298
KA	5873	9451	33
KB6	9044	14555	348
KC4	4640	7467	270
KC4A	10498	16894	174
KC4A	9774	15729	180
KC6	8144	13106	35
KG4	4518	7271	272
KG6	7477	12033	36
KH3	7658	12324	11
KH6	7238	11648	338
KH6 Kure	6930	11152	2
KL7	4554	7329	344
KM6	7072	11381	356
KP4	4212	6778	264
KP6	8509	13694	337
KS6	9854	15858	344
KV4	4131	6648	262
KW6	7474	12028	14
KX6	8181	13166	15
LA	699	1125	29
LU	6994	11255	226
LX	287	462	115
LZ	1336	2150	106
M1	772	1242	128
N0	4681	7533	307
N1	3462	5571	287
N2	3460	5568	289
N3	3629	5840	288
N4	4182	6730	290
N5	4650	7483	293
N6	5449	8769	314
N7	5010	8063	313
N8	3729	6001	296
N9	3882	6247	298
OA	6325	10179	253
OD	2130	3428	111
OE	662	1065	106
OH	1088	1751	48
OH0	963	1550	45
OJ0	963	1550	45
OK	708	1139	95
ON	169	272	100
OR	169	272	100
OX	1673	2692	316
OY	773	1244	342
OZ	475	764	46
P29	8642	13908	45
P5	5323	8566	40
PA	252	406	80
PJ2	4660	7499	261
PJ3	4714	7586	262
PJ4	4635	7459	260
PJ5	4129	6645	260
PJ6	4150	6679	260
PJ7	4103	6603	260
PP	5333	8582	225
PT	5333	8582	225
PY	5333	8582	225
PY0 St Pl	3910	6292	216
PZ	4530	7290	245
RA0	4039	6500	31
RA1	1488	2395	53
RA2	887	1427	68
RA3	1555	2502	64
RA4	1987	3198	68
RA6	2076	3341	85
RA9	2881	4636	55
RB5	1554	2501	80
RC2	1169	1881	71
RD6	2350	3782	89
RF6	2067	3326	87
RG6	2536	4081	80
RH8	2904	4673	82
RI8	3115	5013	78
RJ8	3451	5554	75
RL7	2934	4722	63
RM7	3401	5473	69
RN1	1540	2478	40
RO5	1327	2136	92
RP2	1009	1624	67
RQ2	1027	1653	59
RR2	1131	1820	53
S2	4980	8014	75
S7	5093	8196	121
S8	6111	9834	155
S9	3518	5662	170
SK	804	1294	50

SL	804	1294	50
SM	804	1294	50
SP	846	1361	80
ST	3058	4921	132
SU	2127	3423	123
SV	1410	2269	123
T3	8720	14033	8
T5	4108	6611	125
TA	1626	2617	107
TF	1177	1894	326
TG	5416	8716	280
TI	5384	8664	271
TI9	5803	9339	271
TJ	3261	5248	167
TL	3242	5217	153
TN	3690	5938	159
TR	3603	5798	168
TT	2707	4356	152
TU	3157	5081	187
TY	3008	4841	177
TZ	2389	3845	185
UA0	4039	6500	31
UA1	1488	2395	53
UA1 F J L	2320	3734	15
UA2	887	1427	68
UA3	1555	2502	64
UA4	1987	3198	68
UA6	2076	3341	85
UA9	2881	4636	55
UB5	1554	2501	80
UC2	1169	1881	71
UD6	2350	3782	89
UF6	2067	3326	87
UG6	2536	4081	80
UH8	2904	4673	82
UI8	3115	5013	78
UJ8	3451	5554	75
UK0	4039	6500	31
UK1	1488	2395	53
UK2	887	1427	68
UK3	1555	2502	64
UK4	1987	3198	68
UK5	1554	2501	80
UK6	2076	3341	85
UK7	2934	4722	63
UK8	3115	5013	78
UK9	2881	4636	55
UL7	2934	4722	63
UM8	3401	5473	69
UN1	1540	2478	40
UO5	1327	2136	92
UP2	1009	1624	67
UQ2	1027	1653	59
UR2	1131	1820	53
UT5	1554	2501	80
UW1	1488	2395	53
UW2	887	1427	68
UW3	1555	2502	64
UW4	1987	3198	68
UW6	2076	3341	85
V3A	5225	8409	279
VE1	4046	6511	330
VE2	3135	5045	294
VE3	3410	5488	295
VE4	3946	6350	311
VE5	4168	6708	314
VE6	4219	6790	320
VE7	4721	7598	326
VE8	4218	6788	338
VK0	9273	14923	152
VK0 Heard	8364	13460	138
VK1	10543	16967	66
VK2	10529	16944	59
VK2 L H I	10697	17215	48
VK3	10453	16822	73
VK4	10011	16111	52
VK5	10069	16204	78
VK6	9079	14611	93
VK7	10788	17361	82
VK8	8611	13858	62
VK9N	8503	13684	96
VK9X	9519	15319	18
VK9Y	7129	11473	92
VK9Z	9503	15293	46
VO1	2322	3737	283
VO2	2471	3977	297
VP2	4220	6791	257
VP5	4259	6854	270
VP8	7931	12763	217
VP9	3464	5575	274

YB	7542	12137	73
YI	2528	4068	103
YJ	9993	16082	24
YK	2190	3524	109
YN	5449	8769	275
YO	1280	2060	102
YS	5454	8777	278
YU	1052	1693	112
YV	4617	7430	256
Z2	5127	8251	151
ZA	1186	1909	120
ZB	1097	1765	196
ZC4	1980	3186	112
ZD7	4677	7527	186
ZD8	4195	6751	196
ZD9	6226	10020	190
ZF1	4826	7766	276
ZK1	9981	16062	335
ZK2	10159	16349	343
ZL	11715	18853	26
ZM2	9513	15309	346
ZP	6315	10163	230
ZS	5999	9654	161
ZS3	5360	8626	166
ZV	5362	8629	236
ZW	5362	8629	236
ZY	5362	8629	236
ZZ	5362	8629	236
VQ9 Chagos	5775	9294	108
VR1	9014	14506	0
VR3	8358	13451	337
VR6	9261	14904	286
VR7	8911	14340	33
VS5	6982	11236	67
VS6	6015	9680	57
VS9	4015	6461	112
VU	4867	7832	90
VU7 Andamn	5798	9331	82
VU7 Lacadv	4812	7744	96
W0	4681	7533	307
W1	3462	5571	287
W2	3460	5568	289
W3	3629	5840	288
W4	4182	6730	290
W5	4650	7483	293
W6	5449	8769	314
W7	5010	8063	313
W8	3729	6001	296
W9	3882	6247	298
XE	5563	8953	290
XT	2734	4400	185
XU	6189	9960	71
XV	6144	9888	67
XW	5914	9517	68
XZ	5345	8602	74
Y	570	917	81
YA	3504	5639	81

PX	MILES	KM	BEARING
1A0KM	1429	2300	306
3A	1763	2837	311
3B6	3229	5196	151
3B8	3883	6249	154
3B9	4026	6479	146
3C	2660	4281	223
3D2	9931	15982	77
3D6	4051	6519	187
3V8	1466	2359	286
3X	3363	5412	254
3Y	6275	10098	197
4N	1178	1896	320
4S7	3199	5148	103
4U1ITU	1833	2950	311
4U1UN	5685	9149	314
4W	1223	1968	151
4X	0	0	0
4Z	0	0	0
5A	1202	1934	269
5B4	255	410	331
5H	5520	8883	256
5L	3371	5425	249
5N	2591	4170	235
5R	3662	5893	166
5T	3078	4953	267
5U	2226	3582	246
5V	2810	4522	238
5W	10372	16692	59
5X	2199	3539	183
5Z	2204	3547	175
6O	1957	3149	159
6W	3180	5118	260
6Y	6783	10916	297
7J	5552	8935	65
7O	1440	2317	152
7P	4295	6912	187
7Q	1230	1979	181
7X	1749	2815	286
7Z	804	1294	130
8P	5987	9635	283
8Q	3116	5015	118
8R	6251	10060	276
8Z	671	1080	104

PX	MILES	KM	BEARING
9G	2916	4693	242
9H	1235	1987	289
9J	3131	5039	190
9K	1047	1685	96
9L	3497	5628	253
9M2	4771	7678	99
9M6	5576	8973	89
9M8	5410	8706	94
9N	2978	4792	82
9Q	2647	4260	198
9U5	2428	3907	189
9V1	4931	7935	100
9X	2359	3796	189
9Y4	6175	9937	283
A2	3981	6407	191
A3	9859	15866	84
A4	1520	2446	108
A5	3280	5279	81
A6	1289	2074	106
A7	1062	1709	108
A9	1021	1643	109
AP	2058	3312	84
BV	4504	7248	66
BY	5125	8248	70
C2	8706	14011	64
C3	1984	3193	302
C5	3423	5509	260
C6	6487	10440	303
C9	3248	5227	174
CE	8326	13399	245
CE0A	10241	16481	271
CE0X	8580	13808	258
CE0Z	8744	14072	250
CE9	8335	13414	209
CE9	8309	13372	212
CN	2341	3767	287
CO	6743	10852	302
CP	7431	11959	260
CR3	3459	5567	258
CR9	4770	7676	76
CT1	2457	3954	296
CT2	3496	5626	297
CT3	2734	4400	284

Fig. B-6. Direction distance table for G3ZCZ/4X located at 31.8 35.2.

PX	MILES	KM	BEARING
CX	7410	11925	240
CZ6	6252	10061	341
D2	3181	5119	208
D4	3938	6337	268
D6	3041	4894	168
DA	1840	2961	321
DB	1840	2961	321
DF	1840	2961	321
DJ	1840	2961	321
DL	1840	2961	321
DM	1772	2852	328
DT	1788	2877	328
DU	5469	8801	80
EA	2246	3614	296
EA6	1869	3008	294
EA8	3107	5000	279
EA9	3001	4830	274
EI	2529	4070	318
EL	3371	5425	249
EP	943	1518	70
ET	1199	1930	164
F	2016	3244	309
FB8W	5509	8866	168
FB8X	5998	9653	158
FB8Y	8616	13866	154
FB8Z	5537	8911	147
FC	1601	2576	304
FG7	6019	9686	287
FH8	3128	5034	166
FK8	9267	14913	89
FM7	6063	9757	286
FO8	11402	18349	18
FO8 Clprtn	9643	15518	294
FP8	4716	7589	313
FR7	3887	6255	157
FS7	6043	9725	289
FW8	10004	16099	67
FY7	5871	9448	273
G	2282	3672	316
GD	2481	3993	320
GI	2544	4094	320
GJ	2268	3650	313
GM	2478	3988	324
GU	2293	3690	313

PX	MILES	KM	BEARING
GW	2399	3861	318
H4	8291	13343	78
HA	1322	2127	327
HB	1735	2792	315
HB0	1715	2760	316
HC	7606	12240	282
HC8	8243	13265	290
HH	6545	10533	295
HI	6411	10317	293
HK	7202	11590	285
HK0	7279	11714	296
HL	5012	8066	56
HP	7319	11778	292
HR	7427	11952	301
HR0	7168	11535	301
HS	4208	6772	90
HT	7488	12050	298
HV	1429	2300	306
HZ	856	1378	120
I	1361	2190	307
IC	1312	2111	303
IH	1363	2193	290
IS	1564	2517	299
IT	1264	2034	294
J28	1488	2395	159
J3	6174	9936	283
JA	5591	8998	53
JD1	5742	9241	61
JE	5618	9041	48
JF	5629	9059	54
JG	5533	8904	55
JH	5494	8841	59
JI	5545	8924	58
JJ	5607	9023	50
JL	5591	8998	53
JR	5629	9059	54
JR6	5428	8735	66
JT1	3895	6268	57
JW	3299	5309	356
JX	3167	5097	342
JY	49	79	74
K0	6858	11037	330
K1	5679	9139	313
K2	5681	9142	314

PX	MILES	KM	BEARING
OH	2015	3243	349
OH0	2069	3330	345
OJ0	2069	3330	345
OK	1574	2533	327
ON	2064	3322	319
OR	2064	3322	319
OX	3822	6151	329
OY	2795	4498	331
OZ	2094	3370	331
P29	7559	12165	84
P5	4887	7865	54
PA	2024	3257	322
PJ2	6568	10570	288
PJ3	6642	10689	288
PJ4	6534	10515	287
PJ5	6047	9731	289
PJ6	6075	9776	289
PJ7	6028	9701	290
PP	6203	9982	252
PT	6203	9982	252
PY	6203	9982	252
PY0 St Pl	4697	7559	256
PZ	6105	9825	275
RA0	4168	6708	35
RA1	1948	3135	1
RA2	1716	2762	340
RA3	1677	2699	4
RA4	1598	2572	20
RA6	1053	1695	27
RA9	2433	3915	38
RB5	1258	2024	2
RC2	1574	2533	349
RD6	952	1532	45
RF6	968	1558	27
RG6	1363	2193	46
RH8	1492	2401	61
RI8	1751	2818	63
RJ8	2115	3404	66
RL7	2179	3507	45
RM7	2292	3689	59
RN1	2293	3690	360
RO5	1100	1770	344
RP2	1693	2725	344
RQ2	1821	2931	346

PX	MILES	KM	BEARING
K3	5849	9413	313
K4	6406	10309	315
K5	6878	11069	317
K6	7546	12144	339
K7	7120	11458	337
K8	5956	9585	320
K9	6105	9825	322
KA	5591	8998	53
KB6	9815	15795	46
KC4	6702	10786	296
KC4A	8868	14271	169
KC4A	8413	13539	180
KC6	7461	12007	72
KG4	6601	10623	298
KG6	6833	10996	68
KH3	7904	12720	49
KH6	8698	13998	14
KH6 Kure	7624	12269	35
KL7	6085	9793	3
KM6	7972	12829	30
KP4	6192	9965	292
KP6	9805	15779	27
KS6	10467	16845	60
KV4	6088	9797	291
KW6	7657	12322	49
KX6	8185	13172	57
LA	2263	3642	337
LU	7697	12387	241
LX	1941	3124	318
LZ	922	1484	330
M1	1490	2398	311
N0	6858	11037	330
N1	5679	9139	313
N2	5681	9142	314
N3	5849	9413	313
N4	6406	10309	315
N5	6878	11069	317
N6	7546	12144	339
N7	7120	11458	337
N8	5956	9585	320
N9	6105	9825	322
OA	7948	12791	271
OD	153	246	6
OE	1575	2535	321

PX	MILES	KM	BEARING
UK1	1948	3135	1
UK2	1716	2762	340
UK3	1677	2699	4
UK4	1598	2572	20
UK5	1258	2024	2
UK6	1053	1695	27
UK7	2179	3507	45
UK8	1751	2818	63
UK9	2433	3915	38
UL7	2179	3507	45
UM8	2292	3689	59
UN1	2293	3690	360
UO5	1100	1770	344
UP2	1693	2725	344
UQ2	1821	2931	346
UR2	1927	3101	350
UT5	1258	2024	2
UW1	1948	3135	1
UW2	1716	2762	340
UW3	1677	2699	4
UW4	1598	2572	20
UW6	1053	1695	27
V3A	7385	11885	303
VE1	5937	9554	348
VE2	5363	8631	318
VE3	5638	9073	319
VE4	6099	9815	332
VE5	6280	10106	336
VE6	6252	10061	341
VE7	6654	10708	347
VE8	5931	9545	355
VK0	7376	11870	165
VK0 Heard	6290	10122	158
VK1	8691	13986	113
VK2	8805	14170	109
VK2 L H I	9149	14723	106
VK3	8507	13690	116
VK4	8501	13681	101
VK5	8093	13024	116
VK6	6954	11191	122
VK7	8715	14025	123
VK8	7046	11339	97
VK9N	6358	10232	123
VK9X	9089	14627	75

PX	MILES	KM	BEARING
RR2	1927	3101	350
S2	3380	5439	86
S7	2887	4646	148
S8	4489	7224	185
S9	2769	4456	225
SK	2002	3222	340
SL	2002	3222	340
SM	2002	3222	340
SP	1592	2562	336
ST	1139	1833	189
SU	332	534	249
SV	860	1384	303
T3	8867	14270	57
T5	1957	3149	159
TA	636	1024	334
TF	3282	5282	330
TG	7576	12192	303
TI	7439	11972	294
TI9	7841	12619	292
TJ	2466	3969	227
TL	1943	3127	213
TN	2498	4020	212
TR	2745	4418	222
TT	1588	2556	227
TU	3144	5060	244
TY	2666	4290	239
TZ	2594	4175	256
UA0	4168	6708	35
UA1	1948	3135	1
UA1 F J L	3375	5431	4
UA2	1716	2762	340
UA3	1677	2699	4
UA4	1598	2572	20
UA6	1053	1695	27
UA9	2433	3915	38
UB5	1258	2024	2
UC2	1574	2533	349
UD6	952	1532	45
UF6	968	1558	27
UG6	1363	2193	46
UH8	1492	2401	61
UI8	1751	2818	63
UJ8	2115	3404	66
UK0	4168	6708	35

PX	MILES	KM	BEARING
VK9Y	5058	8140	116
VK9Z	8250	13277	92
VO1	4529	7289	312
VO2	4697	7559	320
VP2	6084	9791	286
VP5	6331	10188	297
VP8	8110	13051	224
VP9	5593	9001	303
VQ9 Chagos	3561	5731	130
VR1	9355	15055	55
VR3	9701	15612	24
VR6	11423	18383	294
VR7	8138	13096	78
VS5	5436	8748	91
VS6	4856	7815	76
VS9	1789	2879	134
VU	2898	4664	104
VU7 Andamn	3971	6391	99
VU7 Lacadv	2733	4398	111
W0	6858	11037	330
W1	5679	9139	313
W2	5681	9142	314
W3	5849	9413	313
W4	6406	10309	315
W5	6878	11069	317
W6	7546	12144	339
W7	7120	11458	337
W8	5956	9585	320
W9	6105	9825	322
XE	7787	12532	314
XT	2782	4477	249
XU	4602	7406	90
XV	4678	7528	86

PX	LATITUDE	LONGITUDE (E)	(W)
XW	4437	7140	85
XZ	3758	6048	87
Y	1772	2852	328
YA	1952	3141	74
YB	5804	9340	100
YI	520	837	78
YJ	9236	14863	85
YK	184	296	34
YN	7562	12170	299
YO	1008	1622	333
YS	7596	12224	301
YU	1178	1896	320
YV	6441	10366	284
Z2	3457	5563	186
ZA	1055	1698	311
ZB	2323	3738	289
ZC4	255	410	331
ZD7	4283	6893	226
ZD8	4250	6840	239
ZD9	5698	9170	216
ZF1	6951	11186	300
ZK1	10854	17467	58
ZK2	10655	17147	68
ZL	10076	16215	119
ZM2	10228	16460	52
ZP	7226	11629	249
ZS	4546	7316	191
ZS3	4143	6667	201
ZV	6591	10607	262
ZW	6591	10607	262
ZY	6591	10607	262
ZZ	6591	10607	262

1A0KM	9930	15980	292	9H	9742	15678	283
3A	10204	16421	298	9J	7060	11362	243
3B6	5719	9204	262	9K	7536	12128	292
3B8	5223	8405	256	9L	9958	16025	219
3B9	4889	7868	260	9M2	3945	6349	307
3C	8639	13903	240	9M6	3488	5613	324
3D2	2414	3885	65	9M8	3454	5559	316
3D6	6383	10272	233	9N	5949	9574	310
3V8	9948	16009	280	9Q	7640	12295	246
3X	10040	16157	223	9U5	7428	11954	252
3Y	5621	9046	201	9V1	3757	6046	306
4N	9583	15422	297	9X	7474	12028	252
4S7	5397	8685	297	9Y4	9981	16062	132
4U1ITU	10269	16526	299	A2	6609	10636	232
4U1UN	10365	16680	72	A3	2122	3415	66
4W	7480	12038	276	A4	7019	11296	290
4X	8540	13743	287	A5	5718	9202	313
4Z	8540	13743	287	A6	7253	11672	290
5A	9551	15370	275	A7	7472	12025	289
5B4	8734	14056	289	A9	7511	12087	289
5H	9397	15123	174	AP	6730	10831	302
5L	9792	15758	223	BV	5218	8397	333
5N	9151	14727	240	BY	4598	7400	337
5R	5795	9326	249	C2	2977	4791	36
5T	10630	17107	233	C3	10495	16890	292
5U	9533	15341	248	C5	10324	16614	221
5V	9262	14905	235	C6	9829	15818	97
5W	3149	5068	71	C9	6370	10251	249
5X	7379	11875	256	CE	6874	11062	150
5Z	7139	11489	259	CE0A	5928	9540	120
6D	6973	11222	268	CE0X	7131	11476	139
6W	10323	16613	228	CE0Z	6735	10839	143
6Y	9560	15385	107	CE9	5205	8376	169
7J	4513	7263	345	CE9	5335	8586	169
7O	7305	11756	274	CN	10788	17361	272
7P	6233	10031	231	CO	9577	15412	99
7Q	7891	12699	269	CP	8145	13108	147
7X	10218	16444	277	CR3	10265	16519	220
7Z	7736	12450	284	CR9	4619	7433	328
8P	10172	16370	133	CT1	10979	17669	284
8Q	5408	8703	286	CT2	12019	19342	284
8R	9727	15654	141	CT3	11094	17854	262
8Z	7868	12662	289	CX	7310	11764	163
9G	9447	15203	233	CZ6	8746	14075	47

Fig. B-7. Direction distance table for G3ZCZ/VK3 located at −37.8 145.

D2	7818	12582	235
D4	10744	17290	204
D6	6344	10209	254
DA	10126	16296	306
DB	10126	16296	306
DF	10126	16296	306
DJ	10126	16296	306
DL	10126	16296	306
DM	9922	15967	309
DT	9943	16001	309
DU	3902	6279	332
EA	10768	17329	285
EA6	10389	16719	284
EA8	11192	18011	242
EA9	10918	17570	240
EI	10716	17245	317
EL	9792	15758	223
EP	7848	12630	298
ET	7658	12324	273
F	10468	16846	299
FB8W	4518	7271	229
FB8X	3697	5950	231
FB8Y	2095	3371	184
FB8Z	3592	5781	247
FC	10109	16268	291
FG7	10227	16458	126
FH8	6222	10013	254
FK8	1665	2679	51
FM7	10158	16347	128
FO8	4159	6693	90
FO8 Clprtn	6697	10777	108
FP8	11180	17992	53
FR7	5314	8552	253
FS7	10240	16479	122
FW8	2796	4500	64
FY7	9927	15976	151
G	10568	17007	310
GD	10606	17068	318
GI	10651	17141	320
GJ	10630	17107	306
GM	10498	16894	321
GU	10642	17126	307
GW	10615	17083	314
H4	2322	3737	22
HA	9617	15477	301
HB	10133	16307	300
HB0	10105	16262	301
HC	8603	13845	124
HC8	8078	13000	113
HH	9793	15760	111
HI	9919	15963	113
HK	9050	14564	122
HK0	9063	14585	108
HL	5326	8571	345
HP	9005	14492	114
HR	8906	14332	102
HR0	9165	14749	102
HS	4669	7514	312
HT	8854	14249	105
HV	9930	15980	292
HZ	7668	12340	286
I	9856	15861	292
IC	9825	15811	290
IH	9873	15889	283
IS	10087	16233	288
IT	9785	15747	285
J2B	7353	11833	272
J3	9996	16087	131
JA	5117	8235	354
JD1	4591	7388	350
JE	5377	8653	358
JF	5048	8124	354
JG	5061	8145	352
JH	4883	7858	349
JI	4873	7842	350
JJ	5245	8441	356
JL	5117	8235	354
JR	5048	8124	354
JR6	4576	7364	343
JT1	6069	9767	333
JW	9243	14875	348
JX	9958	16025	346
JY	8488	13660	287
K0	8768	14110	64
K1	10400	16737	74
K2	10357	16668	72
K3	10231	16465	75
K4	9680	15578	78
K5	9175	14765	79
K6	7880	12681	63

K7	8284	13331	61
K8	9917	15959	66
K9	9718	15639	65
KA	5117	8235	354
KB6	3671	5908	60
KC4	9640	15514	108
KC4A	2900	4667	174
KC4A	3606	5803	180
KC6	2974	4786	7
KG4	9741	15676	106
KG6	3532	5684	360
KH3	4106	6608	27
KH6	5535	8907	53
KH6 Kure	5031	8096	30
KL7	7690	12376	29
KM6	5112	8227	37
KP4	10127	16297	116
KP6	4450	7161	64
KS6	3157	5081	73
KV4	10219	16445	118
KW6	4186	6737	23
KX6	3531	5682	28
LA	9954	16019	323
LU	7156	11516	159
LX	10262	16515	305
LZ	9275	14926	297
M1	9955	16021	295
N0	8768	14110	64
N1	10400	16737	74
N2	10357	16668	72
N3	10231	16465	75
N4	9680	15578	78
N5	9175	14765	79
N6	7880	12681	63
N7	8284	13331	61
N8	9917	15959	66
N9	9718	15639	65
OA	8050	12955	133
OD	8575	13800	289
OE	9902	15935	302
OH	9480	15256	321
OH0	9617	15477	321
OJ0	9617	15477	321
OK	9815	15795	305
ON	10335	16632	308

OR	10335	16632	308
OX	10611	17076	5
OY	10382	16708	334
OZ	10063	16194	316
P29	2266	3647	0
P5	5507	8862	345
PA	10238	16476	310
PJ2	9702	15613	122
PJ3	9641	15515	120
PJ4	9730	15658	122
PJ5	10235	16471	122
PJ6	10212	16434	121
PJ7	10259	16510	121
PP	8729	14048	168
PT	8729	14048	168
PY	8729	14048	168
PY0 St Pl	9828	15816	190
PZ	9810	15787	145
RA0	6904	11111	348
RA1	9066	14590	321
RA2	9606	15459	313
RA3	8951	14405	316
RA4	8509	13694	314
RA6	8429	13565	305
RA9	7682	12363	324
RB5	8939	14386	308
RC2	9322	15002	312
RD6	8174	13154	302
RF6	8447	13594	303
RG6	7957	12805	307
RH8	7595	12223	306
RI8	7376	11870	308
RJ8	7038	11326	311
RL7	7581	12200	319
RM7	7096	11420	315
RN1	9121	14678	328
RO5	9200	14806	303
RP2	9487	15267	313
RQ2	9488	15269	316
RR2	9413	15148	319
S2	5509	8866	310
S7	5951	9577	267
S8	6004	9662	230
S9	8707	14012	238
SK	9742	15678	318

SL	9742	15678	318
SM	9742	15678	318
SP	9645	15522	309
ST	8087	13014	269
SU	8742	14069	282
SV	9377	15090	289
T3	3315	5335	41
T5	6973	11222	268
TA	9016	14509	294
TF	10545	16970	348
TG	8745	14073	100
TI	8899	14321	110
TI9	8490	13663	112
TJ	8813	14183	243
TL	8424	13557	253
TN	8227	13240	244
TR	8563	13780	238
TT	8853	14247	259
TU	9541	15354	228
TY	9319	14997	238
TZ	10021	16127	242
UA0	6904	11111	348
UA1	9066	14590	321
UA1 F J L	8781	14131	347
UA2	9606	15459	313
UA3	8951	14405	316
UA4	8509	13694	314
UA6	8429	13565	305
UA9	7682	12363	324
UB5	8939	14386	308
UC2	9322	15002	312
UD6	8174	13154	302
UF6	8447	13594	303
UG6	7957	12805	307
UH8	7595	12223	306
UI8	7376	11870	308
UJ8	7038	11326	311
UK0	6904	11111	348
UK1	9066	14590	321
UK2	9606	15459	313
UK3	8951	14405	316
UK4	8509	13694	314
UK5	8939	14386	308
UK6	8429	13565	305
UK7	7581	12200	319
UK8	7376	11870	308
UK9	7682	12363	324
UL7	7581	12200	319
UM8	7096	11420	315
UN1	9121	14678	328
UO5	9200	14806	303
UP2	9487	15267	313
UQ2	9488	15269	316
UR2	9413	15148	319
UT5	8939	14386	308
UW1	9066	14590	321
UW2	9606	15459	313
UW3	8951	14405	316
UW4	8509	13694	314
UW6	8429	13565	305
V3A	8934	14377	99
VE1	8594	13830	37
VE2	10463	16838	59
VE3	10227	16458	63
VE4	9286	14944	53
VE5	8982	14455	52
VE6	8746	14075	47
VE7	8151	13117	47
VE8	8183	13169	32
VK0	3202	5153	205
VK0 Heard	3473	5589	228
VK1	273	439	56
VK2	515	829	52
VK2 L H I	862	1387	66
VK3	55	89	0
VK4	932	1500	20
VK5	434	698	294
VK6	1605	2583	273
VK7	374	602	164
VK8	1954	3145	332
VK9N	2203	3545	275
VK9X	2364	3804	43
VK9Y	3466	5578	289
VK9Z	1502	2417	13
VO1	11362	18285	48
VO2	10844	17451	41
VP2	10154	16341	126
VP5	10011	16111	106
VP8	6082	9788	165
VP9	10713	17240	94

VQ9 Chagos	5036	8104	276
VR1	3320	5343	50
VR3	4596	7396	63
VR6	4918	7915	108
VR7	2351	3783	18
VS5	3528	5678	320
VS6	4567	7350	330
VS9	6797	10938	279
VU	5687	9152	296
VU7 Andamn	4705	7572	303
VU7 Lacadv	5800	9334	290
W0	8768	14110	64
W1	10400	16737	74
W2	10357	16668	72
W3	10231	16465	75
W4	9680	15578	78
W5	9175	14765	79
W6	7880	12681	63
W7	8284	13331	61
W8	9917	15959	66
W9	9718	15639	65
XE	8397	13513	88
XT	9775	15731	236
XU	4304	6926	315
XV	4365	7025	319
XW	4589	7385	318
XZ	5144	8278	312
Y	9922	15967	309
YA	6993	11254	306

YB	2948	4744	313
YI	8109	13050	292
YJ	1865	3001	49
YK	8498	13676	290
YN	8779	14128	105
YO	9301	14968	299
YS	8737	14060	102
YU	9583	15422	297
YV	9759	15705	128
Z2	6725	10823	240
ZA	9538	15350	292
ZB	10796	17374	275
ZC4	8734	14056	289
ZD7	8268	13306	212
ZD8	8992	14471	208
ZD9	6996	11259	198
ZF1	9384	15102	102
ZK1	3298	5307	80
ZK2	2971	4781	78
ZL	1556	2504	110
ZM2	3381	5441	67
ZP	7866	12659	158
ZS	6285	10114	225
ZS3	7019	11296	225
ZV	8887	14302	154
ZW	8887	14302	154
ZY	8887	14302	154
ZZ	8887	14302	154

1A0KM	5249	8447	356
3A	5546	8925	352
3B6	2876	4628	65
3B8	2606	4194	79
3B9	3036	4886	82
3C	2504	4030	346
3D2	8607	13851	156
3D6	794	1278	53
3V8	4789	7707	353
3X	3626	5835	321
3Y	1612	2594	201
4N	5382	8661	1
4S7	5261	8467	64
4U1ITU	5583	8985	352
4U1UN	7788	12533	305
4W	3839	6178	30
4X	4680	7532	15
4Z	4680	7532	15
5A	4419	7111	357
5B4	4853	7810	13
5H	4125	6638	284
5L	3378	5436	321
5N	2956	4757	340
5R	2001	3220	69
5T	4255	6848	326
5U	3489	5615	343
5V	2967	4775	335
5W	9068	14593	168
5X	2540	4088	26
5Z	2667	4292	33
6D	3202	5153	38
6W	3922	6312	324
6Y	7197	11582	282
7J	8382	13489	78
7O	3684	5929	33
7P	629	1012	67
7Q	3486	5610	21
7X	4836	7783	348
7Z	4367	7028	27
8P	6009	9670	288
8Q	4513	7263	65
8R	5660	9109	282
8Z	4707	7575	26
9G	3108	5002	332
9H	4828	7770	356
9J	1567	2522	25
9K	4885	7861	31
9L	3546	5707	319
9M2	6068	9765	84
9M6	7003	11270	90
9M8	6582	10592	91
9N	6083	9789	55
9Q	2029	3265	12
9U5	2264	3643	22
9V1	6020	9688	87
9X	2329	3748	21
9Y4	6039	9719	285
A2	735	1183	35
A3	8330	13405	155
A4	4717	7591	39
A5	6300	10139	58
A6	4728	7609	36
A7	4675	7523	32
A9	4652	7486	32
AP	5554	8938	44
BV	7697	12387	65
BY	7877	12676	75
C2	9370	15079	135
C3	5397	8685	347
C5	3909	6291	320
C6	7444	11980	289
C9	1880	3025	51
CE	4943	7955	240
CE0A	7044	11336	227
CE0X	5699	9171	244
CE0Z	5431	8740	238
CE9	3784	6090	210
CE9	3800	6115	213
CN	5002	8050	340
CO	7523	12107	285
CP	5175	8328	259
CR3	3852	6199	319
CR9	7347	11824	74
CT1	5373	8647	340
CT2	5825	9374	326
CT3	5007	8058	334
CX	4100	6598	247
CZ6	9685	15586	314

Fig. B-8. Direction distance table for G3ZCZ/ZS located at −33.9 18.3.

HB	5618	9041	354
HB0	5624	9051	354
HC	6539	10523	266
HC8	7260	11684	260
HH	6946	11178	285
HI	6808	10956	286
HK	6598	10618	272
HK0	7280	11716	275
HL	8524	13718	64
HP	7036	11323	273
HR	7657	12322	274
HR0	7571	12184	278
HS	6330	10187	73
HT	7509	12084	273
HV	5249	8447	356
HZ	4483	7214	29
I	5250	8449	357
IC	5146	8281	357
IH	4886	7863	355
IS	5139	8270	353
IT	4939	7948	356
J28	3531	5682	32
J3	6075	9776	285
JA	9072	14600	69
JD1	8719	14031	78
JE	9348	15044	65
JF	9052	14567	71
JG	8944	14394	70
JH	8712	14020	73
JI	8767	14109	73
JJ	9214	14828	67
JL	9072	14600	69
JR	9052	14567	71
JR6	8285	13333	77
JT1	7702	12395	52
JW	7798	12549	0
JX	7369	11859	351
JY	4695	7556	16
K0	9327	15010	295
K1	7729	12438	304
K2	7807	12564	305
K3	7842	12620	302
K4	8181	13166	295
K5	8527	13723	289
K6	10074	16212	284
D2	1664	2678	352
D4	4417	7108	312
D6	2217	3568	52
DA	5823	9371	354
DB	5823	9371	354
DF	5823	9371	354
DJ	5823	9371	354
DL	5823	9371	354
DM	5940	9559	357
DT	5942	9562	357
DU	7472	12025	85
EA	5299	8528	343
EA6	5130	8256	348
EA8	4859	7820	327
EA9	4587	7382	328
EI	6185	9954	345
EL	3378	5436	321
EP	5238	8430	27
ET	3667	5901	27
F	5612	9031	349
FB8W	1960	3154	127
FB8X	2788	4487	129
FB8Y	4786	7702	160
FB8Z	3266	5256	113
FC	5276	8491	353
FG7	6266	10084	289
FH8	2236	3598	55
FK8	8054	12961	145
FM7	6195	9970	288
FO8	8793	14151	195
FO8 Clprtn	7789	12535	239
FP8	7248	11664	317
FR7	2432	3914	78
FS7	6410	10316	289
FW8	8955	14411	159
FY7	5298	8526	285
G	5987	9635	348
GD	6235	10034	347
GI	6275	10098	346
GJ	5873	9451	347
GM	6335	10195	348
GU	5902	9498	347
GW	6111	9834	347
H4	8578	13805	127
HA	5589	8994	1

OR	5928	9540	351
OX	7464	12012	337
OY	6773	10900	348
OZ	6234	10032	355
P29	8044	12945	117
P5	8530	13727	61
PA	5980	9624	352
PJ2	6502	10464	282
PJ3	6584	10596	281
PJ4	6468	10409	282
PJ5	6406	10309	289
PJ6	6434	10354	289
PJ7	6426	10341	290
PP	4123	6635	272
PT	4123	6635	272
PY	4123	6635	272
PY0 St Pl	3874	6234	298
PZ	5520	8883	283
RA0	8603	13845	36
RA1	6564	10563	9
RA2	6108	9830	2
RA3	6317	10166	11
RA4	6262	10077	17
RA6	5701	9175	19
RA9	6940	11169	28
RB5	5896	9488	11
RC2	6099	9815	6
RD6	5507	8862	22
RF6	5618	9041	18
RG6	5855	9422	26
RH8	5755	9262	31
RI8	5909	9509	34
RJ8	6094	9807	39
RL7	6599	10620	30
RM7	6403	10304	38
RN1	6891	11090	7
RO5	5627	9056	7
RP2	6150	9897	3
RQ2	6287	10118	3
RR2	6432	10351	4
S2	6130	9865	61
S7	3132	5040	59
S8	677	1089	88
S9	2502	4026	342
SK	6348	10216	359
K7	9851	15853	293
K8	8286	13335	304
K9	8477	13642	303
KA	9072	14600	69
KB6	9813	15792	165
KC4	7095	11418	283
KC4A	4522	7277	173
KC4A	3875	6236	180
KC6	8757	14093	112
KG4	7165	11531	285
KG6	8805	14170	100
KH3	10303	16581	116
KH6	11497	18502	199
KH6 Kure	11170	17976	103
KL7	10554	16985	347
KM6	11469	18457	121
KP4	6645	10694	288
KP6	10298	16573	181
KS6	9011	14501	169
KV4	6542	10528	289
KW6	10243	16484	109
KX6	9806	15781	124
LA	6503	10465	356
LU	4335	6976	244
LX	5817	9361	352
LZ	5334	8584	6
M1	5387	8669	356
N0	9327	15010	295
N1	7729	12438	304
N2	7807	12564	305
N3	7842	12620	302
N4	8181	13166	295
N5	8527	13723	289
N6	10074	16212	284
N7	9851	15853	293
N8	8286	13335	304
N9	8477	13642	303
OA	6054	9743	257
OD	4821	7758	15
OE	5663	9113	357
OH	6494	10451	3
OH0	6487	10440	1
OJ0	6487	10440	1
OK	5762	9273	358
ON	5928	9540	351

SL	6348	10216	359
SM	6348	10216	359
SP	5934	9550	1
ST	3538	5694	18
SU	4479	7208	11
SV	4972	8001	3
T3	9729	15657	139
T5	3202	5153	38
TA	5159	8302	9
TF	7110	11442	343
TG	7828	12598	273
TI	7245	11659	272
TI9	7257	11679	266
TJ	2741	4411	347
TL	2862	4606	3
TN	2343	3771	358
TR	2403	3867	345
TT	3378	5436	360
TU	3149	5068	327
TY	3074	4947	337
TZ	3785	6091	335
UA0	8603	13845	36
UA1	6564	10563	9
UA1 F J L	7992	12862	7
UA2	6108	9830	2
UA3	6317	10166	11
UA4	6262	10077	17
UA6	5701	9175	19
UA9	6940	11169	28
UB5	5896	9488	11
UC2	6099	9815	6
UD6	5507	8862	22
UF6	5618	9041	18
UG6	5855	9422	26
UH8	5755	9262	31
UI8	5909	9509	34
UJ8	6094	9807	39
UK0	8603	13845	36
UK1	6564	10563	9
UK2	6108	9830	2
UK3	6317	10166	11
UK4	6262	10077	17
UK5	5896	9488	11
UK6	5701	9175	19
UK7	6599	10620	30
UK8	5909	9509	34
UK9	6940	11169	28
UL7	6599	10620	30
UM8	6403	10304	38
UN1	6891	11090	7
UO5	5627	9056	7
UP2	6150	9897	3
UQ2	6287	10118	3
UR2	6432	10351	4
UT5	5896	9488	11
UW1	6564	10563	9
UW2	6108	9830	2
UW3	6317	10166	11
UW4	6262	10077	17
UW6	5701	9175	19
V3A	7784	12527	276
VE1	9840	15836	327
VE2	7892	12701	312
VE3	8058	12968	308
VE4	9088	14625	310
VE5	9399	15126	309
VE6	9685	15586	314
VE7	10275	16536	312
VE8	10167	16362	337
VK0	3371	5425	156
VK0 Heard	2968	4776	134
VK1	6680	10750	142
VK2	6921	11138	142
VK2 L H I	7197	11582	146
VK3	6456	10390	140
VK4	7277	11711	135
VK5	6284	10113	135
VK6	5426	8732	122
VK7	6213	9999	145
VK8	7055	11354	114
VK9N	5117	8235	115
VK9X	8778	14126	142
VK9Y	5132	8259	93
VK9Z	7721	12425	130
VO1	7083	11399	319
VO2	7607	12242	321
VP2	6253	10063	288
VP5	7039	11328	289
VP8	3953	6362	226
VP9	7069	11376	301

VQ9 Chagos	3926	6318	73
VR1	9683	15583	151
VR3	10434	16791	184
VR6	7858	12646	211
VR7	8536	13737	125
VS5	6793	10932	90
VS6	7417	11936	75
VS9	3971	6391	42
VU	5150	8288	59
VU7 Andamn	5745	9245	73
VU7 Lacadv	4763	7665	58
W0	9327	15010	295
W1	7729	12438	304
W2	7807	12564	305
W3	7842	12620	302
W4	8181	13166	295
W5	8527	13723	289
W6	10074	16212	284
W7	9851	15853	293
W8	8286	13335	304
W9	8477	13642	303
XE	8571	13793	272
XT	3466	5578	332
XU	6491	10446	78
XV	6773	10900	77
XW	6682	10753	74
XZ	6261	10076	66
Y	5940	9559	357

YA	5777	9297	40
YB	6406	10309	98
YI	4913	7906	23
YJ	8265	13301	144
YK	4844	7795	16
YN	7548	12147	272
YO	5437	8750	6
YS	7717	12419	272
YU	5382	8661	1
YV	6222	10013	282
Z2	1315	2116	36
ZA	5175	8328	1
ZB	5069	8158	340
ZC4	4853	7810	13
ZD7	1951	3140	303
ZD8	2724	4384	304
ZD9	1710	2752	252
ZF1	7463	12010	281
ZK1	8842	14229	176
ZK2	8704	14007	170
ZL	6969	11215	162
ZM2	9352	15050	168
ZP	4472	7197	256
ZS	278	447	78
ZS3	646	1040	341
ZV	4878	7850	271
ZW	4878	7850	271
ZY	4878	7850	271
ZZ	4878	7850	271

Index

Index